Catalyst Surface:
Physical Methods
of Studying

Advances
in
Science
and
Technology
in
the USSR

Chemistry Series

ADVANCES IN
SCIENCE AND TECHNOLOGY
IN THE USSR

Catalyst Surface: Physical Methods of Studying

Edited by KH. M. MINACHEV and E. S. SHPIRO
Translated from the Russian by G. Leib

Mir Publishers
Moscow

CRC Press
Boca Raton   Ann Arbor   Boston

Library of Congress Cataloging-in-Publication Data

Catalyst surface: physical methods of studying / edited by Kh. M. Minachev and E. S. Shpiro.
    p. cm.--(Advances in science and technology in the USSR. Chemistry series)
    Translated from the Russian.
    Includes bibliographical references and index.
    ISBN 0-8493-7532-0
    1. Catalysts. 2. Surfaces (Technology) I. Minachev, Kh. M. (Khabib Minachevich) II.
Shpiro, E. S. III. Series.
QD505.C3866 1990
541.3′95—dc20   90-46126
                                                 CIP

© Mir Publishers. 1990

    This edition published by CRC Press, Inc., 2000 Corporate Blvd. N.W. , Boca Raton,
Florida.

International Standard Book Number 0-8493-7532-0

Library of Congress Card Number 90-46126
Printed in the United States

PREFACE

New production processes involving catalysts can only be created at present when comprehensive specifications of a catalyst are available and the mechanism of their action is understood in detail. This is because the industry now puts stricter requirements on catalysts to ensure their selectivity, ecological cleanliness, and low energy and material costs. Our hopes that these expectations will come true are based on two positive trends in science and engineering, namely, the development of production processes to synthesize novel materials with controllable physicochemical properties, and the development and employment of physical methods (primarily for surface analysis) to completely characterize material at the molecular and atomic levels.

The 1970's saw substantial progress in studying the structure of the surfaces of model catalytic systems (single crystals and thin films) and their interaction with reacting molecules with the aid of physical methods of surface analysis. The progress in studying real catalysts (including industrial ones) has not been so great, and their development remains empirical. In the 1980's, the gap in our understanding the action of simulated and real catalytic systems has begun to shrink owing to the more comprehensive studying of real systems by a set of modern physical techniques that has been developed theoretically and methodologically. For example, at our laboratory alone, over 50 studies involving the analysis of catalyst surfaces by electron emission spectroscopy (EES) and ion spectroscopy (IS) have been completed in recent years. The total number of publications in this field has exceeded 1000. The goals of the studies have changed appreciably. In addition to solving some particular problems, they are aimed at establishing a fundamental relation between surface and catalytic properties.

This is why the reviews of catalyst surface analysis performed by physical techniques in the late 1970's and early 1980's [see, for example, A. W. Czanderna (ed.). *Methods of Surface Analysis.* New York: Elsevier (1975); J. Thomas and R. Lambert (eds.). *Characterization of Catalysts.* Chichester: Wiley (1980); and the monograph: Kh. M. Minachev, G. V. Antoshin, and E. S. Shpiro. *Fotoelektronnaya spektroskopiya i ee primenenie v katalize* (Photoelectron Spectroscopy and Its Use in Catalysis). Moscow: Nauka (1981)] no longer reflect the present state of this impetously developing field of science. All this prompted us to attempt a newer and deeper analysis and overview of studies of catalyst surfaces by EES and IS, chiefly on the basis of our own results, and also of data that appeared in Soviet and foreign publications after 1980.

In addition to new results, the present monograph briefly sets out the fundamentals of the techniques used and the methodology of studying real catalytic systems. We hope our foreign readers will be interested in the work of Soviet authors and be able to appraise their contribution to the development of this important field of surface and catalysis physical chemistry.

Kh. Minachev, E. Shpiro

CONTENTS

8   Contents

LIST OF ABBREVIATIONS

AES	Auger electron spectroscopy
ARUPS	Angle-resolved ultraviolet photoelectron spectroscopy
ARXPS	Angle-resolved X-ray photoelectron spectroscopy
CAT	Constant analyzer transmission
CHA	Concentric hemispherical analyzer
CMA	Cylindrical mirror analyzer
CNDO	Complete neglect of differential overlapping
CRR	Constant relative resolution
DV	Discrete variation
EES	Electron emission spectroscopy
EHM	Extended Hückel method
ELS	Electron loss spectroscopy
EPR	Electron paramagnetic resonance
ESCA	Electron spectroscopy for chemical analysis
EXAFS	Extended X-ray absorption fine-structure spectroscopy
FABMS	Fast atom bombardment mass spectroscopy
FAT	Fixed analyzer transmission
FEM	Field emission microscopy
FRR	Fixed relative resolution
FWHM	Full width at half maximum peak height
HREELS	High resolution electron energy loss spectroscopy
HRTEM	High resolution transmission electron microscopy
IPS	Inverse photoemission spectroscopy
IR	Infrared spectroscopy
IS	Ion spectroscopy
ISS	Ion scattering spectroscopy
LCAO	Linear combination of atomic orbitals
LEED	Low-energy electron diffraction
LEELS	Low-energy electron loss spectroscopy
MASNMR	Magic angle spinning nuclear magnetic resonance
Me	Metal
MO	Molecular orbital
MSD	Multi-channel solid detector
NEXAFS	Near extended X-ray absorption fine-structure spectroscopy
NMR	Nuclear magnetic resonance
PAX	Photoemission of absorbed xenon

PZC	Point zero charge
RED	Radial electron distribution
REM	Rare earth metal(s)
SAM	Scanning Auger spectroscopy (microprobe)
SCF	Self consistent field
SEXAFS	Surface-sensitive extended X-ray absorption fine-structure spectroscopy
SIMS	Secondary ion mass spectrometry
SMSI	Strong metal-support interaction
SNIMS	Secondary neutral ion mass spectrometry
SW	Scattered wave
TDS	Thermal desorption spectroscopy
TEM	Transmission electron microscopy
TOF	Turnover frequency
UHV	Ultrahigh vacuum
UPS	Ultraviolet photoelectron spectroscopy
XAES	X-ray Auger electron spectroscopy
XANES	X-ray absorption near-edge spectroscopy
XAS	X-ray absorption spectroscopy
XPD	X-ray photoelectron diffraction
XPS	X-ray photoelectron spectroscopy
XRD	X-ray diffraction analysis

INTRODUCTION

The 1970's are customarily considered the beginning of a new stage in the fundamental studies of catalysis and in the prediction of the properties of various catalytic systems. This stage is characterized by the development of new ways of synthesizing catalysts ("heterogenized" complexes, materials with the structure of zeolites, etc.) and also by the application of experimental and theoretical techniques enabling one to describe the structure of a surface and the nature of gas-solid interaction. The use of techniques such as low-energy electron diffraction (LEED), Auger electron spectroscopy (AES), and low-energy electron loss spectroscopy (LEELS) resulted in appreciable progress in understanding the catalytic action of atomically pure surfaces of metal single crystals [1, 2]. In particular, it became possible to explain the cause of the different structural sensitivity of reactions, establish the substantial contribution of the reaction medium to the formation of the "working" surface of a catalyst, and confirm the validity of a local approach when describing processes of adsorption and catalysis. At present, the entire set of techniques of the surface science is employed for this class of catalysts including the most up-to-date ones such as scanning tunneling microscopy. These techniques have a high surface sensitivity and provide information of the fundamental characteristics of solid surfaces, namely, the crystallographic and electron structure, the chemical state and local environment of elements, and the nature and geometry of the bond of the gas or liquid molecules to the solid. At present, we can list over fifty techniques and methods that were or are being used to study surfaces.

At the same time, to achieve real progress in the development of catalysts, it is essential to study the formation and nature of the activity of real catalytic systems with the aid of these techniques. Owing to fundamental restrictions, only some surface analysis techniques can be used for such investigations. The most effective techniques that have already recommended themselves in studying the surface of heterogeneous catalysts include electron emission spectroscopies (EES), X-ray photoelectron spectroscopy (XPS), Auger electron spectroscopy (AES), and ultra-violet photoelectron spectroscopy (UPS) (Table 1). They were first applied to catalysis at the beginning of the 1970's. Auger electron spectroscopy, especially during the first period, was employed for a qualitative and semi-quantitative analysis of the composition of metal and alloy catalyst surfaces. X-ray photoelectron spectroscopy [or electron spectroscopy for chemical analysis (ESCA)], developed in the fundamental and

Table 1. Characteristics of Electron Emission and Ion Spectroscopies

Technique	Basic information	Average sensitivity		Depth of analyzed layer, nm	Local resolution, nm	Applicability to specimens with various electrophysical and structural characteristics
		concentration, %	monolayer fractions			
X-ray photoelectron spectrosopy (XPS)	Valence state, electron structure, surface composition	0.5-1.0	10^{-1}	2-4	10^4	Any specimens
Ultra-violet photoelectron spectroscopy (UPS)	Electron structure of surface, nature and geometry of adsorption bond	—	10^{-1}	1-2	—	Conductors, semiconductors
Auger electron spectroscopy (static) (AES)	Electron structure, surface composition, depth profiles	0.05-0.1	10^{-2}	1-3	—	Ditto
Scanning Auger spectroscopy (microprobe) (SAM)	Surface composition of local surface areas	0.1	10^{-1}	1-3	20-50	Ditto

Method	Information obtained				
Secondary ion mass spectrometry (SIMS)	Fragments and local structure, composition of surface (semi-quantitative analysis), depth profiles	0.01	10^{-3}	0.5^a-100^b	50-100	Any specimens
Ion scattering spectroscopy (ISS)	Composition and structure of upper surface layers, depth profiles	0.1	10^{-2}	0.2-0.3^c	100	Ditto
Low energy electron loss spectroscopy (LEELS)	Vibrational states of absorbed atoms and molecules, nature and geometry of adsorption bond	—	10^{-3}-10^{-2}	—	—	Single crystals, conductors and semiconductors

[a] Static mode.
[b] Dynamic mode, SIMS.
[c] Monolayer.

applied works of Siegbahn's school [3-5], was found to be more universal. Owing to these works, the first high-resolution spectrometers were developed. A high sensitivity of XPS of the core levels was revealed when studying the electron structure of substances and chemical transformations. Although attention was also given earlier to the surface sensitivity of the technique, thorough investigations in this field began in the 1970's [6-28].

Let us consider the basic features of photoelectron spectroscopy that are especially important when studying adsorption and catalysis.

1. The effective escape depth of photoelectrons is determined by the mean free path up to inelastic collisions with a lattice (λ) and is within 2-4 nm. The attenuation of the signal of the photoelectrons is of an exponential nature. This suggests that the thickness of the layer being analyzed does not exceed 3λ, while about 60% of the intensity falls to the share of layer λ. Hence, surface layers less than 10 nm thick are analyzed by using XPS.

2. The energetic position of the inner level lines (or the chemical shift) characterizes the valence and coordination state of elements, the degree of ionicity of a bond or the effective charge, while the position, shape, and intensity of the valence bands characterize the energy of the valence orbitals and the density of the occupied states in the valence band.

3. The energy and angular distribution of the photoelectrons emitted from adsorbates reflect the type and geometry of an adsorption bond.

4. The intensity of the lines is the basis for a quantitative analysis of a surface.

5. The technique can be used to study all the elements of the periodic table except hydrogen.

In the late 1960's, the first publications also appeared describing the use in catalysis of techniques based on the interaction of 0.5-10-keV ions with a surface. The best known of them include secondary ion mass spectrometry (SIMS) and low-energy ion scattering spectroscopy (ISS) [6-8, 29, 30]. In both techniques, the surface of a specimen is bombarded by ions, most often of inert gases. In SIMS, the sputtered substance in the form of ions is analyzed, and in ISS — the primary ions that have spent their energy on elastic collisions with the substance. The main merits of SIMS is its high elemental sensitivity (Table 1) and the possibility of procuring data on the composition of the top and deeper layers. The main feature of ISS is its extremely high sensitivity to a surface — under definite conditions information is received from only the first layer of a specimen. It should be noted that owing to the difficulties involved in running experiments and interpreting their results, SIMS and ISS have meanwhile not been embodied to such an extent in catalysis as the electron spectroscopy techniques.

The main object of the present monograph is to consider the fundamentally new information that is given by EES and IS or will most likely be given when these techniques are further developed for understanding the mechanism of formation of active surfaces of various types of catalysts and the nature of their activity. Our readers can acquaint themselves with more particular data independently by turning to the references. A number of problems remain in each of the considered techniques whose solution requires their further theoretical and methodological development or the use of additional techniques. These problems include a quantitative description of the electron structure, quantitative analysis of the surface and near-surface layers of heterogeneous systems, improvement of the accuracy of the spectral parameters and their standardization for non-conducting specimens, and the development of procedures wherein the pretreatments of the specimens in situ and the simulation of a catalytic process are closer to real conditions. All these matters are dealt with in the present monograph.

We considered primarily the works including investigations of model and real objects, and employing not only EES and IS, but also such effective techniques as EXAFS, NMR in solids, and varieties of electron microscopy. It is just such works, in our opinion, that will allow one to understand more deeply the contribution of the collective and local properties of a surface to the proceeding of an elementary catalytic event and will be the foundation for building up a modern theory of the selection of optimum catalysts.

The present monograph has been conceived to be helpful not only for specialists, but also for a broad circle of scientists working in the field of the physical chemistry of surfaces and catalysis. Chapter 1 briefly treats the physical fundamentals of the EES and IS techniques that were used to obtain the main results. Since these questions were repeatedly considered in special monographs and reviews [3, 5, 8, 10, 30-34], the main attention in this chapter is devoted to the specific features of the analysis and interpretation of the spectra of complicated systems, a comparison of the possibilities of individual EES and IS techniques, as well as to other related techniques of surface analysis. Chapter 2 describes the procedure of an experiment with a view to the latest developments and discusses the problems of standardizing spectral data. Chapters 3-5 contain the results of studying catalysts by EES and IS. Here Chap. 3 on the example of the latest achievements reveals the possibilities of the techniques in studying the fundamental properties of a surface: the electron structure, chemical state of the components, and the composition, local structure, and morphology of the surface. It also discusses the most important theoretical and methodological problems such as the electron structure of small metal clusters and the quantitative analysis of heterogeneous surfaces. Chapter 4 discusses in detail data on various types of catalysts, namely, metals and alloys, oxides, zeolites, and mono- and polymetallic supported catalysts. Chapter 5 gives examples of studying the formation of surfaces and the nature of the active centers of new and existing catalysts used in the most important processes of petroleum refining, petrochemistry, organic, and inorganic syntheses.

1

Physical Fundamentals of Electron Emission and Ion Spectroscopies

1.1　Photoelectron Spectroscopy

Photoelectron spectroscopy is based on the phenomenon of the photoelectric effect, and the main equation describing the latter is the Einstein relation

$$h\nu = E_b + E_k + E_{rec} \tag{1.1}$$

where $h\nu$ is the energy of a quantum (generally mono- or nonmonochromatic radiation is used, Al $K\alpha$ and Mg $K\alpha$ radiation in XPS, and He^I and He^{II} radiation in UPS), E_b is the binding energy of an electron in an atom determined as the energy needed to remove an electron to infinity with $E = 0$, E_k is the kinetic energy of a photoelectron, and E_{rec} is the recoil energy; at $h\nu < 1500\,eV$, the recoil energy $E_{rec} \leqslant 0.1\,eV$ [3] and it may be disregarded in the following. For molecules in the gas phase, Eq. (1.1) can be written in the form

$$h\nu = E_k' + E_{b,v} \tag{1.2}$$

where E_k' is the kinetic energy of a photoelectron, and $E_{b,v}$ is the binding energy of an electron relative to the vacuum level.

For solid conducting specimens, we have

$$h\nu = E_k + E_{b,F} + \varphi_{sp} \tag{1.3}$$

where $E_{b,F}$ is the binding energy of an electron determined relative to the Fermi level of the spectrometer material, and φ_{sp} is the spectrometer work function. The latter quantity can be found when there is a sharp boundary in the change in photoemission intensity at the Fermi level. The determination of E_b for semiconductors and isolators is difficult because of the absence of an electric contact between a specimen and the spectrometer and the ambiguity in the position of the Fermi level. As a result of photoemission, an uncompensated charge appears on such specimens. In the general case, it can be either positive or negative, e.g. because of the emission of secondary electrons from the chamber walls, and this charge must be taken into account when determining E_b:

$$h\nu = E_k + E_{b,F} + \varphi_{sp} \pm \varphi_{ch} \tag{1.4}$$

where φ_{ch} is a correction for charging of a specimen. How it is determined is considered in Chap. 2.

1.1.1 Chemical Shifts and Width of Core Level Lines. Shape of Valence Bands

The shifts of E_b of the core levels are the main source of information on the nature of a chemical bond. The calculations of E_b are performed by various methods (Hartree-Fock-Slater-Dirac, MO LCAO SCF) in the approximation of "frozen" orbitals by the Koopmans theorem, i.e. of the equivalence of the wave functions for an atom in its initial and final states. The latter assumption is approximate and takes no account of the correlation changes associated with ionization of the atoms [3, 31, 35]. Such calculations, and also the notions of the electrostatic model of point charges [3-5, 9] reveal that for an atom A, we have

$$E_{b,A} = kq_A + \sum_{i \neq A} \frac{q_i}{R_i} + L \tag{1.5}$$

where k is the constant of the Coulomb interaction between the core and valence electrons, q_A is the effective charge of atom A, q_i is the effective charge of the surrounding atoms, R_i is the interatomic distance, and L is the reference level; if $L = 0$, Eq. (1.5) can be written as

$$\Delta E_{b,A} = \Delta E(q) - \Delta V \tag{1.6}$$

The first term shows how the chemical shift depends on the charge, while the second one shows how it depends on the potential of the Madelung interaction (ΔV) with the neighboring atoms of the lattice; for an ionic crystal [3, 5, 9, 22], we have

$$\Delta E_b = q \left(\frac{1}{r} - \frac{\alpha}{R} \right) \tag{1.7}$$

where q is the effective charge of an atom, r is the average radius of the valence shell, α is the constant of the Madelung interaction, and R is the approximate interatomic distance.

The most clearly expressed correlations include the direct dependence of the chemical shift on the effective charge that is evaluated in various ways, including the simplest method of electronegativities. Such a relation is not obvious because even to a first approximation the chemical shift is a small difference of two large quantities [Eq. (1.6)]. If we take into account, on the other hand, that the final state of a photoionized atom differs from the initial one, a relaxation term has to be introduced into Eq. (1.6):

$$E_b = E_{in} - E_{fin} \tag{1.8}$$

$$\Delta E_b = \Delta E(q) - \Delta V - \Delta E_R \tag{1.8a}$$

$$\Delta E_R = \Delta E_{IR} + \Delta E_{ER} \tag{1.9}$$

Here E_R (ΔE_R), the relaxation energy, is connected with the change in the form of the wave functions in photoionization; it is spent on redistribution of the electron density in screening of the hole potential at the inner level and includes two components: E_{IR} — intra-atomic relaxation describing the interaction of the electrons of a given atom with a hole, and E_{ER} — extra-atomic relaxation, i.e. the energy of interaction of the electrons of the neighboring atoms and the hole in atom A. The values of E_{IR} amount to scores of electron-volts, while those of E_{ER} never exceed 10 eV. The extra-atomic relaxation E_{ER} is of a substantial significance for a condensed state and therefore changes greatly in a transition, for instance, from the gaseous state to an adsorbed one or from an atomic state to a crystalline one. The matters pertaining to the appraisal of the contribution of E_R for these cases are treated in the following sections. The calculated values of the chemical shifts when applying a unit charge to an atom are of the order of magnitude of scores of electron-volts [9, 31, 32], the experimental values of the shifts do not exceed 5-6 eV and in most cases are 1-2 eV. When the second and third terms of Eq. (1.8a) are taken into consideration, the convergence of the experimental and theoretical data noticeably improves [10, 31, 35-38].

Also found were the dependences of the chemical shift on the oxidation state of the elements, the electron-donor and electron-acceptor nature of the ligands, the reciprocal of the interatomic distance for compounds of one type ($1/R_{AB}$) [5, 9, 34, 39, 40], and also on the thermodynamic characteristics [41]. Correlations between the chemical shifts and the heat of the relevant reactions have been found for many substances [34]. The same approach can be used to estimate the energy of adsorption of molecules on a surface [9]. As regards the dependence of ΔE on $1/R$, the most definite correlation has been obtained between the shift of the O $1s$ line and the interatomic distance for oxides of the type Me_2O_3 [40].

An analysis of how ΔE_b depends on the oxidation state is important for identifying the chemical state of elements. In the simplest case, an increase in the oxidation state of an element by unity in a series of compounds with similar surroundings is attended by a change in the shift of about 1 eV. This rule is not always observed, however (see the data for the oxides of Co, Pd, and Cu in [9, 27]). An increase in the ionicity of a bond (with the same formal oxidation state) causes a positive shift, while an increase in the covalent nature of the bond causes a negative shift. An analysis conducted by Carlson [32] shows that the chemical shift with the same variation of the oxidation state grows to the right within the periods of the periodic table and diminishes down the groups.

Wagner *et al.* [39] proposed a way of identifying chemical compounds based on determining the Auger parameter

$$\alpha = E_b + E_k \tag{1.10}$$

where E_b is the binding energy of a photoelectron, and E_k is the kinetic energy of an Auger electron in a transition including the same level. Wagner *et al.* plotted two-dimensional diagrams with $E_{k,i}$ and α laid off along the ordinate axis and E_b along the axis of abscissas. Since the contribution of the relaxation processes in photoionization and Auger ionization is different, such diagrams make it possible to single out the "chemical" part of ΔE_b more precisely and identify the state in the absence of shifts or with anomalous shifts in the X-ray photoelectron spectra. The most complete data on chemical shifts are given in a reference book by Nefedov [34]. Table 2 presents the average values of the chemical shifts observed in a change in the state of transition elements — the main components of heterogeneous catalysts.

In addition to apparatus factors, the width of the core level lines is determined by the intrinsic width of the levels that depends on the lifetime of the hole states, namely, on the Coster-Kronig transitions, and on the Auger and radiation transitions [3, 31]. Since some of them follow an interatomic mechanism, the width of the lines depends on the type of chemical compound. Moreover, the change in E_{ER} contributes to it, which is reflected in broadening of the levels of small particles.

Table 2. Chemical Shifts Attending a Change in the Oxidation State of Elements in Oxides and Zeolites[*]

Element	Compound	Level	ΔE_b	Element	Compound	Level	ΔE_b
Ti	TiO	$2p_{3/2}$	2.8	Sn	SnO_2	$3d_{5/2}$	1.7
	Ti_2O_3	"	3.9	Mo	MoO_2	"	1.4
	TiO_2	"	5.4		MoO_{3-x}		3.1
Cr	CrO_3		5.4		MoO_3		4.5
	Cr_2O_3		2.5	Ru	RuO_2		0.7
	$Cr^{3+}Y$		2.9		RuO_3		2.5
Fe	FeO		2.6		$Ru^{3+}Y$		3.7
	$Fe^{2+}Y$		3.5	Rh	Rh_2O_3		1.5
	$\alpha\text{-}Fe_2O_3$		3.1		$Rh^{3+}Y$		3.0
	$Fe^{3+}Y$		4.5	Pd	PdO		1.6
Co	CoO		2.0		$Pd^{2+}Y$		3.4
	$Co^{2+}Y$		4.0	Ag	Ag_2O		− 0.4
	Co_3O_4		2.5		AgO		− 0.7
	$CoAl_2O_4$		3.0	W	WO_2	$4f_{7/2}$	2.2
Ni	NiO		2.0		WO_{3-x}	"	3.4
	$Ni^{2+}Y$		4.1		WO_3		4.4
	$Ni(OH)_2$		3.0	Re	ReO_2		2.6
Cu	CuO		1.6		RcO_3		3.7
	$Cu^{2+}Y$		2.0		Re_2O_7		5.2
	Cu_2O		0	Pt	PtO		2.7
Zn	ZnO		0		$Pt^{2+}Y$		2.9
Sn	SnO	$3d_{5/2}$	1.7		PtO_2		3.9

[*] Relative to the metal.

The energy structure of the inner levels is of a discrete nature, i.e. $E_i = f(Z)$ (Fig. 1), whereas the spectrum of the valence levels of the condensed state is quasicontinuous. At large values of $h\nu$ (XPS), the spectra of the valence band reflect the density of the occupied states of the valence electrons. At low values of $h\nu$ (UPS), the shape of a spectrum depends greatly on the emission energy and carries information on the initial and final states: the bulk energy structure and the occupied surface states [10, 11, 31] or the vacant ones (inverse photoemission) [42].

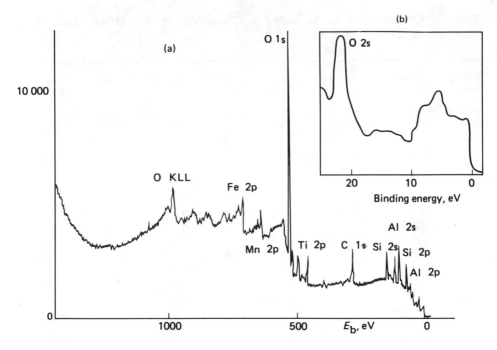

Fig. 1. Survey X-ray photoelectron spectra of inner levels (a) and valence band (b): a — zeolite with additives of Ti, Mn, and Fe oxides; b — Rh/TiO$_2$

1.1.2   Structure of the Core Level Spectra

Many spectra of photoelectrons are not simple singlets or spin doublets, but have a varying specific structure. The latter may include peaks of discrete energy losses on inelastic collisions or produced by X-ray satellites, the multiplet splitting of lines, and satellites due to many-electron processes (of the shake-up or shake-off type). Multiplet splitting is observed for elements in the paramagnetic state [11, 31, 43, 44], and it is caused by the exchange interaction of an escaping photoelectron with an unpaired d or f electron of the outer shell. The magnitude of the multiplet splitting is maximal for a sublevel with the same quantum number n as the d and f level (e.g. $3d$-$3s$), but it is also observed for other levels. For instance, the multiplet (exchange) interaction in paramagnetic compounds of Co and Cr [44] increases the splitting between the components of the relevant spin doublets.

Satellites of the shake-up type (excitation of an electron to a higher state) or shake-off type (excitation of an electron to a continuum state) appear because of the shaking of the unstable photoionized state of an atom [45-52]. They have been interpreted as a result of (a) an intra-atomic transition of a second electron from the $3d$ state to an unoccupied $4s$ state [45], and (b) monopole transitions with charge transfer of the ligand-metal type [46, 47]. The second mechanism explains the dependence of the satellite intensity and splitting of the satellite peak (ΔE_{sat}) on the ligand electronegativity. When the covalent nature of a bond grows, the intensity diminishes, and ΔE_{sat} grows. Moreover, a relation was noted between these parameters and the paramagnetism [45-47], and also between them and the unpaired spin density on a transition ion [50]. The model of the appearance of a shake-up satellite as a result of the interaction of a core photoelectron with an unpaired d electron leading to inversion of the spin of the latter has been proposed on this basis. In addition to the $2p$ levels of the $3d$ elements, intensive satellites are observed in the $3d$ and $4d$ spectra of the lanthanides and actinides [51, 52]. Recently both low-energy and high-energy satellite peaks have been detected. Their intensity may exceed that of the main lines. The configurational interaction of various final states is one explanation of this phenomenon [51].

The change in the satellite structure sometimes enables one to judge on the chemical and spin state of an ion. For example, the vanishing of a satellite points to the transitions $Cu^{2+} \rightleftarrows Cu^{+}$ (Fig. 2) and $Co^{2+} \rightleftarrows Co^{3+}$ in the relevant oxides. But concise criteria relating the parameters of the satellite structure to the degree of covalency of a bond, the coordination number, or the

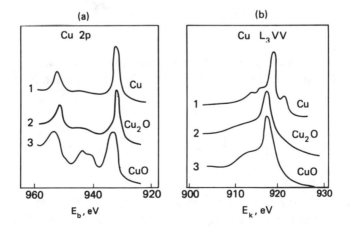

Fig. 2. Photoelectron (Cu $2p$) (a) and Auger electron (Cu L_3VV) (b) spectra for Cu (1), Cu_2O (2), and CuO (3). Spectrum 3 in Fig. 2a contains a shake-up satellite

spin density have not been found in the general case. This is explained by the difficulty of describing many-body processes. As a consequence, a satellite structure is mainly used as "fingerprints" of a definite state.

Peaks of inelastic energy losses associated with the excitation and ionization of electrons of a solid are observed in the spectra of metals. Their analysis provides information on the density of the states in the valence band [53] and also on the escape depths of photoelectrons. The latter circumstance enables one to use such spectra for quantitative analysis [54]. A more intensive structure of the inelastic losses is observed in the spectra of broad-band semiconductors. These losses are due to inner (hole) and outer (photoelectron) collective plasmon excitations, and also to less considerable band-band type excitations. An analysis of their intensities and the distance between the loss-induced and main peaks is useful when studying the morphology of thin specimens of Al_2O_3/Al and SiO_2/Si [54]. The peaks of inelastic losses were found to have the greatest intensity with submonolayer coatings.

1.1.3   Intensities of Photoelectron Spectra

We have already noted that only the photoelectrons which experienced no inelastic collisions contribute to the intensity of the main spectrum lines. The probability of an electron escaping from layer "x" of a flat specimen is [55-57]:

$$P = \exp \frac{-x}{\lambda \sin \theta} \qquad (1.11)$$

where λ is the escape depth of the electrons equal to the inelastic mean free path, and θ is the angle of photoemission. The integral intensity for element X (level i) is:

$$I_i = SK_i \frac{1}{\sin \theta} \int_0^\infty c_x \sigma_i \exp \frac{-x}{\lambda_i \sin \theta} \, dx \qquad (1.12)$$

where K_i is a multiplier taking into account a number of experimental and theoretical factors, c_x is the bulk concentration, S is the area of the specimen irradiated by the beam of X rays, and σ_i is the photoionization cross section. At $\theta = 90°$, about 60% of the signal is received from a layer whose thickness is $x = \lambda$, and 95% from one for which $x = 3\lambda$ (Fig. 3). The intensity is deter-

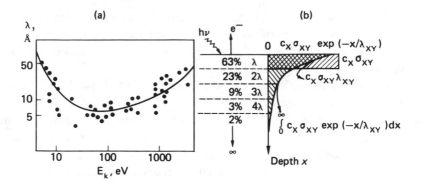

Fig. 3. A "universal curve" (a) and the contribution of various layers of thickness x to the total intensity at $\theta = 90°$ (b). The weakening of the intensity with the depth is shown at the right (element X, level Y)

mined in the integral form as follows [31, 55, 56]:

$$I_i = FGS_iD\sigma_iP_i(\beta)\lambda_iC_ic_i \tag{1.13}$$

Here F is the intensity of the flux of X-ray photons, G is a geometric factor taking into consideration the electrons registered by the spectrometer, S_i is the sensitivity of the analyzer to electrons of a given kinetic energy (E_k), D is the sensitivity of the detector, $P_i(\beta)$ is an asymmetry factor taking into consideration the angular dependence of the photoionization cross section, C_i is a factor taking into consideration the attenuation of the intensity when the electrons pass through the layer of surface contaminants, and c_i is the bulk concentration. Expressions (1.12) and (1.13) imply that the absorption of X rays in the layer being analyzed may be disregarded.

Since not all the parameters contained in Eqs. (1.12) and (1.13) are known in the general case, in practice one determines the ratio of the intensities for two selected elements (one of which is considered to be a standard element). For the pair X and Y, we have:

$$R_x = \frac{I_{X_i}}{I_{Y_i}} \tag{1.14}$$

or

$$R_{X,Y} = R_{P(\beta)}R_{S_i}\frac{\sigma_{X_i}}{\sigma_{Y_i}}\frac{c_X}{c_Y}R_\lambda R_C \tag{1.15}$$

where R is the ratio of the relevant parameters in Eq. (1.13), and c_X/c_Y is the ratio of the concentrations of the elements X and Y (for a homogeneous specimen it is the ratio of the bulk concentrations). Let us see how the factors in Eqs. (1.13)-(1.15) are determined.

Asymmetry Factor $P(\beta)$: if the angle between the direction of incidence of the X rays and the escape of the photoelectrons $\beta = 54°44'$, then $P(\beta) = 1$. Theoretically, β varies from 2 (an s shell) to 0.5 (all other shells). In most commercial instruments, $\beta = 90°$, and the maximum error introduced into $R_{P(\beta)}$ is about 20% (in most cases it does not exceed 10%). Calculated values of $P(\beta)$ [58] can be used for more accurate measurements. The **transmission coefficient of an analyzer** $T_i = S_i D$ depends on the kinetic energy of the electrons (see Chap. 2). The **photoionization cross sections** σ_i have been found for most elements of the periodic table by measuring the relative intensities of the elements in various reference compounds. These values are close to those calculated by the Hartree-Fock-Slater method in the approximation of a free atom [59-63]. The determination of the **escape depth (free mean path) of electrons** λ is to date the most involved problem in quantitative analysis. It is usually assumed that λ is determined by the inelastic collisions of the photoelectrons with lattice elements (mainly electron-electron interaction). Here the dependence of the intensity attenuation on the layer depth follows the Lambert-Beer law, and the experimental value of λ (or, more precisely, the length of signal attenuation) is determined by measuring the relative intensities of the signal from the substrate and a film of known thickness in one of the following two ways [55, 64, 65]:

$$I_1 = I_0 \exp - \frac{d}{\lambda_i \sin \theta} \tag{1.16}$$

or

$$I_2 = I_\infty \left\{ 1 - \exp \left(-\frac{d}{\lambda_i \sin \theta} \right) \right\} \tag{1.17}$$

By combining these equations, we obtain

$$\frac{I_2}{I_1} = \frac{I_\infty}{I_0} \frac{1 - \exp(-d/\lambda_i \sin \theta)}{\exp(-d/\lambda_i \sin \theta)} \tag{1.17a}$$

where I_1, I_2, I_0, and I_∞ are the intensities of the signal (peak) from the substrate coated with a film of thickness d, from the thin film, of the same signal for the pure substrate, and of the peak (I_2) in a thick film (a massive specimen), respectively.

The quantity λ depends on a number of factors, namely, the closeness of packing of the substance [55], the photoemission angle [57], and the kinetic energy E_k [55, 66-68]. With fixed angles θ, the escape depth of the electrons is an approximate function of E_k. This is illustrated by what is called a "universal curve" (see Fig. 3) on which the experimental and theoretical values of λ for various specimens are arranged with a certain scatter [8, 22, 68]. A number of authors have found analytical expressions relating λ, E_k, and the characteristics of the specimens (for Eq. (1.18) see [64], and for (1.19) see [9, 67]):

$$\lambda = CE^n \quad (n = 0.5 \text{ to } 0.7) \tag{1.18}$$

$$\lambda(E) = \frac{E_k}{\log E_k - b} \tag{1.19}$$

where b is a constant.

Equation (1.19) has been proposed by Penn for calculating λ in inorganic compounds. Another way of writing this equation is

$$\lambda(E) = \frac{E_k}{P(\log E + Q)} \tag{1.19a}$$

where P and Q are characteristics of the substance being studied. This procedure takes into account the specific features of the structure of a substance, but only in the approximation of an electron gas. Expression (1.19a) holds for $E_k > 200$ eV. For low kinetic energies, we have [69]

$$\lambda \propto \frac{1}{\sqrt{E_k}} \tag{1.20}$$

which is indeed confirmed by the minimum on the universal curve. Penn's procedure ensures an accuracy of 5% for the representative elements and 40% for the transition ones. The ratio of the values of λ for two values of E_k is described by the expression

$$\frac{\lambda(E_1)}{\lambda(E_2)} = \frac{E_1 \, (\ln E_2 - 2.3)}{E_2 \, (\ln E_1 - 2.3)} \tag{1.21}$$

Chang [70] took λ approximately as $\lambda \approx 0.2E_k^{0.5}\alpha$, where α is the thickness of a monolayer and is assumed to be 0.25 nm.

By generalizing the data for over 350 specimens, Seah and Dench [68] used a statistical approach to the appraisal of λ and proposed various analytic expressions for its determination. When estimating λ as the number of monolayers, the following equation is used:

$$\lambda_m = K_1 E_k^{-2} + K_2(\alpha E_k^{1/2}) \tag{1.22}$$

where K_1 and K_2 are constants equal to 2170 and 0.72, respectively, for inorganic compounds, and α is the thickness of a monolayer set by the expression

$$\alpha^3 = \frac{M}{\varrho n} N \times 10^{24} \tag{1.23}$$

where M is the molecular mass, ϱ is the volumetric density, n is the number of atoms in a molecule, and N is Avogadro's constant. When E_k exceeds 150 eV, we have

$$\lambda_i = K_2(\alpha E_k^{1/2}) \tag{1.23a}$$

As expected, Eqs. (1.22)-(1.23a) are observed the most accurately for a definite class of compounds, namely, metals, oxides, halides, and organic compounds. Tokutaki et al. [71] compared the accuracy of the procedures proposed by Penn [67], Seah and Dench [68] and concluded that the latter is more acceptable. They proposed, in turn, a precised way of calculating λ for Auger electrons that takes into account the attenuation of the primary electron beam. Later Tanuma et al. [72] proposed a general formula for calculating λ and appraising its dependence on E_p (for a given material) or on the kind of material (for a given E_p)

$$\lambda_i = \frac{E}{E_p^2 \beta \ln(\gamma E)} \quad (\text{Å}) \tag{1.24}$$

$$E_p = 28.8 \left(\frac{\varrho N_v}{A}\right)^{3/2} \quad (\text{eV}) \tag{1.24a}$$

Here N_v is the total number of valence electrons per atom, ϱ is the volumetric density, g/cm^3, A is the atomic or molecular mass, β and γ are empirical coefficients determined by the equations

$$\beta = -2.52 \times 10^{-3} + \frac{1.05}{(E_p^2 + E_g^2)^{1/2}} + 8.1 \times 10^{-4} P \tag{1.25}$$

$$\gamma = 0.151 \beta^{-0.49} \tag{1.26}$$

where E_g is the width of the forbidden gap (band) for nonconducting materials. The difference between individual values of λ and those calculated by Eqs. (1.24)-(1.26) for 31 compounds (metals, oxides, carbon) averaged 12%.

It should be noted that by Monte Carlo calculations [73] the escape depth of electrons is determined not only by inelastic, but also by elastic collisions with lattice elements, especially with glancing photoemission angles. Here tunnelling of the electrons occurs owing to the anisotropy of the photoelectron diffraction in a crystal [74]. But if quantitative analysis is conducted without a reference specimen [75, 76] or if the line intensities are measured for electrons with close kinetic energies [73], the elastic scattering will not affect the accuracy of the measurements. Jablonski *et al.* [77, 78] proposed a way of calculating λ based on measuring the intensity of the peak of elastically scattered electrons in AES. Satisfactory agreement (20-30%) of the experiments and calculations taking the elastic scattering cross section into account was found for a number of elements. Attention is also attracted to the fact that when no account is taken of the elastic collisions, the value of λ may be overstated by 30%.

Hence, the developed procedures now enable one to determine λ with a greater accuracy, and this, in turn, enables one to count on an increased accuracy of quantitative analysis by XPS.

Equation (1.15) can be transformed to

$$R_{X,Y} = T_{X,Y} \frac{\sigma_{X_i}}{\sigma_{Y_i}} \tag{1.27}$$

where $T_{X,Y}$ is a multiplier taking into consideration all the factors treated above; $T_{X,Y} \to 1$ when $E_k > 1000\,\text{eV}$ and $\Delta E_{X-Y} < 400\,\text{eV}$. Hence

$$R_{X,Y} = \frac{\sigma_{X_i}}{\sigma_{Y_i}} \frac{c_X}{c_Y} \tag{1.28}$$

The accuracy obtained when using this equation is generally not below 20%. The concentrations can be evaluated with a smaller error by using the method of graduation curves, in other words by plotting the relative integral intensities for two elements against their atomic ratio in homogeneous specimens with a stoichiometric composition [9]. A linear form of these relations points to the absence of appreciable matrix effects in XPS. Serious problems are associated with the correct determination of the integral intensities of lines into which complications have been introduced by shake-up satellites, peaks of inelastic losses, etc. In such cases, the accuracy of analysis diminishes [79]. The various ways of taking the background into account in integration are treated in Chap. 2.

1.1.4   UPS: Spectra of the Adsorbed State.
Angle-Resolved Photoemission

In ultraviolet photoelectron spectroscopy, the sources are He^I and He^{II} resonance radiation, and also synchrotron radiation. The varieties of this method include angle-resolved UPS (ARUPS), UPS with polarization of the radiation, and UPS using a varying wavelength (synchrotron). The main tasks relegated to UPS are (a) studying the energy of the valence orbitals of adsorbates depending on the electron and crystallographic structure of the surface, (b) determining the localization sites of the adsorbed molecules, as well as the nature and geometry of the adsorption bond, (c) identifying the bulk band structure, and (d) identifying the occupied and vacant surface states.

The change in the energy position of a molecule's electron levels in adsorption is determined by the expression:

$$E_i = E_{i,\text{gas}} - E_{i,\text{ads}} \tag{1.29}$$

where $E_{i,\text{gas}}$ is the binding energy of the i-th level of a molecule in the gas phase, and $E_{i,\text{ads}}$ is the binding energy of the i-th level in an adsorbed molecule. Since $E_{i,\text{gas}}$ is determined relative to a vacuum level, the magnitude of the work function must be added to $E_{i,\text{ads}}$ for a specimen coated with an adsorbate. Large changes occur in E_{ER} in adsorption, therefore Eqs. (1.8)-(1.9) must be used to determine the chemical shifts. Demuth and Eastman [80] have proposed a method of separating chemical and relaxation shifts. In particular, they assume that $\Delta E = \Delta E_{\text{ER}}$ for nonbonding orbitals. Another difficulty in determining the chemical shifts in adsorbate spectra consists in the change in the work function in adsorption. Because of the difficulty of determining $\varphi = f(\theta)$, in many cases φ is taken for a clean surface or for one completely covered by the adsorbate. The change in the spectrometer work function φ_{sp} can be found by the cut-off of the secondary electrons reaching the analyzer [81]. An interesting procedure for determining the local work function of individual areas of a surface is based on measuring the position of the $5p$ level of physically adsorbed xenon [82]. It has been shown both theoretically and experimentally that the vacuum level is the reference one for xenon. Moreover, the fine structure of the ultraviolet electron spectrum of Xe is a probe for determining the adsorption centers of various structures. It is quite probable that in the joint adsorption of Xe and another molecules, this procedure can also be used for determining φ in the presence of an adsorbate.

The most powerful tool for studying the structure of a surface and the nature of an adsorption bond is angle-resolved photoelectron spectroscopy

[83]. When using polarized ultraviolet radiation, one can study how the intensity of the peaks of valence orbitals having different symmetry depends on the azimuthal (θ) and polar (Φ) angles of emission. For instance, the dependence of the position of 4σ CO on Φ shows that on the Ni(110) face the angle between the axis of CO and a normal to the surface is 26°; no inclination is observed for Pt(111) [84]. Another procedure for determining the orientation of a molecule consists in finding the resonance of peaks of definite orbitals (5σ and 4σ CO). The intensity grows sharply when the directions of the polarized radiation and the emission of electrons coincide with the axis of a molecule. For example, it has been proved that CO is at right angles to the Ni(100) surface [85]. Synchrotron radiation with a varying wavelength ($20\,\text{eV} < h\nu < 45\,\text{eV}$) was used in these experiments.

Still another possibility of studying the nature of an adsorption bond with the aid of photoelectron spectroscopy using synchrotron radiation is associated with the manifestation of what is known as the Cooper minimum of the photoionization cross section of the $4d$ and $5d$ levels at appropriate values of $h\nu$ [83]. Here the relative intensity of the adsorbate spectrum grows, which enables one to study its structure in greater detail. In the spectra of CO adsorbed on Pt at $h\nu = 150\,\text{eV}$, one can observe a peak of the 3σ orbital, which is impossible when using UPS.

Angle-resolved photoelectron spectroscopy is employed for determining the intrinsic surface states of metals or the differences in the density of the states on different faces of single crystals [86]. For example, two surface resonance levels have been found for Mo. They are displaced relative to the Fermi level by 0.5 eV (face [100]) and 4.5 eV (face [110]). The technique of inverse photoemission (IPS) developed recently [87, 88] and based on measuring the continuum of synchronous radiation is used to identify the unoccupied surface states of metals or the vacant levels of adsorbates. Unoccupied surface states of the Shockly type in the sp-hybridized gap above the Fermi level have been discovered on the surface Ag(110) and the low-index copper faces (111), (001), and (110) [89]. The combination of this technique with XPS and UPS opens up new possibilities for revealing the role of the surface states of metals in adsorption and catalysis.

1.2   Auger Electron Spectroscopy

The transition of a photoionized atom from an excited to the ground state can occur in two ways [3-5]. For instance, the vacancy in the K shell can be filled at the expense of an X-ray-induced transition $K \rightarrow L_3$

or because of a radiationless transition KLL. In the latter case, two vacancies form in the L shell (L_1L_1, L_1L_3) and an Auger electron is emitted. The energy of Auger electrons is determined by the binding energy of the electrons participating in the Auger transition, consequently the Auger spectra, like the XPS ones, provide chemical information. For a KLL transition [3, 31]:

$$E_{KL_1L_2}(Z) \approx E_K(Z) - E_{L_1}(Z) - E_{LL}(Z + 1) - \varphi_{sp} \qquad (1.30)$$

where E_K (or E_L)(Z) is the energy of level K (L) of the given element (Z), and $E_{LL}(Z + 1)$ is the energy of level L of the element $Z + 1$ (twofold ionization of an atom is taken into account). The energy of an Auger electron does not depend on $h\nu$, which makes it possible to separate the lines of the Auger and photoelectrons by using X rays of various energies (e.g. Mg K_α and Al K_α). In the independent technique of Auger electron spectroscopy, the spectra are excited with the aid of primary electrons having an energy of 1-10 keV. Each of the techniques, i.e. XAES and AES, has its merits. In the first case, it is simpler to obtain spectra with a high resolution, while in the second with an optimum ratio E_p/E_{Aug} the probability of the emission of Auger electrons is higher. This makes it possible to obtain low-energy electron spectra (up to 100 eV) and thus increase the surface sensitivity. Moreover, the electron beam can be focussed to a very small size (50-100 nm) and scanned over the surface. This method has been named scanning Auger microprobe (SAM).

We can single out three fields of AES application [6-8, 32, 90]: (a) the qualitative and semiquantitative analysis of the surface composition, (b) studying of the chemical state of elements and their electron structure, and (c) a more accurate quantitative analysis of a surface, depth profiling, and determination of the structure of a surface. The last two fields have become popular in the last 5-10 years in connection with the improvement in the experimental procedures and the quantitative calculations of the lineshape and intensity of the Auger spectra. The shift in the kinetic energy E_k of the Auger electrons participating in a KL_1L_2 transition can be written as follows:

$$\Delta E_{KL_1L_2} = E_K - E_{L_1} - E_{L_2} - (E_K + \Delta E_K - E_{L_1} - \Delta E_{L_1} - E_{L_2}$$
$$- \Delta E_{L_2}) = \Delta E_K + \Delta E_{L_1} + \Delta E_{L_2} \qquad (1.31)$$

where ΔE_K, ΔE_{L_1}, and ΔE_{L_2} characterize the changes in the binding energies of the core levels determined in a photoelectron process. They already include the changes in E_{ER} also called the dynamic relaxation [35-38]. For a more accurate approximation, the quantity $\Delta E(L_2)$ must be corrected with a view to the energy of interaction of two holes in the final multiplet state θ, i.e. to $F(L_1, L_2, \theta)$. It is also necessary to take into account the total relaxation energy $R_{tot} = R_s + \Delta E_{ER}$ in which the static relaxation R_s is added to the dynamic component. The potential taking both types of interaction into consideration, i.e. U_{ef}, is also called the Lang-Williams Auger parameter [35], and unlike the Wagner Auger parameter [39] $\alpha = E_b(K) + E_k(KL_1L_2, \theta)$, it is determined as follows:

$$U_{ef} = \xi(L_1L_2) = E_b(K) - E_b(L_1) - E_b(L_2) - E_k(KL_1, L_2, \theta)$$
$$= F(L_1, L_2, \theta) - R_{tot}(L_1, L_2) \qquad (1.32)$$

or

$$\Delta E_{KL_1L_2} = -\Delta E_K + \Delta E_{L_1} + \Delta E_{L_2} - U_{ef} \qquad (1.33)$$

The relaxation mechanism and the procedure for appraising the relaxation term from photoelectron and Auger electron spectra will be considered in greater detail when describing the spectra of metallic clusters (Chap. 3).

In a very simple model, the difference in the contribution of the chemical and relaxation terms of photoelectrons to ΔE_b and of Auger electrons to ΔE_k is expressed as follows:

$$\Delta E_b = \Delta E - \Delta E_{ER} \qquad (1.34)$$
$$\Delta E_k = -\Delta E + 3\Delta E_{ER} \qquad (1.35)$$

The changes in the energy of Auger electrons in the oxidation of metals are generally greater than the shifts in the photoelectron spectra. For example, the shift Zn LMM in the transition Zn $\rightarrow$ ZnO is 4.7 eV, whereas the peak for Zn $2p_{3/2}$ does not practically change its position [90]. The data obtained for thin-layer specimens suggest the influence of the screening potential on ΔE_k. When Zn is oxidized in thin films, the shift drops to 2.6 eV [90.] The same effect is observed when Ar condenses on Ag; the shift of E_k for Ar $L_{23}MM$ changes from 7.1 ± 0.5 eV in comparison with the gaseous state for submonolayer coatings to 2.5 ± 0.5 eV in the formation of a thick film [90]. It is assumed that in the first case the potential of hole screening by electrons of the metal is higher than in solid Ar, which is an insulator. It should be noted that the opposite effects are observed when thin metal films condense on an insulator.

To obtain chemical information from Auger spectra, one must have a theoretical description of the complete structure, namely, the intensities of individual transitions and the width of the lines. Such data have been obtained only for atoms and some simple compounds [91]]. The calculation of N (KVV) for NH_4Cl by the X_α-SW method agrees well with an experimentally obtained spectrum of polycrystalline NH_4Cl [92]. A number of interesting conclusions on the nature of the bond in a molecule can be made from the structure of the spectrum, e.g. on the degree of delocalization and covalence of the bond N—H ... Cl. Information on the interaction of various elements is carried by the interatomic Auger transitions, for instance with the participation of Me and O. The width of the Auger peaks is another parameter used to identify the chemical state. Much narrower lines are observed for metals than for insulators or semiconductors. The spectra of Cu L_3VV for Cu, Cu_2O, and CuO [93] are a typical example (see Fig. 2b). Oxidation of Cu to Cu_2O is attended not only by a shift of the main peak, but also by broadening of the lines.

A more detailed electron structure of compounds can be established in the quantitative analysis of the shape of Auger lines. It includes mathematical processing of the spectra (removal of the overall background of scattered electrons, the background of inelastic scattering, and instrument broadening) and theoretical calculations. If Auger spectra reflect core level-valence band transitions (of the CVV or CCV type), the local density of the occupied states of the metallic and nonmetallic components can be determined after their processing. Such information substantially supplements the data on the integral density of states in the valence band (XPS). Within the scope of the Cini-Sawatsky theory [94-97], the shape of an Auger electron spectrum depends appreciably on the relation between the values of the intra-atomic interaction of two holes (U_{ef}) and the interatomic interaction when a hole jumps over to another atom (W). An Auger spectrum of the band type is observed when $U_{ef} \ll W$, whereas the spectrum consists of narrow components of the atomic type with a typical multiplet structure when $U_{ef} \gg W$ (Fig. 4).

An intermediate case is often observed in practice. When these approximations are taken into account and when using the relevant processing procedure, the shape of an Auger electron spectrum can be used to establish the predominating type of bond and determine the partial magnitudes of the charges on the atoms in a compound. The carbides and nitrides of the Group IV-VI metals have been used as an example to show how the shape of the Auger electron spectra depends on the stoichiometry, phase composition, and the presence of vacancies in the lattice of these compounds [99-101].

The intensity of the Auger electron signal (the ABC transition) is described by the equation [32]:

$$dI = FN_i\sigma_iB(E_p)P_{ABC}\lambda_i \, d\Omega T_i \tag{1.36}$$

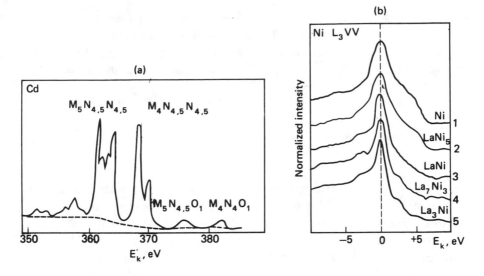

Fig. 4. Auger electron spectra of quasiatomic (a) [7] and band (b) [98] types:
a — metallic Cd MNN spectrum; b — L_3VV spectra of nickel and its alloys with La

where F is the flux of primary electrons, N_i is the number of atoms being analyzed, σ_i is the ionization cross section of level A, $B(E_p)$ is the backscattering factor, P is the probability of the Auger transition ABC, Ω is a geometric factor, T_i is the transmission coefficient of the spectrometer, and λ_i is the escape depth of the electrons. The first two and last three quantities in Eq. (1.36) are the same as in Eq. (1.13). The probability P_{ABC} at the energies employed in AES differs only slightly from unity. The values of $\sigma_i(E_p, E_k)$ are determined chiefly by calculations [102]; the values of σ_i usually grow beginning from the threshold energy E_p. But it is very difficult to use Eq. (1.36) for quantitative analysis. It is more convenient and reliable to analyze the ratios of the intensities that are independent of the primary current. One of the critical factors determining the accuracy of quantitative analysis in AES is the secondary electron backscattering factor, because the secondary electrons excite additional Auger electrons. $B(E_p)$ also depends on the matrix. Sometimes, e.g. when analyzing an adsorbate-adsorbent system, the quantity B can be excluded by calibrations [103]:

$$\frac{n_\alpha}{n_\alpha^0} = \frac{I_{\alpha,ABC}}{I_{\alpha,ABC}^0} \frac{1 + r_\alpha^0(E_p, E_A)}{1 + r_\alpha^\beta(E_p, E_A)} \tag{1.37}$$

where $I_{\alpha,\text{ABC}}$ and $I^0_{\alpha,\text{ABC}}$ are the intensities of the Auger electron peaks for an element in the matrix being studied and a standard specimen (α is the coverage by the adsorbate within the limits of a monolayer), r^0_α are factors taking into account σ_A and $B(E_p)$ for a standard specimen, and r^β_α are the same for the matrix β.

When analyzing binary alloys (X, Y), the ratio of the intensities is [104]:

$$\frac{I(X)}{I(Y)} = \frac{c_X}{c_Y} \frac{I_X}{I_Y} S_{X,Y} \tag{1.38}$$

Here I_X and I_Y are the intensities for pure metals, and $S_{X,Y}$ are empirical factors of elemental sensitivity. They are known for a number of elements [104, 105], but in the majority of cases such an analysis without additional calibrations leads to an error of 50-100%. To take the matrix effect into consideration, complete information on the coefficients in Eq. (1.36) is needed. A method has been proposed for the semi-empirical determination of the elemental sensitivity based on calculations of σ_i and λ_i [106]. The accuracy of quantitative analysis in AES is improved substantially by Monte Carlo calculations of the backscattering factors [107, 108]. If, moreover, it is possible to appraise B experimentally, e.g. by analyzing homogeneous specimens with a known stoichiometry, this can sharply increase the accuracy of analyzing specimens of the same type. Most of such examples meanwhile relate to the analysis of quite simple binary systems such as bimetal alloys.

By analyzing them *ab initio* [107, 108], we have:

$$\frac{I(X)}{I(Y)} = \frac{c_X}{c_Y} M_{X,Y}(c) S_{X,Y} \tag{1.39}$$

where $M_{X,Y}(c)$ is a correction for the matrix effect:

$$M_{X,Y}(c) = \frac{B(X)\lambda(X)}{B_X\lambda_X(\varrho/A)_X} \bigg/ \frac{B(Y)\lambda(Y)}{B_Y\lambda_Y(\varrho/A)_Y}$$

$$= B_M(X,\ Y)\lambda_M(X,\ Y)\frac{(\varrho/A)_Y}{(\varrho/A)_X} \tag{1.40}$$

Here B is the backscattering factor, B_M is a correction for the backscattering factor in the matrix, i.e. for the same elements in the alloy, λ is the escape depth of the Auger electrons for the pure elements, λ_M is a correction for λ in the matrix, ϱ is the density, A is the atomic mass, X and Y are indices (subscripts) signifying the pure metals, and (X, Y) are the same for the alloy. McHung and Sheffield [108] have thoroughly analyzed the factors of Eq. (1.39) for homogeneous Pd-Cu alloys. In particular, the relations $B = f(E_p)$ obtained by Monte Carlo calculations and experimentally were compared. As a result, the accuracy of analyzing the surface composition of such specimens was improved up to an error of only 5%.

Hence, an analysis of the data contained in the latest publications points to an appreciable improvement in the accuracy of quantitative analysis by AES, although this analysis remains more involved than when using XPS.

1.3　　Ion Spectroscopies

Secondary ion mass spectrometry (SIMS) was developed in the middle of the 1960's [6, 8, 109, 110]. A substantial contribution to the development of SIMS was made by the Soviet scientist Ya. Fogel. He was the first to give attention to the effectiveness of the method of secondary ion emission for studying processes on a surface such as adsorption, catalysis, and corrosion (see, for example, [111, 112]). The widespread use of SIMS for these purposes began in the middle of the 1970's, which is explained by a number of factors, namely, (a) improvement of the apparatus, (b) the development of more dependable theoretical models for describing the interaction of ions with a substance, and (c) the combination of SIMS with other techniques including

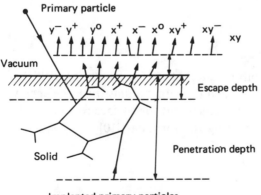

Fig. 5. Interaction of ions with a solid and the emission of secondary neutral, excited, and ionized (+ and −) particles [113]

XPS, AES, and TDS. The latter factor is evidently the most significant one because the complicated processes attending measurements in SIMS make it difficult to interpret the results. As we have already indicated, SIMS is based on ion bombardment resulting in the sputtering of the target in the form of neutral and charged particles (Fig. 5). The latter are analyzed by mass spectrometry and provide information on the qualitative and quantitative composition of a surface. SIMS in a general case is a destructive technique, and therefore we must give greater attention to the processes on a surface attending ion bombardment.

1.3.1   Ion Sputtering

In addition to SIMS, this procedure is used for cleaning a surface, in depth profiling, and in ISS. In the theory of multiple collisions [114], the sputtering yield (coefficient), i.e. the number of atoms and ions ejected per incident ion is:

$$Y(E) = \lambda F(x, E) \tag{1.41}$$

where λ is a matrix constant inversely proportional to the surface bond energy U_0, i.e. $\lambda = 0.042/NU_0$, N is the atom density (Å^{-3}); U_0 can readily be approximated by the sublimation energy, and $F(x, E)$ is an energy function (x is the distance to the surface).

The sputtering yield is related to the surface bond energy:

$$Y(E) = \frac{\beta}{U_0} \tag{1.42}$$

where β is a quite smooth function of Z. The sputtering yield Y also depends on the mass of an ion (m_1) and the target atoms (m_2), the energy of the ions, the electron structure of the target, and the surface roughness. The sputtering rate s (Å/min) is evaluated by the formula [110]

$$s = \frac{0.06 Y I_p^+ A}{\varrho} \tag{1.43}$$

where I_p^+ is the primary ion flux density, $\mu\text{A/cm}^2$, A is the atomic or molecular mass, g/mol, and ϱ is the density of the target material, g/cm^3. The sputtering yields have been found for many one-component targets at various values of

Table 3. Sputtering Yields (Y) of Selected Materials [115]

Element[a]	Y	Specimen[b]	Y_1 (oxide)	Y_2(metal)
Be	1.1	Al_2O_3	1.5	3.2
Al	1.94	SiO_2	3.6	2.1
Si	1.0	MoO_3	9.6	2.8
Ti	1.13	TiO_2	1.6	2.1
Fe	1.34	V_2O_5	12.7	2.3
Ni	1.86	WO_3	9.2	2.6
Cu	3.64	Cu	—	8.0
Mo	1.14	Ag	—	15.0
Pd	3.06	Au	—	15.0
Ag	3.8 ⎫ 4.7 ⎭			
Cd	11.2			
Ta	0.91			
W	1.10			
Pt	2.0			
Au	3.08 ⎫ 4.02 ⎭			

[a] Ar^+, 1 keV.
[b] Kr^+, 10 keV.

the instrument parameters [8, 109, 113, 115] (Table 3). The rate of sputtering a specimen (s_2) can be appraised if we know the rate of sputtering of a reference specimen (s_1) and the ratio of the yields from the expression $s_2/s_1 = Y_2/Y_1$. The yield is related to the masses of the target and an ion as follows:

$$Y \propto \frac{m_1 m_2}{m_1 + m_2} \tag{1.44}$$

where m_1 is the mass of an ion, and m_2 is that of a target atom. Since the yield grows with increasing m_2, it is good practice to use gases such as krypton or xenon for sputtering especially heavy atoms.

In the bombardment of two- and multicomponent targets, a problem comes to the forefront that is associated with the nonuniform (preferential) sputtering of one of the components. Therefore, in the initial sputtering stage, AES and XPS will provide information on the composition of the altered layer. Next the processes of enrichment of the surface in the component sputtered with difficulty and of diffusion of the second metal will compensate each other. A steady state will set in which an equilibrium composition corresponds to. The phenomenon of preferential sputtering has been studied in binary alloys by AES, XPS, SIMS, and ISS [8, 9, 116-118]. The following relation is observed between the bulk concentrations ($c_{b,1}$, $c_{b,2}$) of a binary alloy and the surface concentrations ($\overline{c}_{s,1}$, $\overline{c}_{s,2}$) in a state of equilibrium (provided that only the surface layer is sputtered):

$$\frac{\overline{c}_{s,1}}{\overline{c}_{s,2}} = \frac{c_{b,1}}{c_{b,2}} \frac{Y_2}{Y_1} \tag{1.45}$$

The method of graduation curves [117] has been proposed for the depth profiling of sputtered alloys by XPS. The ratio of the photoelectron intensities is

$$\frac{I_1}{I_2} \propto \frac{\sigma_1 \lambda_1}{\sigma_2 \lambda_2} \frac{c_{b,1}}{c_{b,2}} \frac{Y_1}{Y_2} \tag{1.46}$$

If Y_2/Y_1 is constant, the ratio of the intensities is determined by the slope of the graduation curve. In the general case, the values of Y differ for alloys and pure metals [9, 119]. To find the true profile of concentrations, it is important to assess the thickness of the altered layer, which in various models [9] ranges from 10 to 50 Å.

The main processes attending ion etching (see Fig. 5) have been treated in detail in [118-121]. They include (a) preferential sputtering, (b) cascade collisions due to the penetration of ions into the bulk of a sample, (c) the thermodiffusion of defects and their annihilation, (d) radiation-induced segregation, and (e) Gibbs adsorption. Ion bombardment causes chemical changes in the surface layers and, primarily, reduces oxides and other compounds to a metallic state. The probability of such reduction grows when the energy of formation of these compounds G^* diminishes [122]. Hence, ion etching or depth profiling enables one to procure information on the distribution of elements over the depth of a layer, but here a number of factors must be taken into consideration. The effect of preferential sputtering can be lowered by increasing E_p and at $\theta \approx 90°$. But this is attended by an increase in destruction and in the effect of light atom impinging. We can note, nevertheless, that the sputtering yields for many metals are within one order of magnitude (see Table 3). This makes it possible to obtain semiquantitative information on the distribution of the elements in the upper layers even for multicomponent samples with an irregular surface. More accurate analysis is possible when using several techniques, namely, AES, XPS, and ISS.

1.3.2 Secondary Ion Mass Spectrometry

Since only secondary charged particles are registered in SIMS[*], the positive or negative ion yield, i.e. the total number of positive or negative secondary ions sputtered from a specimen per incident primary particle, is one of the most important parameters. This quantity is generally much smaller than unity [109, 110, 113] because, for example, a positive ion is rapidly captured by an electron near the surface. The positive ion yield for nickel, for instance, is 3×10^{-4}. Let us briefly consider the physical fundamentals of SIMS. A number of processes occurring when primary ions collide with a target have been described above. The ejection of secondary particles occurs with a greater probability not as a result of a direct impact, but because of a cascade of collisions. Various authors estimate the escape depth of the sputtered particles to range from 0.6 to 2.0 nm [110-123]. The emitted particles can be in the most diverse states, namely, singly and doubly excited, ionized, neutral, and exciton. This is due to the plurality of processes taking place in multifold collisions of the particles, e.g. resonance ionization, and exchange of charges, electron capture, and auto-ionization with the emission of Auger electrons.

A variety of theories describing the emission of secondary ions have been advanced [8, 113, 115]. Among them are:

(a) the **kinetic model**, assuming that the target atoms are emitted in the form of neutral particles, and that de-excitation with their transformation into ions occurs outside the surface owing to the Auger effect;

(b) the **auto-ionization model**, considering the excitation of an electron in an inner shell of an atom to be the initial stage, followed by the formation of a secondary ion and the ejection of an Auger electron;

(c) the **chemical model of ionization**, considering all substances to be ionic compounds in which the breaking of a bond causes the formation of positive and negative ions;

---

[*] Fast atom bombardment mass spectrometry (FABMS) and the mass spectrometry of secondary particles completely ionized by an additional source are related techniques.

(d) the **thermodynamic model**, presuming that a plasma forms in ion bombardment, and it is in local thermal equilibrium. The ion yield is constant during the entire process (neutralization is not taken into consideration);

(e) the **cascade model of collisions**, considering that sputtering and ionization are a result of cascades of collisions induced by the implantation of primary particles. It enables one to find the relation between the atomic and molecular ions by taking the dissociation energy into account.

It should be noted that the thermodynamic model [113, 124], notwithstanding a number of vague physical points, has become the most popular one for practical analysis in SIMS. The formation of ions obeys the law of mass conservation, i.e. $X^0 \rightleftarrows X^+ + e^-$. The equilibrium constant, which is found by an equation of statistical mechanics, is:

$$K_{n^+} = \frac{n_{X^+} n_{e^-}}{n_{X^0}} \tag{1.47}$$

In this model, the dependence of the ion yield on the ionization energy has an exponential form, and the influence of the matrix is taken into consideration. Werner [113] relates the formation of molecular ions of a definite structure to the valence of the metal oxides. This allows one to obtain chemical information by SIMS. It has been pointed out [8, 9, 110, 113] that the theoretical models of the secondary ion emission are not related to one another, but only describe some laws of the change in the ion emission on the basis of definite (sometimes close) assumptions. This is why an empirical approach is meanwhile the main one when analyzing and interpreting the data of SIMS.

When determining the absolute or relative concentration of an element, one must have information on a number of physical and instrumental parameters. The secondary ion current (I_s) is related to the concentration (c) by the following equation [113]:

$$I_s = I_p Y \beta^+ c f = I_p Y^+ c f \tag{1.48}$$

where I_p is the primary ion current, Y is the sputtering yield equal to the total number of sputtered particles related to the number of primary ions that experienced a collision, β^+ is the degree of ionization equal to the ratio of the number of particles sputtered in the form of positive ions to their total number. The product $Y\beta^+ = Y^+$ is the positive ion yield, and f is the transmission coefficient of the mass analyzer. The thickness of the layer sputtered in unit time is [113]:

$$\dot{z} = \frac{3.6 \times 10^{-4} M}{\varrho i_p Y} \quad \mu\text{m/h} \tag{1.49}$$

where M is the molecular mass, ϱ is the target density, g/cm^3, and i_p is the primary ion current density, mA/cm^2.

Depending on the value of $\dot{z}$, two modes of SIMS are distinguished — static and dynamic. The former was proposed for studying surfaces by Benninghoven [29]. In the static mode, the current density is very low (up to 1 nA/cm^2), and the destruction of a specimen caused by sputtering is minimal. Consequently, the main information is received from the upper layers of a substance (the sputtering rate in this mode is about 1 Å/h. The dynamic mode, conversely, is intended for analyzing the volume of a specimen. Here the current density is $i_p \approx 1$ mA/cm^2, and the sputtering rate reaches 10-100 nm/min. Intermediate variants of the technique generally have to be used in practice.

If a standard specimen is available, the concentration of a metal can be appraised:

$$\frac{I_m}{I_{st}} = \frac{Y_m^+ c_m f_m}{Y_{st}^+ c_{st} f_{st}} \tag{1.50}$$

If $f_m = f_{st}$, we have

$$\frac{I_m}{I_{st}} = I_{m,rel} = \frac{Y_m^+}{Y_{st}^+} \frac{c_m}{c_{st}} = Y_{rel}^+ c_{m,rel} \tag{1.51}$$

Data on $Y^{\pm}$ and $Y_{rel}^{\pm}$ for pure metals are given in diagrams by various authors, but these quantities may differ greatly (by an order of magnitude) in a matrix. Each parameter in Eq. (1.48) depends on many instrumental factors and on the properties of the matrix. The value of $Y^{\pm}$ depends (a) on the primary current density and energy, and it generally increases with a growth in the energy after a threshold energy of 30-80 eV, (b) on the target temperature; at comparatively low temperatures, Y^+ drops with elevation of the temperature, while it does not change at high temperatures, and (c) on the sputtering and emission angle — the angular relations are of a periodic nature due to channeling of the primary ions (a minimum) and disordering of a crystal (maximum). The orientation effects are used in scanning SIMS when obtaining an image in secondary ions. The value of $Y^{\pm}$ grows in polycrystals with an increase in θ to about 90°.

The energy distribution of the secondary ions depends on the nature of the metal. Silver and cobalt have a narrow distribution, while vanadium and manganese have the highest average energies. The average and maximum energies of ions diminish as the $3d$ shell is filled [8]. The resolution is improved in practice by discriminating the ions with a narrow interval of low energies (several eV).

The dependences of Y and Y^+ on z are often opposite in shape. The explanation is that Y is inversely proportional to the surface bond energy, whereas $Y^+ \propto E_b^{5/2}$ [125]. Such a relation explains the growth in the sputtering yield in the surface oxidation of metals, which can be performed by introducing a small amount of O_2 into the Ar stream or by bombarding the surface with O_2^+ ions. The formation on a surface of a stronger ionic bond of the type $Me^+{-}O^-$ instead of a metallic bond increases the probability of chemical ionization of the surface [8, 113, 118]. The yield of negative ions is increased by using an element with a low ionization potential, e.g. cesium. This leads to the formation of a similar pair: $Me^-{-}Cs^+$. The observed phenomenon has both positive and negative consequences: (a) the yields of secondary ions during oxidation not only grow, but also level out, thus reducing the matrix effect, and (b) when analyzing the concentration profile in a nonuniformly oxidized layer, the change in $Y^{\pm}$ in the course of depth profiling must be taken into consideration.

Let us now consider some features of the emission of atomic and molecular (cluster) ions of interest for studying the local structure of a surface. It follows from models of secondary ion emission that the escape depth of polyatomic or cluster ions is much smaller than the sputtering zone and is usually restricted to two or three atomic layers. Moreover, the ratio Me_n^+/Me_{n-1}^+ depends on the discrimination energy of the ions. The peaks of the energy distribution of molecular ions are shifted in the direction of low energies in comparison with atomic ions [113]. As a whole, the ratio Me_n^+/Me_{n-1}^+ is less than unity and drops with a growth in n [113]. Of great interest are mixed cluster ions whose analysis can illustrate the location of atoms on a surface. Numerous investigations of various systems [8, 29, 110, 113] have revealed that the cluster ions analyzed in spectra are indeed surface fragments and do not form in the gas phase. But this requires additional confirmation in each specific case, namely, a comparison of different specimens measured under similar conditions, and variation of E_p, E_s, and θ.

Another matter of importance is the elemental sensitivity of SIMS, including the threshold concentrations. As a whole, they are higher than in XPS or AES, but depend greatly on the instrumental factors, including the transmission coefficient of the mass analyzer. At $f \approx 10^{-3}$, the minimal detectable concentrations for elements such as Cu, Ni, and Al range from 1.5×10^{-3} to 1×10^{-4} at. %; for their oxides, accordingly, the detection limit grows by another one or two orders of magnitude. But when using the static or quasistatic mode of SIMS, which is the most suitable for studying a surface, the sensitivity of the technique diminishes. Approximate estimations of the threshold concentrations can be made by using nomograms [113]; they also enable one to appraise the relative elemental sensitivity, although involved calibrations are needed for conducting quantitative analysis.

In summarizing our brief exposition of the fundamentals of SIMS, we can conclude that to date this technique remains a highly sensitive qualitative or semiquantitative analytical one. The following features of SIMS are significant for investigating the surface of a catalyst:

(a) a high sensitivity and the possibility of detecting different isotopes. The latter circumstance is of interest for determining the localization sites of active centers and adsorbate molecules;

(b) the high surface sensitivity reached in static SIMS;

(c) the possibility of analyzing the elemental composition and local environment of atoms on a surface;

(d) depth profiling.

The following shortcomings and difficulties appear when SIMS is used:

(a) the destruction and ion-stimulated effects caused by bombardment (a change in the composition and chemical state);

(b) the difficulty of quantitative analysis due to the matrix effect;

(c) the charging of insulators and the associated drop in resolution, background distortion, etc. The latter effect is eliminated partly when using an emitter of low-energy electrons and manifests itself to a smaller extent when a surface is bombarded by fast atoms (FABMS), which do not bring along a charge to the surface, instead of by ions. This technique has attracted the attention of investigators in the field of catalysis owing to the smaller destruction produced by the fast atoms [29, 126].

New possibilities of SIMS are also associated with an increase in its sensitivity due to the ionization in a plasma of sputtered neutral particles (SNIMS), and also with the obtaining of an image of the secondary ions reflecting the surface topography. The minimal beam diameter is 1-0.1 μm in modern instruments, which makes it possible to use SIMS as a microprobe technique for obtaining a two- and three-dimensional pattern. These possibilities of SIMS can be employed only for a limited range of model catalysts. The most typical trend is to use SIMS together with other surface analysis techniques, which makes the interpretation of the results less ambiguous. It is sufficient to note that a comparison of the concentration profile obtained by analysis of the removed layer (SIMS) and of the surface (AES, XPS) allows one to single out the effect of preferential sputtering.

1.3.3   Ion Scattering Spectroscopy

Investigators have been studying the fundamental laws of low-energy ion scattering on solid surfaces for more than 25 years [127], and a major contribution was made by Soviet scientists [128, 129]. The possibility of using this technique for analyzing a surface was first shown by Smith [130] who studied beams of noble gases with energies of 0.5-3 keV sputtered on Ni and Mo. The potential of ISS for studying surfaces has been treated in greater detail in quite a number of monographs and reviews [6, 10, 24, 30, 127, 131]. A special review has been published recently [132] devoted to the studying of catalysts by ion scattering spectroscopy.

The model of low-energy (slow) ion scattering is based on the postulate that the process can be considered as a simple elastic binary collision of an incident ion with a target atom (the billiard ball model). It has been proved experimentally that such an assumption is quite justified. Proceeding from the binary collision model, the energy losses in scattering are determined by the law of kinetic energy and momentum conservation:

$$\frac{E_1}{E_0} = \frac{m_1^2}{(m_1 + m_2)^2} \left\{ \cos \theta \pm \left(\frac{m_2^2}{m_1^2} - \sin^2 \theta \right)^{1/2} \right\}^2 \tag{1.52}$$

where E_0 is the energy of an incident ion, E_1 is the energy of a reflected ion, m_1 is the mass of an ion, m_2 is the mass of a target atom, and θ is the scattering angle. If $\theta = 90°$, we have

$$\frac{E_2}{E_0} = \frac{m_2 - m_1}{m_2 + m_1} \tag{1.53}$$

i.e. the mass of an element is easily identified from the scale of energies.

The quantitative analysis of the composition is a more involved matter [127, 133]. The intensity of a scattered beam is proportional to the differential scattering cross section in the given solid angle and to the quantity $(1 - P_n)$, where P_n is the probability of neutralization $(1 - P_n \approx 10^{-3})$, and also to the coefficient of screening of one atom by another one (S) [10]:

$$I_{sc} \propto \sigma_{sc}(1 - P_n)S \tag{1.54}$$

The differential scattering cross section σ_{sc} is a function of the potential of interaction between an ion and an atom, which for low energies has not been determined exactly and is selected empirically [127, 133]. The ambiguity in appraising σ_{sc} is introduced by neutralization of the ions. The effectiveness of the ejection of an ion without neutralization $(1 - P_n)$ depends on E_0, the target material, and the nature of the adsorbed atoms. In general, the elemental sensitivity grows with an increase in z and m_2.

Factors such as the matrix effect, preferential sputtering, surface roughness, and altering of the background also introduce complications into the quantitative analysis of polyatomic targets and especially of catalysts [134, 135]. This is why the elemental sensitivities have been determined strictly only for simple systems in particular for binary alloys [136]. In other cases, graphical plots of the scattering yields against the atomic number of an element can be used for semiquantitative analysis [137], or additional calibrations are required.

Although the neutralization and screening (shadowing) effects make quantitative determinations more involved, it is exactly because of their features that ISS has a high surface sensitivity. Indeed, although the ions penetrate into the near-surface layers several nanometers deep, their neutralization in the bulk takes place more effectively than on the surface, and the escape depth of singly scattered ions is limited to one or two monolayers [127, 133]. Calculations show that the scattering of low-energy ions is attended by the appearance of a shadow cone whose width is close to the interatomic distance (1.5 Å for He$^+$). That is, with a definite direction of scattering, the ions "fail to see" the atoms of the second layer. The correctness of this has been proved experimentally when measuring the intensity of scattering from opposite faces of crystals [138], and also in the formation of ordered surface structures in adsorption or segregation [139, 140]. The second case is illustrated in Fig. 6 [140]. In the ordered segregation of sulfur forming the structure c(2 × 2) on Ni(001), the intensity of the peaks of He$^+$ scattered from Ni/S changes de-

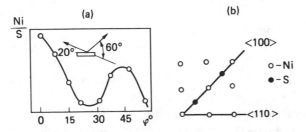

Fig. 6. Intensity of scattering He$^+$ ions from a surface Ni(001) coated with sulfur against the azimuthal scattering angle (a), and (b) structure of a surface monolayer of Ni(001) + S[c(2 × 2)] [140]

pending on the azimuthal scattering angle. In accordance with the arrangement of the sulfur atoms, screening occurs in the direction ⟨100⟩ — the signal from Ni/S is halved.

The analysis of the multifold scattering of ions and of how the intensity of single scattering depends on E_0 is significant for procuring information about a surface. Multiple scattering leads to the appearance of steps on the peaks at the high-energy side. The absence of single scattering spectra at the initial instant is due to the fact that first only multiple scattering occurs, and the typical ISS spectra appear only after the formation of a sufficient number of defects. Hence, by studying multiple scattering, one can obtain information on surface defects, e.g. on stepped surfaces of single crystals [127].

Another interesting phenomenon observed in the scattering of low-energy ions is the periodic change in the intensity (oscillation) depending on the energy E_0 of an incident ion. These changes are the most pronounced for $4d$ elements when they collide with He^+ ions and are explained by the quasiresonance exchange of the charge [133], namely, the $4d$ levels of these elements are close in their energy to the $1s$ level of He^+. Although the oscillations hinder quantitative analysis, they are "fingerprints" on definite surface atoms, which is important when analyzing elements with close masses. Moreover, the oscillations also depend on the chemical state and close environment of elements.

From the viewpoint of studying catalysts, the merits of ISS are as follows:

(a) an extremely high surface sensitivity;

(b) the possibility of studying various catalytic systems including metals, semiconductors, and insulators. In the latter case, charging is compensated by the emission of low-energy electrons;

(c) semiquantitative analysis of the composition of a surface;

(d) structural information, which is especially valuable for single crystals and for studying adsorption.

The shortcomings of ISS include:

(a) a low resolution; the magnitude of the resolution in spectrometers with a hemispherical analyzer is determined exclusively by the ratio $S = m_2/m_1$ and the collection angle θ. When the ratio S decreases, the resolution grows, but if ions heavier than He^+ are used at $\theta \geqslant 90°$, there is no possibility of detecting elements with light masses;

(b) ion scattering is attended by sputtering of the target and, consequently, by all the effects observed in bombardment by ions. But unlike Ar^+, the destructive effect and sputtering rate of He^+ are appreciably (by an order of magnitude) lower. As in SIMS, the authenticity of the results of ISS can be improved by combining it with XPS, AES, or SIMS.

1.4   Other Surface Science Techniques

High Resolution Low-Energy Electron Loss Spectroscopy (HREELS). This technique is based on the following principle. The surface of a metal coated by an adsorbate is irradiated by a highly monochromatized source of electrons with an energy of $E = 2\text{-}5\,eV$. The electrons experience inelastic collisions, and the energy of the electrons scattered by the adsorbate layer is measured by an analyzer, as in photoelectron spectroscopy [36]. Since the mechanisms of energy losses by low-energy electrons and of the appearance of surface vibrations are interrelated, while the selection rules for HREELS and IR (infrared spectroscopy) are the same, the direct and quantitative comparison of the results of the two techniques is possible. HREELS is inferior to IR in resolution (the resolution of modern instruments is 5-10 meV or 40-80 cm^{-1} and is achieved only for single crystals). But this technique has a number of merits, namely, (1) its sensitivity exceeds 0.1% of the adsorbate monolayer on a metal; (2) rapid scanning is possible within a broad range of energies from 5 meV (400 cm^{-1}) to 1 eV (8000 cm^{-1}); and (3) under the same experimental conditions, the data of HREELS can be compared with the results obtained by other techniques such as ARUPS, IPS (inverse photoemission spectroscopy), and XPS.

The results obtained for the adsoption of unsaturated hydrocarbons on metals [1, 141] are striking examples of the effectiveness of HREELS in determining the geometry of an adsorption bond and the structure of adsorption complexes. When ethylene or acetylene is adsorbed, a new structure of the ethylidene type CH_3—$C\lessgtr$ is detected. It plays a very important role as an intermediate in hydrogenation-dehydrogenation reactions on metals. The very low frequencies of CO vibrations on pure metals registered in the loss spectra of low-energy electrons were a direct proof of the formation of structures joined to the metal by pi (C— and O—) bonds. These structures are intermediates in the dissociation of CO [141].

Extended X-ray Absorption Fine-Structure Spectroscopy (EXAFS) is widely used in studying catalytic processes [142], and some results of these investigations are treated in Chaps. 3-5. There is a surface-sensitive variant of this technique — SEXAFS, in which the Auger electrons formed when filling a photoelectron vacancy are registered. The yield of the Auger electrons is assumed to be approximately proportional to the coefficient of surface X-ray absorption. SEXAFS can be employed for determining the geometry of an adsorbate [143]. Another variant — NEXAFS[*] — is also widely used to find

---

[*] All these techniques are based on employing a synchrotron source.

the geometry of an adsorption bond. In this technique, the oscillations that appear near the absorption edge are registered. They are due to transitions from core levels to unoccupied ones above and below the continuum [144]. For illustrative purposes, we can give data on the formation of formate (HCO_2) and methoxy (CH_3O) groups on Cu(100). In all cases, sigma resonance from a C—O bond is observed, but its position depends on the nature of the adsorption complex. It follows from theoretical calculations that the energy of sharp resonance is proportional to the square of the interatomic distance, hence the data of NEXAFS enable one to determine the interatomic distance in C—O, which is 1.25 Å in the formate group. Recently, correlations have been found between the energy of sharp resonance and the interatomic distances for a large number of molecules, consequently NEXAFS can be used effectively to procure information not only on the orientation of a molecule on a surface, but also on the bond length. Information on the structure can also be obtained in principle without a synchrotron source by employing X-ray photoelectron diffraction (XPD) [145]. This technique is used to analyze the intensity arising when the wave functions of the directly emitted electrons and of those elastically scattered on the surrounding atoms are superposed. It enables us, for instance, to determine the angle of inclination of a CO molecule on Cu, Ni, or Pt [145]. These data agree well with the results of NEXAFS.

2 Instrumentation and Experimental Procedures

From the time when the first serial spectrometers appeared (1969), their design and outfitting with additional facilities have constantly changed. At the same time, the functioning principle of spectrometers and their main components (sources, energy analyzers) have remained virtually the same as in the first instruments. Modernization is mainly aimed at improving the output parameters such as the sensitivity and resolution, and also at creating the greatest possible set of facilities enabling one to conduct various measurements of practically any solid specimens, as well as of liquids and gases when special attachments are available. These conclusions do not relate to specially developed types of instruments in which there have been created unique possibilities of measurements using, e.g., synchrotron radiation, polarized UV radiation, highly monochromatized X-ray sources, and Auger and SIMS microscopes with local space resolution.

The main investigations of catalysts (including those performed by the authors), however, have been conducted with the help of serial instruments, and their features are just what will be described in this chapter. In addition to a brief description of the experimental equipment, more detailed information on which can be found in special publications [7, 8, 9, 113, 146], this chapter includes material on the procedure of measuring and analyzing the basic spectral characteristics. It also deals with the problems arising in the calibration of the spectra of nonconducting specimens, the effects of their destruction when acted upon by ionizing radiations, in determining the instrument coefficients needed for quantitative analysis, etc. Special attention is given to the mathematical processing of spectra, and also to the procedures (including those developed by the authors) for preparing specimens for analysis, transferring them into a spectrometer without contact with the atmosphere and without processing that models the conditions of activation of the catalysts and the catalytic reaction.

2.1   Instrumentation

The existing spectrometers for surface analysis can be conditionally divided into two types:

(a) one- or two-module spectrometers allowing one or two techniques to be used, e.g. XPS-AES, XPS-UPS, AES-LEED, SIMS-ISS, and UPS-HREELS;

(b) multiple-module instruments-laboratories for surface science using a set of techniques, e.g. XPS-SAM-UPS-SIMS-ISS-LEED-HREELS, etc. The first type of instruments prevailed up to the beginning of the 1970's, while the second type is the most popular at present. Each type has its merits and drawbacks, but when dealing with the studying of catalysts, preference must be given to the multiple-module instruments because only a combination of techniques can ensure detailed information on these exceedingly complicated objects. In our investigations, we employed a single-module XPS spectrometer ES 200B and a surface analyzing system XSAM 800 (XPS-UPS-SAM-SIMS-ISS) of the firm "Kratos". These instruments in addition to spectrometers produced by the firms "VG", "Leybold-Heraeus", and "Perkin-Elmer" are the most widespread, and a description of their characteristics can give a quite complete idea of the procedure of an experiment. In the following, we shall consider in greater detail the techniques of electron emission spectroscopy, which are the main ones used in studying catalysts. It should also be mentioned that EES, SIMS, and ISS employ common components, namely, sources, analyzers, and the vacuum system.

2.1.1   Electron Spectrometers

One of the main requirements which an electron spectrometer must meet is the provision of a high vacuum essential for the excitation and analysis of electrons and the ensuring of surface purity. Depending on what problems are to be solved, either a high (10^{-6}-10^{-8} torr) or ultrahigh vacuum (UHV) (10^{-9}-10^{-11} torr) is needed. The difference in the maximum vacuum achieved generally depends not on the type of vacuum pumps, but on their speed and also on the system of sealing the main components and the provision of means for the differential outgassing of the individual units. The vacuum is maintained in an ES 200B spectrometer by a system consisting of rotary and vapor-oil diffusion pumps for outgassing the fast insertion lock for the rapid introduction of the specimens, an X-ray source, an analysis chamber, and an energy analyzer space. As in most other instruments, the space of the

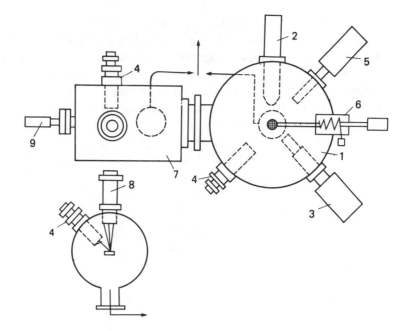

Fig. 7. Main units and their arrangement in an XSAM 800 spectrometer (Kratos Analytical)

1 — analysis chamber; *2* — X-ray tube; *3* — electron gun; *4* — ion gun; *5* — UV lamp; *6* — X-Y-Z-θ manipulator; *7* — preparation chamber; *8* — quadrupole mass analyzer; *9* — specimen insertion lock

X-ray source is separated from the measuring chamber by a thin Al filter, which makes it possible to maintain a vacuum difference in the two parts of an instrument of about two orders of magnitude. The maximum vacuum in the source chamber is 5×10^{-8} torr, and in the specimen chamber one of 5×10^{-10} torr can be reached. But catalyst specimens can also be analyzed at a lower vacuum of 1×10^{-7} torr, which was usually achieved in 30 minutes after insertion of a specimen.

A UHV is constantly maintained in an XSAM 800 spectrometer. This is achieved by using high-speed vapor-oil diffusion, turbomolecular, or ion pumps, improved sealing of the main units assembled on metal seals, multistage outgassing of the specimen insertion system, and also differential outgassing of the main sources (Fig. 7). The XSAM 800 spectrometer consists of a spherical analysis chamber accommodating the X-ray, UV, electron and ion sources, an energy analyzer, an X-Y-Z-θ manipulator ensuring exact positioning of a specimen and allowing it to be moved in three directions and rotated through the specified angle. The second part of the instrument is the preparation chamber accommodating a secondary ion mass spectrometer, fast insertion locks, and sputtering ion guns. This chamber has a sealed connection to a specially designed microreactor for pretreating catalysts under atmospheric pressure and with a gas admission system (see Fig. 7).

Principle of Electron Spectrometer Operation (Fig. 8). A schematic diagram showing how photoelectrons are excited and analyzed in ES 200B and XSAM 800 spectrometers is presented in Fig. 8. Most frequently, Al or Mg $K\alpha$ radiation ($h\nu$ = 1253.6 eV and 1486.6 eV, respectively) is used as sources.

(a)

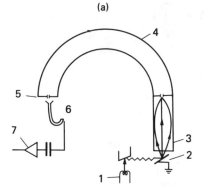

(b)

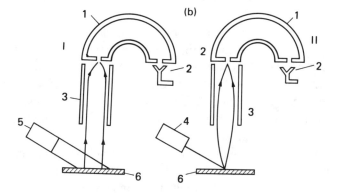

Fig. 8. Schematic diagram of an electron spectrometer:

a — general scheme of excitation and registration of photo and Auger electrons: 1 — X-ray tube; 2 — specimen; 3 — entrance slit; 4 — analyzer; 5 — collector slit; 6 — detector; 7 — amplifier; b — two operating modes of an XSAM 800 spectrometer: I — low magnification; II — high magnification; 1 — analyzer; 2 — detector; 3 — lens system; 4 — electron gun; 5 — X-ray tube; 6 — specimen

These sources provide a high intensity and relatively small width of the $K\alpha_{1,2}$ doublets, namely, 0.83 eV (Al) and 0.68 eV (Mg). Moreover, it is possible to use a combination of high-energy and low-energy X-ray anodes, the greatest interest among which is aroused by the radiations Ti $K\alpha$ ($h\nu$ = 4510 eV, FWHM = 2.0 eV), Cu $K\alpha$ ($h\nu$ = 8048 eV, FWHM = 2.6 eV), Y $M\xi$ ($h\nu$ = 132.3 eV, FWHM = 0.47 eV), and Zr $M\xi$ ($h\nu$ = 151.4 eV, FWHM = 0.77 eV). The combining of sources with different energies enables one to vary the depth of the layer being analyzed from 5 to 100 Å, i.e. obtain a non-destructive profile. In XSAM 800 spectrometers, an X-ray tube with four anodes — Al, Mg, Ti, and Ag — is employed. The use of sources such as Zr $M\xi$ with low $h\nu$ values also sharply increases the sensitivity with respect to the low-intensity lines of the subvalence and valence band because of the higher photoionization cross sections of these levels [147, 148]. A drawback of a zirconium anode is the rapid contamination of its surface, and also the high attenuation of the radiation when it passes through an Al filter.

In addition to nonmonochromatized sources, serial spectrometers are provided with a set of single crystal monochromators that increase the resolution to 0.4-0.5 eV by cutting out one of the lines of the exciting radiation, e.g. Al $K\alpha$. But here the photon flux sharply attenuates, which can be compensated by an increase in the anode area. In practice, this has been achieved only in a specially developed instrument in Uppsala [149].

The flux of X-ray photons impinges on the specimen at a definite angle (45° in many cases), and the emitted photoelectrons pass via the inlet slit into the energy analyzer of the electrostatic type, where they are separated by energies (see Fig. 8). Two types of analyzers are in greatest favor for analyzing electrons, namely, the concentric hemispherical analyzer (CHA) and the cylindrical mirror analyzer (CMA). The latter can be of a single- or double-pass design. The first serial instrument VIEE-15 used an analyzer consisting of a sphere and a cylinder. Since the CMA has a high transmission, but a poorer resolution, it is used more often in Auger spectrometers, whereas the CHA with its higher energetic resolution is more suitable for XPS. The modern XSAM 800 and VG ESCA LAB 5 instruments are featured by a system of electron-optical lenses for compensating aberration. It is arranged between the specimen and the entrance slit of the analyzer (see Fig. 8) and enables the ratio between the transmission and the resolution of a spectrometer to be varied within broad limits. Owing to their high collecting ability, the lenses can "extract" electrons from a small area ($d \approx 1$ mm), which is essential in SAM or ISS. The system of lenses made it possible to obtain good parameters for XPS and AES when using the CHA, although the double-pass CMA is more effective for an Auger microscope with a high local resolution and a low-current electron beam.

There are two modes for passing electrons through an analyzer: (a) at ΔE = const (CAT or FAT — constant or fixed analyzer transmission), when the electrons retarded at the inlet are transmitted at a constant potential across the analyzer poles (E = 5-200 eV); here a constant absolute resolution is maintained; and (b) at $\Delta E/E$ = const (CRR or FRR — constant or fixed relative resolution), when the preliminarily retarded electrons are transmitted at a variable potential depending on E_k; here the relative resolution is constant, and the signal/noise ratio is higher at low values of E_k. The FRR mode is the main one used in the spectrometers of the firm "Kratos".

Recently, apart from the single-channel electron multiplier with $K_{amp} \approx 10^8\text{-}10^9$ (channeltron), multi-channel solid detectors (MSD) controlled by a computer began to be used. The three-channel detector is one of their varieties. The merits of the MSD consist in a higher counting rate that can compensate the losses of sensitivity associated with the use of a monochromator, and, which is even more significant, in that these detectors are position-sensitive, i.e. can pick up an image of a specimen in photoelectrons. The first "ESCA microscopes" with local resolution were created on this basis. With a view to the high characterizing ability of XPS with respect to chemical states and its universal nature, the appearance of a possibility of local analysis even with a rough resolution is of appreciable interest, for example, when studying industrial catalysts. The developed models of the "ESCA microscope" [150-153] employ various principles for picking up photoelectrons from a small area and obtaining an image of individual sections, one of them being the use of position-sensitive MSD. The spatial resolution in the electron spectrometers modified for this purpose is 150-250 μm and can be increased to 10 μm [153].

2.1.2   Resolution and Sensitivity of Electron Spectrometers

The absolute energy resolution (the FWHM of a peak being registered) is determined by the expression [7]:

$$(\Delta E)^2 \approx (\Delta E_r)^2 + (\Delta E_{lev})^2 + (\Delta E_{in})^2 \tag{2.1}$$

where ΔE_r is the width of the exciting radiation line, ΔE_{lev} is the natural width of the level, and ΔE_{in} is the instrument broadening. In the excitation of type Al $K\alpha$ or Mg $K\alpha$ radiation spectra, the main contribution is made by ΔE_r.

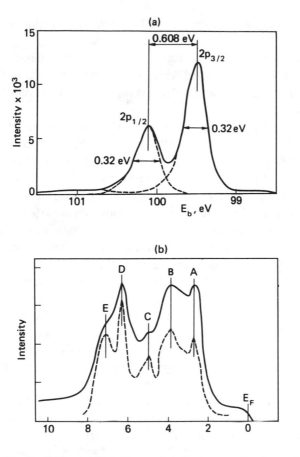

Fig. 9. High-resolution X-ray photoelectron spectra (monochromatized) of core Si $2p_{1/2-3/2}$ level of Si(III) (a) and valence $5d$ band of Au (b) [149]. The dashed line shows the theoretical density of the valence states [134]

The maximum resolution at the minimum broadenings produced by ΔE_{in} and measured as FWAM for the Ag $3d_{5/2}$ peak is about 0.8 eV. Monochromatization of the Al $K\alpha_{1,2}$ radiation can result in ΔE being 0.4-0.5 eV, while when fine focussing is used, it may approach 0.2 eV. Figure 9 presents spectra of the Si $2p$ core level in a single crystal of Si(111) and of the Au valence band obtained in an instrument developed in Uppsala [149]. For Si $2p$, a clear spin doublet is observed (without a monochromator, the line looks like a symmetric singlet), while the resolution of the valence band spectra is so high that they reflect virtually the entire structure of the theoretically calculated density of the valence states [154].

The relative resolution $\Delta E/E_0 \times 100\%$ depends on the energy of the electrons and, as we have already noted, may be constant or variable depending on the mode of electron retardation. The average value of the relative resolution is 0.1-1%.

The sensitivity in XPS can be expressed by the counting rate at the maximum of the standard peaks. At a resolution of 0.8 eV, the counting rate for Ag $3d_{5/2}$ is 5×10^4 cps, while with diminishing of the resolution to 1.1 and 1.5 eV it grows to 2×10^5 and 2×10^6 cps, respectively. When an MSD is used, the counting rate will increase at least by an order of magnitude. The signal contrast $C = S/B$ is another more universal characteristic of the sensitivity. It is determined as the ratio of the signal (S) to the background (B), which reaches 30:1 for the best instruments. Although the minimum concentrations determined by XPS depend on many factors, they can be estimated to be 0.5-1% of the magnitude of the monolayer, or 0.1-1% by mass.

The CMA is also used to analyze photoelectrons excited by UV radiation and Auger electrons excited by electron impacts. In the first case, the source is a gas-discharge lamp providing radiation of the resonance lines of He^I ($h\nu = 21.2$ eV) or a mixture of He^I and He^{II} ($h\nu = 40.8$ eV). The intrinsic width of the He^I lines is 0.003 eV, and the spectrometer resolution is 0.09-0.18 eV. The counting rate at the maximum of the Au $5d$ peak is $3-5 \times 10^4$ cps. An ultraviolet source is outfitted with a two-stage system of differential outgassing; as a result, when helium flows into the discharge region and lowers the vacuum to 1×10^{-8} torr, a vacuum of 5×10^{-10} torr is retained in the analysis chamber.

An electron gun for exciting Auger electrons is provided with beam fine focussing and raster means; as a result, the spatial resolution in the standard version of SAM is 0.2-1 μm, while in the most up-to-date Auger microscopes it reaches 50-200 Å. Monte Carlo calculations [155] show that with an infinitely small size of the beam, the maximum resolution which can be achieved in SAM at $E = 2-3$ keV will be 30 Å. This value is comparable with the resolution of scanning electron microscopy. But SAM has meanwhile failed to come into favour for analyzing catalysts, which have an extremely nonuniform surface and are often insulators. The static variant of AES with a wide defocussed electron beam is more popular. When using the CHA with system of lenses or the double-pass CMA, the maximum resolution achieved in AES is close to that in XPS (about 1%). The absolute resolution does not depend on the line width of the exciting radiation, but is determined by the second and third terms of Eq. (2.1). The signal contrast in the differential form, in which the sensitivity in AES is expressed, is 200-500 at a low resolution, while at a high resolution it exceeds 100 (analysis of the Cu L_3VV peak). The minimum concentrations determined by AES are higher than in XPS and reach 10^{-2}-$10^{-1}\%$ of the magnitude of the monolayer.

2.1.3 Excitation and Registration of Scattered Ion Spectra and Secondary Ion Mass Spectra

The same sources of ions can be employed in ISS and SIMS. The chief requirement for ISS is stability of the set energy (E_0). This requirement is met by many types of sources, including ones with ionization by a Penning or high-voltage discharge of a gas concentrated in the restricted space of an ion gun. An example of such a source is the Kratos firm's Mini Beam gun producing an ion current from 5×10^{-9} to 10^{-6} A with $E_p = 0.05$-5 keV. The beam diameter varies from 0.05 to 1 mm, the spot area is 10×10 mm, and the inleakage is 10^{-4}-10^{-7} torr. With differential outgassing, the inleakage can be reduced to 10^{-9} torr. Two types of analyzers are used to analyze the scattered ions, namely, a CHA of electrons that operates with the reverse polarity on its hemispheres, or a CMA with a coaxially built in ion gun, which ensures collection of the ions within a large solid angle and, consequently, a high transmission (Fig. 10). The second type of analyzer is employed in spe-

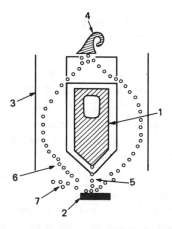

Fig. 10. Schematic view of an ISS spectrometer with a CMA (3M, USA):

1 — ion source; *2* — specimen; *3* — CMA; *4* — detector; *5* — primary ions; *6* — scattered ions; *7* — sputtered ions

cial instruments of type ISS-SIMS 3M or SIMS 800 "Kratos". The first type of analysis is used in the multiple-module instruments XSAM 800. A combined spectrometer has also been developed. It combines the better resolution of the CHA and sensitivity of the CMA and makes it possible to measure beams of scattered ions resolved by angles and energy [156]. Shimizu and Kurokawa [157] provided an Auger microprobe JAMP-3 with a source of floating high voltage, which enabled them to obtain AES and ISS spectra consecutively using the same CMA. In our experiments, we used a combination of a Mini Beam 1 and a CHA to obtain an ion scattering spectrum. We selected the following parameters for the measurements: $^4He^+$ ions, $E_p = 0.7\text{-}2$ keV, $I_p = 5 \times 10^{-8}\text{-}5 \times 10^{-7}$ A, and $p = 5 \times 10^{-8}$ torr. The maximum rate of counting the peaks scattered from metals (Pd, Cu) was 10^3 cps. Since the spectra have a very low background in the region of $E/E_0 \geqslant 0.5$, it was possible to register very weak signals (10^2 cps) at currents below 50 nA. Under these conditions, the rate of scattering the He^+ ions was very low [115]. This made it possible to obtain information on the uppermost surface layers. The resolution, which was appraised along the "valley" between equivalent peaks of Ni and Cu, was 75%.

SIMS. At present, there are diverse variants of SIMS enabling one to analyze predominately not only the surface or bulk (static and dynamic mode) of an integral region of specimens, but also to probe with local resolution, and to obtain an image of the region being studied in positive or negative ions ("chemical maps"). The various sources and mass analyzers employed in SIMS have been described in monographs [9, 110, 113]. Mention must be made of the latest types of sources ensuring a high brightness that allow one to obtain very narrow ion beams. The latter provide currents sufficient for analysis. A field emission liquid-metal (Ga, Cs) source marketed by the Kratos firm makes it possible to obtain a spatial resolution up to 500 Å. Here the image in secondary ions produced at currents of 2×10^{-9} A reflects the distribution of the elements on the very surface. A duoplasmotron source allows both inert and active gases to be ionized and highly stable ion beams to be obtained with variation of the current from 3×10^{-9} to 1×10^{-5} A and a beam diameter from 5 to 1000 μm. The development of this technique is very prospective, seeing that SIMS has the highest sensitivity with respect to the elements among all the techniques of electron emission and ion spectroscopies, and also that it can distinguish different isotopes, among other features [158, 159].

Various types of mass analyzers have been proposed. The analyzer of the quadrupole type is the most popular among them. It ensures a high sensitivity and an acceptable resolution by masses. In most catalyst studies, favor is given to systems consisting of a discharge ion source and a quadrupole mass analyzer provided with special energy filters for singling out the secondary ions within a narrow interval of energies. Depending on the modes, the parameters determining the ion currents [Eq. (1.48)] are varied within the following limits: $I_p = 10^{-5}\text{-}10^{-11}$ A, $i_p = 10^{-2}\text{-}10^{-9}$A/cm^2, $d = 10^{-4}\text{-}10^{-1}$ cm, $Y = 1\text{-}10$, $Y^\pm = 10^{-5}\text{-}10^{-1}$, $f = 10^{-3}\text{-}10^{-4}$, and the sputtering rate s (at $Y = 3$) $= 10^{-4}\text{-}10\ \mu$m/h $= 0.1\text{-}2 \times 10^4$ Å/min. The lower limits correspond to the static mode, and the upper ones, to the dynamic mode. In practice, an intermediate mode is often realized in serial instruments. In our experiments, we used a Mini Beam I ion gun, a quadrupole type mass analyzer $3M$, $M = 1\text{-}300$, $\Delta M/M = 300$, $E_p = 1\text{-}5$ keV, $E_{\text{sec}} \geqslant 10$ eV, and a mass analyzer transmission mode of $f \approx 1/M$. For operation in a mode close to the static one, the ion beam with $E_p = 1$ keV was rastered to an area of 0.25-0.5 cm^2, and the current density did not exceed 5×10^{-6} A/cm^2; the sputtering rate under these conditions did not exceed 5 Å/min [110, 113]. To eliminate the errors associated with crater formation [110, 113], the secondary ions were registered from the inner part (50%) of the sputtered region.

When measuring insulators by ISS and SIMS, an electron emitter with $E = 1\text{-}15$ eV (of the Flood Gun type) was used to compensate the charging that impaired the resolution and distorted the background.

2.2 Basic Spectral Characteristics and Calibration of the Spectra

2.2.1 XPS

The basic relations between the kinetic energy of the electrons being measured and E_b are given by Eqs. (1.1)-(1.4). When dealing with conductors, the Fermi levels of the specimen and spectrometer level out, while the potential difference $\varphi_0 - \varphi_{\text{sp}}$ (Fig. 11) sets in owing to the difference in the work function of the two metals at the specimen-vacuum interface. When determining the ionization energy (E_{ion}) relative to the Fermi level, one must find φ_0 [160] because $E_{\text{ion}} = E_b + \varphi_0$ or $E_{\text{ion}} = h\nu - E_k + (\varphi_0 - \varphi_{\text{sp}})$. Hence, to determine E_b for the levels of a conductor, it is sufficient to know the value of φ_{sp}, which is measured along the Fermi edge of Ni or Pd, or according to the position of the core levels of elements such as Ag and Au. Provided that E_b for Ag $3d_{5/2} = 368.2$ eV, the value of φ_{sp} in the instruments ES 200B and XSAM 800 is about 4.5 eV. We have already mentioned that φ_0 is deter-

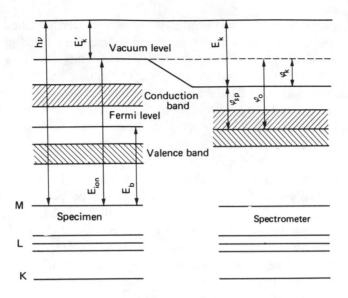

Fig. 11. Diagram of the energy levels for a solid specimen and spectrometer. For the notation, see Eqs. (1.1)-(1.4)

mined by the procedure of measuring the cut-off of the secondary electrons [160].

If a specimen is a semiconductor or insulator, it fails to have a good electrical contact with the spectrometer. Consequently, in photoionization, the specimen acquires a static uncompensated charge, which can be either positive or negative. The latter occurs because of the emission of secondary electrons from the chamber walls. The taking of the charge into account is a serious methodical problem. The value of φ_{ch} [Eq. (1.4)] reaches several eV and depends on the density of ionization, the temperature, and the specimen thickness [9, 17, 161]. The magnitude of the charge for thin films is noticeably smaller, which is confirmed by numerous measurements of the interface of systems of the type $SiO_2/SiO_x/Si$ [23]. Different ways have been proposed for taking the charge into account or compensating it.

1. Compensation of the charge with the aid of electron flood guns with $E < 10$ eV, and also by increasing the emission of secondary electrons from the spectrometer walls by using special coatings lowering the work function [162]. It is impossible to measure insulators without charge compensation when working with X-ray monochromators. The absence of Bremsstrahlung

may cause the lines to shift by scores and hundreds of eV. The problems associated with the charge overcompensation of specimens when a flood gun is used, and also the difficulties involved in compensating a nonuniform charge on the surface of catalysts have been repeatedly discussed in the literature [32, 34, 163, 164]. Swift *et al.* [165] indicate that the accuracy of determining E_b in the spectra of powders when using a flood gun does not exceed ± 0.5 eV. Edgell *et al.* [166] studied in detail the influence of the emitter potential and current on the completeness of charge compensation when using monochromatic Ag K_α radiation. A potential of at least 6 V and an emission current of 0.25 A were shown to be needed for this purpose. Major line broadening was observed at potentials above 6 V for some insulators, e.g. the zeolite NaY. Although in preliminary calibration, emitters may be useful for compensating charge build-up, the ambiguity in the Fermi level shift when such a procedure is followed makes it necessary to employ a standard specimen to determine E_b [166].

2. The use of standard or reference materials whose line position can be considered as the energy datum level. The greatest preference among external standards has been given to the line of C 1s belonging to a layer of hydrocarbons adsorbed on the specimen surface [8, 9, 25, 32, 165, 167]. The residual vapour of the diffusion pump oil, products of destruction of the organic materials which the seals are made from, etc. may be a source of hydrocarbons. When the line of C 1s is used as a reference, it is assumed that (a) the charging of the layer of hydrocarbons and the specimen is identical and electrical equilibrium sets in between them, and (b) the hydrocarbons are adsorbed physically, and E_b of C 1s (most often the values from 284.6 to 285 eV are used) remains constant regardless of the type of specimen and the conditions of its processing. The latter assumption can apparently be adopted only when determining E_b (and not E_{ion}) and in the absence of a chemical reaction on the surface, which may happen when working with catalysts [27]. In the latter case, the lineshape is distorted or two C 1s peaks appear. This means that a different reference is needed for the correct determination of E_b. It should be noted that when measurements are conducted in UHV or after ion etching, E_b of C 1s may differ greatly from 285 eV [9], which also necessitates an additional reference. At the same time, notwithstanding the complications noted above, the results obtained for E_b and ΔE_b of insulators at various laboratories [165, 168-170] indicate a quite good convergence when using the C 1s line as a reference.

3. Another external reference material, namely, sputtered Au [12, 165] is also employed to determine the position of individual lines in the spectra of

the initial specimens. Notwithstanding the quite widespread use of this procedure for calibration, it cannot be considered as a universal one. A film with an effective thickness of 0.6 nm is considered to be optimal for establishing electrical equilibrium between the Au and the specimen. At lower coating thickness, the effects described in Chap. 3 for small clusters appear, i.e. shifting of E_b and a change in the shape and width of the Au $4f$ level [171]. Another shortcoming of this procedure is that it is not suitable when studying the influence of various treatments on the state of a catalyst.

4. The use of internal standards is the most reliable procedure, although it is not always possible. For this purpose, the peaks of matrix elements whose position is assumed to be constant are used in a series of specimens of a single type. In our experiments, we used Si $2p$ in zeolites and on SiO_2, Al $2p$ in zeolites and on Al_2O_3, etc. as such lines. Their binding energies were taken equal to the average values from an analysis of a large array of our own and reference data [27, 34].

5. Still another way of solving the problem of charging (especially when nonuniform) is the determination in addition to E_b of the splitting between individual photoelectron lines, and also of the Auger parameter (α). The latter, when it is possible to register the Auger lines, is a very reliable criterion of the chemical state of elements [39]. But for fine particles, the value of α may depend substantially on the dispersion (see Chap. 3).

When spectra are calibrated using one reference, all the lines are assumed to shift identically in charging regardless of E_k. But the large discrepancy of the results for E_b of lines remote in energy from the reference ones may indicate nonuniform charging [172]. Moreover, this is apparently due to the nonlinearity of the energy scale of the spectrometer. The latter was verified by measuring the difference $E_{k,1} - E_{k,2} = h(\nu_1 - \nu_2)$ when using different radiations. For Al $K\alpha$ and Mg $K\alpha$, the value of $h(\nu_1 - \nu_2) = 233.02$ eV. Such a verification was conducted periodically in an XSAM 800 spectrometer for levels differing in E_k, namely, Cu $2p_{3/2}$, Ag $3d_{5/2}$, and Au $4f_{7/2}$. The deviations do not generally exceed 0.1 eV.

Consequently, the accuracy of determining E_b provided that the energy scale has been calibrated correctly depends on the electrophysical properties of a specimen. At an error of measuring E_k in modern instruments not exceeding 0.02 eV, the value of E_b for the lines of metals is determined with an error of ± 0.1 eV, and for insulators with one of ± 0.2 eV. The latter value grows when measuring lines with a low intensity or broad lines and with nonuniform charging. The reproducibility of E_b in the instruments ES 200B and XSAM 800 for insulators and metals was 0.2 eV and 0.1 eV, respectively.

The problems of standardizing the data of XPS are closely related to those treated above. During the past 10 years, a considerable number of measurements involving series of standard specimens were conducted at various laboratories using instruments of various designs and firms [161, 165, 169, 172-175]. One of them [169] used data obtained by Shpiro et al. We must note the better convergence of data, especially with respect to E_b, in the latest verifying works. This is explained by the more thorough procedure for the adequate calibration of various spectrometers. For example, Powell et al. [174] indicated that the variance of the data on E_b of Cu $2p_{3/2}$ relative to Au $4f_{7/2}$ in different instruments reached 2 eV. This is associated not only with a shift in the zero of the energy scale, but also with its nonlinearity. The convergence of the measurements grows sharply when the absolute calibration procedure was used [161]. The error did not exceed ± 0.04 eV for the narrow lines of Cu, Ag, and Au in the modes $\Delta E = $ const and $\Delta E/E = $ const. Anthony and Seah [161] recommend using the following values of E_b (Table 4) in calibrations. In the spectrometer XSAM 800, values of E_b of 84.0, 368.3, and 932.7 eV were obtained for Au $4f_{7/2}$, Ag $3d_{5/2}$, and Cu $2p_{3/2}$, respectively. The measurements of insulators can meanwhile be performed with a lower accuracy. When determining the binding energy for chemical compounds, in five of nine spectrometers the discrepancies did not exceed 0.3 eV [173]. Considerable attention is given to the standardization of XPS data in the works of Nefedov [9, 34]. He published the most complete summary of data on the binding energy for various elements and compounds [34]. The standard values of E_b for a number of substances are also given in this reference book. Earlier reference books [176-179] contain selected data on E_b, the kinetic energy of Auger electrons,

Table 4. Calibration Values of the Binding Energies (eV) $[E_F(Ni) = 0]$ [161]

	Al $K\alpha$	Mg $K\alpha$
Cu $3p$	75.14 ± 0.02	75.13 ± 0.02
Au $4f_{7/2}$	83.98 ± 0.02	84.0 ± 0.01
Ag $3d_{5/2}$	368.27 ± 0.02	368.29 ± 0.01
Cu L_3MM	567.97 ± 0.02	334.95 ± 0.01
Cu $2p_{3/2}$	932.67 ± 0.02	932.67 ± 0.02
Ag M_4NN	1128.79 ± 0.02	895.76 ± 0.02

the Auger parameter, and also maps illustrating individual spectrum lines for various elements. When using reference data, one must verify them by test checks on one's own spectrometers.

Quantitative Analysis. The relative surface concentrations can be determined by using Eqs. (1.13), (1.15), and (1.28). In most cases, the theoretical values of the photoionization cross sections are used [61]. The values of λ_i are approximated with the aid of the analytical expressions given in Chap. 1, the procedure proposed by Seah and Dench [68] being the most widespread one. Another important factor in determining concentrations is the transmission coefficient of the spectrometer T_i. It is individual for each instrument, but depends the most noticeably on the mode of analyzer operation. To a first approximation in the FAT mode, the value of T_i is proportional to $1/E_k$, and in the FRR mode, to E_k [180]. More accurate measurements yield fractional values of the exponents. For instance, when considering the procedure of quantitative analysis without a reference, Hanke *et al.* [181] measured a large number of lines with various kinetic energies E_k and established the following polynomials for determining T_i for an XSAM 800 spectrometer:
in the FAT mode:

$$T_i = S_i(E_k) = 1.954\,72 - 2.783\,57E_k - 0.859\,84(E_k)^3 \qquad (2.2)$$

in the FRR mode:

$$T_i = S(E_k) = 0.031\,73 + 0.331\,78E_k + 0.968\,22(E_k)^2$$
$$- 0.183\,99(E_k)^3 \qquad (2.3)$$

Our measurements of the spectra of thoroughly cleaned metals yielded the following expressions for T_i in the FRR mode: $T_i \propto E_k^1$ for the ES 200B spectrometer, and $T_i \propto E_k^{3/2}$ for the XSAM 800 spectrometer. Hence, Eq. (1.15) can be reduced to the form:

$$\frac{I_1}{I_2} = \frac{c_1}{c_2}\frac{\sigma_1}{\sigma_2}\left(\frac{E_1}{E_2}\right)^2 \qquad (2.4)$$

(λ_i is approximated as proportional to $E^{1/2}$). When determining the ratio of the intensities of lines with a low ($E_{k,1}$) and high ($E_{k,2}$) kinetic energies ($E_{k,1} < 1000\,\text{eV}$, $\Delta E < 400\,\text{eV}$), the problem arises of taking account of the weakening of the intensity by the layer of contaminant. Here it is also necessary to use the calibration coefficients obtained when analyzing stoichiometric compounds.

The standardization of the XPS intensity values is farther from its completion than that of E_b. In round robin investigations [174], the differences in the relative intensities were 30-40% for lines close in energy, while when ΔE was increased to 1000 eV, these differences reached 100%. In later investigations [182, 183], methods were proposed for the absolute calibration of the intensities that made it possible to compare data obtained in various instruments more correctly. The influence of the method of subtracting the background used (straight line, or Shirley's method) and of the specimen size, position, and surface roughness on the intensity was considered. When all the parameters were taken into account, the error in measuring the intensity dropped to $\pm 2\%$ [184]. This enables us to await the appearance of more accurate and universal data on elemental sensitivities in comparison with published data [185]. The latter can be applied to analyzers in the $\Delta E = const$ mode.

The integral intensities were measured as recommended in [9, 34]. Different methods of background subtraction were used, but preference was given to the straight line one. In the analysis of the $2p$ level lines of $3d$ elements, account was taken of the intensity of the peaks of the shake-up satellites. The error in determining the integral intensities did not generally exceed 5-10%, and in determining the surface concentrations in homogeneous specimens, it did not exceed $\pm 20\%$. Kerkhof's model [185] was used when appraising the surface concentrations in heterogeneous specimens containing deposited particles. For zeolite catalysts, the directed migration of metals to the outer surface or into pores was kept in view.

The problems associated with the features of measuring and analyzing Auger spectra, scattered ion spectra, and secondary ion mass spectra have been partly treated in Chap. 1. Since in our experiments we limited ourselves to a qualitative and semiquantitative use of these methods for studying chiefly conducting specimens, we shall describe this procedure only briefly below. Investigations in which attempts have been made to study the surface of model or real systems more completely by AES, ISS, and SIMS are discussed in Chaps. 3-5.

2.2.2   Other Techniques

AES. The spectra of the Auger electrons obtained by X-ray excitation (XAES) were used for identifying the chemical state of elements. When their position was determined (E_k), a correction for the charging of insulators was taken into consideration. In addition to E_k, we determined the Auger parameter $\alpha = E_k$ (Auger electron) $- E_k$ (photoelectron) or the modified Auger parameter $\alpha' = E_b$ (photoelectron) $+ E_k$ (Auger electron).

AES with electron excitation was employed as an additional technique for appraising the change in the surface composition. In most cases, the ratio of the amplitudes of the relevant signals (from minimum to maximum) was taken as a measure of the composition. Sometimes (alloys) the elemental sensitivity coefficients known from the literature for the same systems were taken into account.

ISS. Elements were identified according to the peaks of the single scattering of ^{4}He$^+$ ions by Eq. (1.50) in which $M_1 = 4$ and $\theta_L = 130°$. The composition was assessed by the ratio of the amplitudes or integral intensities. In some cases, the surface concentrations were determined as follows:

$$\frac{I_1}{I_2} = \frac{c_1}{c_2} \frac{\sigma_{1,sc}}{\sigma_{2,sc}} \tag{2.5}$$

where σ_{sc} are the differential single scattering cross sections, which for $\theta = 138°$ have been calculated by Nelson [137] and are plotted against the atomic number Z. With such an appraisal, the difference in the probabilities of neutralization was disregarded. The measurement accuracy did not exceed 30-50%. ISS was conducted in a mode ensuring a low sputtering rate ($I_p = 50-100$ nA and $E_p = 1$ keV). This is why the data of the first runs (100-200 s) can be related to the top surface layer.

SIMS. The peaks in mass spectra were analyzed with a view to the ratio of the natural isotopes for various elements:

$$I = \frac{A_{m_i}}{\varrho_{m_i}} k \tag{2.6}$$

where I is the intensity of the peak of a given element, A_{m_i} is the amplitude of the peak of the isotope m_i, ϱ_{m_i} is the fraction of the isotope m_i present, and k is an amplification factor. Typical survey mass spectra for metals and insulators are presented in Fig. 12. The taking of the isotopes into account made it possible to identify the peaks of atomic and molecular ions of the main elements, although the comparatively low resolution, especially in the region of m/e exceeding 150, did not allow complete qualitative analysis to be performed. Complications were introduced into this analysis by the presence of impurities of light elements (Li, Na, K), which owing to the high sensitivity (10^{-5}%) were always observed in the spectra of metals, and also

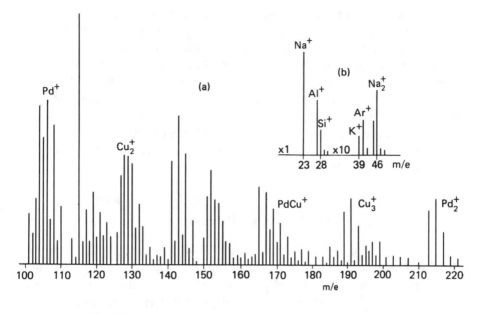

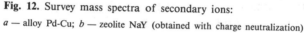

Fig. 12. Survey mass spectra of secondary ions:

a — alloy Pd-Cu; *b* — zeolite NaY (obtained with charge neutralization)

by the presence of peaks due to carbon contaminants, namely, C, C_2, C_2H_x. The resolution, sensitivity, and adjustment were selected on a reference Ti specimen ($m/e = 48$); the measurements were conducted in a transient mode of mass analyzer transmission ensuring an average resolution and transmission within the m/e range of 20-100. Subsequently, the measurements on specimens of one type were conducted at the following constant parameters: (a) the potential on the filters; (b) the argon flow; (c) the secondary ion collection angle (about 90°); (d) the scanning area; and (e) the mode of the ion source. In addition to qualitative analysis (the proportion of ions of the type Me^+, $Me_xO_y^+$, Me_n^+, and $Me_1Me_2^+$), we performed semiquantitative analysis of the surface composition and its alteration along the depth of the layer. For this purpose, the changes in the absolute and relative yields of the secondary ions were plotted against the duration of ion bombardment in various modes. Under the mildest conditions allowing elements to be analyzed with $m/e = 30\text{-}100$ at a concentration of about 1% ($E_p = 0.8$ keV, $i_p = 100$ nA/cm^2), the sputtering rate did not exceed 2-5 Å/min [105].

The composition was appraised by the ratio of the secondary ion yields.

For some elements with close masses [$f_1 = f_2$, Eq. (1.48)], the surface concentrations were determined. The values of Y^+ were taken from the literature.

At present, an attempt is being made to standardize not only the data of electron emission spectroscopy, but also of ISS and SIMS. In the round-robin investigations ASTM-42 being conducted under the auspices of the American Society for Testing and Materials [186, 187], special standards are proposed, and the set of the experimental measuring conditions having the determining significance for reproducing the parameters of ISS and SIMS is indicated. This is why we can expect the appearance in the nearest future of more complete reference data on techniques of ion spectroscopy.

2.3   Spectra Collecting and Processing

Modern instruments are outfitted with microcomputers for performing the automatic registration of spectra, their accumulation, and processing. Although the general programs of data acquisition and analysis are similar, each firm develops its own package of programs. At the same time, there is a large number of instruments in which the spectra are registered manually, and too simplified methods are used to process them. A number of workers have attempted to combine the instruments they use in electron emission spectroscopy with nonspecialized computers such as personal microcomputers. Therefore, to estimate the authenticity and accuracy of the original data obtained by various authors, it will be good to acquaint the reader with the level of solving the tasks imposed on the acquisition and processing of information achieved in modern serial instruments. It should be noted that a number of problems relating to spectrum analysis have not been solved to the end and are the object of special experimental and theoretical research. We shall treat the latter here very briefly and cursorily.

2.3.1   Acquisition and Primary Processing of XPS Data

The firm Kratos has developed several generations of data systems whose possibilities improved with the improvement in the characteristics of the spectrometers and computers themselves. The XSAM 800 instrument employs the systems DS 300 and DS 800 based on the control microcomputers DEC PDP 11. The DS 300 system employs the microprocessor PDP 11/03 L with a memory capacity of 64 kbyte, a magnetic disk store for 512 kbyte, a graphical video terminal, and also a number of other terminal facilities

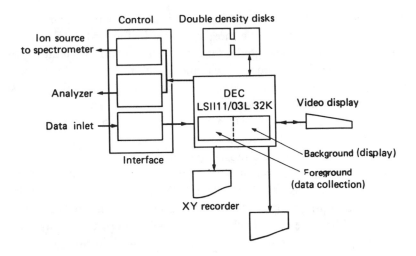

Fig. 13. Structure flow chart of a spectrometer data acquisition and analysis system XSAM 800 (Kratos Analytical)

(Fig. 13). The system operates in a time-sharing mode, namely, when working with a spectrometer, other spectra can be processed at the same time.

The program packages are divided into four groups: (a) data acquisition; (b) data processing on a video display; (c) the input of data from other sources and their output in the form of tables; and (d) reference data. The DS 300 system makes it possible to perform the following control functions: (a) energy scanning in the accumulator mode; (b) control of the analyzer mode of operation; (c) control of the X-ray and UV sources; (d) automatic obtaining of the ion sputtering profile; (e) control of specimen position, etc.

One of the programs sets the parameters for registering the spectra of specimens by individual regions of them in a definite sequence. These data are combined into runs. This program sets the parameters that are common for all the regions, namely, reference data, the X-ray source, the analyzer mode, the mode of profiling and scanning, the magnitude of the work function, and also the parameters that are individual for each region such as the element, energy, number of channels, the step, the dwelling time in a channel, and the number of scans. Up to 10 runs can be set in one experiment, and each of them contains up to 32 regions. This program enables one to conduct experiments for many hours with a set of various specimens.

The spectra are accumulated in four modes: T — true averaging in time; E — exponentially weighted averaging in time; A — scanning in an arithmetical dependence on time; and G — scanning in a geometric dependence on time. The modes T and A are the most popular. In the first mode, the pulses in the channels are summed, and after each scan from all the channels, the intensity with the minimal value in a channel is subtracted. Mode A is used when obtaining ion sputtering profiles or other time relations. In this case, scanning of the present regions is repeated at strictly definite intervals.

After accumulation and storage of the data on magnetic disks, they can be displayed. The display program provides for a substantial volume of spectra processing operations. The most important procedures among them include smoothing by the method of least squares, subtraction of the background, peak synthesis, deconvolution, integration, differentiation, comparison, normalizing, addition, subtraction of spectra, and elimination of the X-ray satellite peaks. There are special subprograms enabling one to procure information on spectra parameters in the digital and graphical forms, plot time dependences or the sequence of the spectra in a third projection (Z plots), etc. Special attention must be given to background subtraction and to procedures of synthesizing peaks and deconvolution.

2.3.2   Determining the Intensity and the True Lineshape

The correctness and accuracy of conducting these procedures determine the accuracy of the quantitative information obtained by XPS on the composition of a surface and the relations between the various chemical states. A number of problems associated with the correct choice of the methods of background subtraction, peak synthesis, and deconvolution remain an object of discussion [180, 188]. The simplest and probably an approximate method of background subtraction is linear approximation. It can be performed in two ways.

1. By subtracting the linear background of the entire spectrum or a limited region:

$$B_i = \frac{I_n - I_m}{n - m} + (i - m)I_m \tag{2.7}$$

where m is the first channel (boundary point), n is the last channel (boundary point), i is the instantaneous channel, I_m is the intensity in the m-th channel, I_n is the intensity in the n-th channel, and B_i is the background in the i-th channel.

2. This method includes (a) subtraction of the minimum value of the intensity from all the channels, and (b) calculation of the background in the i-th channel (beginning from the channel with the lowest intensity). For each point, the background is calculated as

$$B_i = \frac{I_n \sum\limits_{j=m}^{i} I_j}{\sum\limits_{j=m}^{n} I_j} \qquad (2.8)$$

Approximation of the background by a nonlinear function is more accurate. But it is difficult to determine the exact value of this function because in addition to the inelastic scattering of the emitted electrons, the background can also be determined by other processes. The greatest favor has been found by the Shirley approximation [189], by which the background at each point is determined by the inelastically scattered electrons having a higher kinetic energy than the main peak. Consequently, the background at each point is proportional to the integral intensity of the photoelectrons with a higher E_k. A difficulty inherent in this procedure is the accurate choice of the boundary points, especially when analyzing spectra reflecting a mixture of various chemical states of an element. A modification of Shirley's method has been advanced by Bishop [190]. He proposed to approximate a nonlinear background by linear sections. Here the function of the background at the boundaries of each section diminishes when going over from high to low values of E_k. More involved methods of background subtraction have been described by Proctor and Sherwood [188]. They also point out that in linear approximation, which is the most popular in standard programs, the background at the side of high E_b is understated, and at the side of low E_b is overstated.

Instrument broadening is generally taken into account by deconvolution of the peaks by the Fourier transform method [188]. The true function of a spectrum is found by back convolution [188]:

$$f_0 = f_t B \qquad (2.9)$$

where f_0 is the function for the observed spectrum, f_t is the function for the true spectrum, and B is the function of instrument broadening. The deconvolution procedure is analyzed in greater detail by Carly and Joyner [191].

Peak synthesis is a procedure for describing an experimental spectrum with the aid of individual components of a given shape with variation of their number, the ratio of the intensities, and the intrinsic parameters (E_b, FWHM). In the procedure based on the method of least squares, the intensities of an experimental and synthesized spectrum are compared, and the difference is minimized by iterative fitting of the parameters of the individual components. The quality of fitting is determined by evaluating the FIT parameter [188]:

$$\text{FIT} = \sum_{i=m}^{n} \frac{[I_i - F(x_i)]^2}{n - m} \tag{2.10}$$

where I_i is the intensity at the point $x = x_i$, $n - m$ is the total number of points in the spectrum, and $F(x_i)$ is a function describing the shape of a peak. In the simplest model, $F(x_i)$ is a function of a Gaussian or Lorentzian shape, although a mixed shape will be a more accurate approximation.

The nonlinear method of least squares is widely used for optimization of peak synthesis. It takes into account the parameters of the peak tails, namely, the constant tail height, the exponential coefficient of the drop in the intensity on a tail, and the mixing coefficient of these two parameters [188]. An appraisal of peak asymmetry is especially important for metals and alloys because the asymmetry coefficient reflects the degree of interaction in the conduction band. The asymmetry coefficient is appraised for pure metals and alloys by elimination of the instrument broadening (deconvolution) with subsequent analysis of the lineshape using the Doniach-Sunjic function [192, 193]. But when analyzing a mixture of chemical states of the metal-oxide type, the difference in the peak tails of two states can be taken into consideration within the limits of the nonlinear method of least squares. Hughes and Sexton [194] used a variant of this method that had been employed previously quite widely in optical spectroscopy. It was found to be especially accurate for analyzing complicated X-ray photoelectron spectra such as mixtures of various compounds (e.g. TiO_2 and TiC).

The standard variant of the DS 300 system employs a peak synthesis program based on the method of least squares. The operator sets parameters such as the lineshape, the number of components, E_b, FWHM, and I for each of them, the last three parameters being varied by iterations. When analyzing spin doublets, the doublet parameters including the spin-orbital splitting, the FWHM, and the ratio of the intensities of the two components are set additionally and kept constant. It should be noted that the correctness and accuracy of singling out individual components depend on the volume of the known physical information relating to the parameters of the spectra of reference compounds, on the consideration of broadening caused by charging, nonuniformity of the supported particles, on the data relating to the possible number of states procured by independent methods, etc. The presence of ambiguity in such information leads to contradictory interpretation of the ratio of the chemical compounds in the components on the surface of catalysts. Why this happens will be discussed in following chapters. When analyzing singlets with separated peaks, one can expect a high accuracy (an error within 5-10%) in determining the basic parameters of the peaks if using standard programs. For poorly resolved doublets or a mixture of peaks from different elements doped by satellites, etc., the error will generally be within 20-30% if special analysis is not undertaken.

The processing of Auger electron spectra is a still more involved task that is far from being solved. Quite a number of parameters introducing complications into and distorting the form of an Auger spectrum must be taken into account when finding the true lineshape. They include the background produced by the multiply scattered electrons, the background of inelastically scattered electrons, instrument broadening, and, last but not least, the contribution of the final states, which, as shown above, can determine the shape of the Auger spectrum. This problem was solved on a quantitative level only in a few works, examples of which are given in Chap. 3. But they relate to rather simple objects with a high conductance and quite a homogeneous composition of the surface and bulk. The spectra of real catalysts are distorted to a much greater extent because of charging, surface roughness, and the presence of several chemical states. Meanwhile, all these effects can be taken into consideration only qualitatively, and most correctly when comparing Auger and X-ray photoelectron spectra of the core levels.

The DS 300 system enables one to collect data of UPS and ISS, and the DS 800 system also data of SIMS. Moreover, there is a special program for feeding in spectral data from other sources. We used it when processing the spectra obtained in the spectrometer ES 200B. The accumulation of these spectra and their initial processing were performed using a multichannel analyzer of the Nicolet type directly connected to the spectrometer ES 200B.

In addition to the standard programs, we developed algorithms allowing us to conduct quantitative analysis on the basis of the integral intensities of XPS and with a view to the coefficients of Eq. (1.15) on the basis of the procedures described above. When studying supported systems, we also developed a program for appraising the distribution of the components on a surface and their dispersity by various models, including Kerkhof's model [185].

2.4   Specimen Preparation, Transfer and Treatment

The preparation of the surface of specimens for investigation by EES and IS is one of the most important problems in the studying of real catalysts. Spectrometers are generally supplied with standard facilities for preparing clean surfaces of single crystals or films. But many of the relevant procedures cannot be used for preparing specimens of powdered catalysts. For instance, depth profiling may lead to amorphization of the structure and substantial chemical changes. The treatment of catalysts in an atmosphere of various gases under pressures of a score or hundred of torrs in the UHV system of a spectrometer is not convenient because of the possible contamination of the main units and prolonged restoration of the working mode. The achievement of conditions under which treatment is performed in catalytic reactors and the chemical reaction itself is conducted is of still greater importance for establishing a relation between the properties of a surface and the activity of catalysts. Here special procedures and facilities are needed for the preparation of catalysts, their transfer into a spectrometer, and their treatment under *in situ* conditions. Although various firms offer separate devices for these purposes, to date there are no universal attachments on the market that meet all requirements. Moreover, such commercially available facilities are very costly. All this makes research workers themselves develop and fabricate special contrivances and reactors.

Below we briefly describe the procedures for preparing catalyst specimens and attachments for their studying by EES and IS that have been developed at our laboratory. We hope they will also be helpful for other workers in the field, first of all for those using instruments of the same type as we have and working with real catalysts. Moreover, we presume that these data will provide a notion on the scientific methods of studying how active catalyst surfaces form.

2.4.1   Preparation and Insertion of Specimens

The main way of preparing specimens is to press (or rub) them into a metal grid, mounting, or foil. First the specimens are ground into a finely dispersed powder. The specimen support can be made from stainless steel, a nickel-chromium alloy, copper, brass, indium, or lead. This method made it possible to perform heat treatment of the specimens in vacuum or in an atmosphere of various gases. It was also found that the mounting of a specimen in a grid attached to a metallic holder reduced the charging in comparison with its application onto an adhesive tape. The latter procedure was used for the express analysis of specimens requiring no further treatment or when studying specimens whose grinding and pressing may change their chemical state or the composition of the surface, and may also cause amorphization of the surface layer. Alloys were studied in the form of foil 0.05-0.1 mm thick that was in electrical contact with the holder.

Spectrometers are provided with transfer locks intended for the rapid insertion of up to four specimens through the vacuum gate valves while retaining a high vacuum in the analysis chamber. Prior to analysis, the specimens were subjected to diverse treatment including heat treatment in vacuum, treatment with H_2, O_2, CO, or NO, and bombardment by Ar^+ ions. Preparation chambers were used for these purposes. They had a tight vacuum connection to the spectrometer. Special reactors were also used for treatment at high temperatures and atmospheric pressure.

2.4.2   Preparation Chamber of the Spectrometer ES 200B

A preparation chamber was specially developed for treatment in conjunction with the spectrometer ES 200B. It was joined to the spectrometer and the gas admission system (Fig. 14a). The chamber was evacuated by a rotary roughing pump to 10^{-2} torr and an ion magnetic discharge pump providing a maximum vacuum of 5×10^{-9} torr. Also mounted on the chamber were a monopole-type head MX2401 of a mass spectrometer for measuring the residual gases, and an ion gun with $E_p = 1$ keV and $I_p \leqslant 10 \, \mu A$. A specimen was inserted into the preparation chamber via a system of valves similar to the main inlet of the spectrometer ES 200B. The chamber was connected to a gas admission system that was evacuated by a diffusion pump to 5×10^{-7} torr. Treatment was conducted at pressures of 2-5 torr and temperatures up to 600 °C. A special heater was fabricated for this purpose. It was installed in the specimen holder together with a chromel-alumel thermocouple (Fig. 14b). Moreover, the vapour of liquid nitrogen could be fed in through capillaries in the holder framework for cooling a specimen.

The preparation chamber was separated from the analysis chamber by a gate valve, which allowed treatment to be attended by the analysis of other specimens. After the termination of treatment, the preparation chamber was evacuated to 10^{-7}-10^{-8} torr, and the specimen was transferred into the analysis chamber in vacuum. The experimental conditions we selected were close to those used in investigations by IR, EPR, NMR, etc. They are quite suitable for analyzing the surface layers of powdered catalysts and solving problems

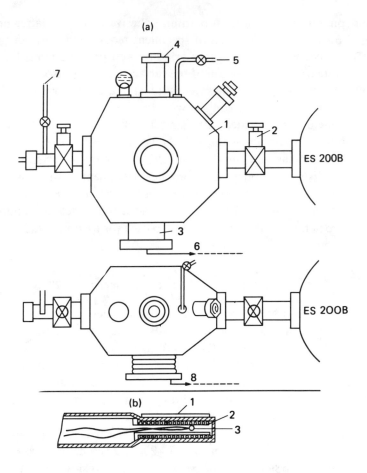

Fig. 14. Preparation chamber of the spectrometer ES 200B (a) and insertion probe (b):

(a): *1* — chamber; *2* — gate valve; *3* — vacuum transducer; *4* — mass spectrometer; *5* — gas inlet; *6* — outlet to sorption pump; *7* — outlet to rotary pump; *8* — outlet to ion pump;

(b): *1* — specimen; *2* — heater; *3* — thermocouple

associated with the determination of the chemical state of elements, the composition of a surface, or the studying of strongly adsorbed complexes. But sometimes other approaches are needed, namely, (a) for studying a surface in close to pure conditions by techniques more sensitive to surface contamination than XPS; and (b) for modelling the conditions of treatment in a catalytic reactor.

2.4.3 A Reactor for High-Temperature Treatment of Catalysts under Atmospheric Pressure

A special arrangement has been developed for treating specimens at temperatures up to 900 °C in a flow of various gases under atmospheric pressure and transferring activated specimens without the access of air into a spectrometer ES 200B [195]. It consists of a reactor with a removable furnace and a transfer unit connected to the spectrometer (Fig.15). To prevent overheating of the Viton seals, the reactor is provided with a jacket cooled by cold water. During treatment, after a holder with a specimen has been inserted into the reactor, a scrubbed gas is circulated through it, flows into the transfer unit, and emerges via a capillary. As a result, an excess pressure is created in the reactor. When transferring a specimen, the transfer unit is attached to the entry device (transfer chamber) of the spectrometer, which is continuously scavenged by purified argon. The tapered adapter on the spectrometer pushes the Teflon plug into the transfer unit, while the seal moves with the stopper onto the adapter. The counterflow of argon prevents air from getting into the

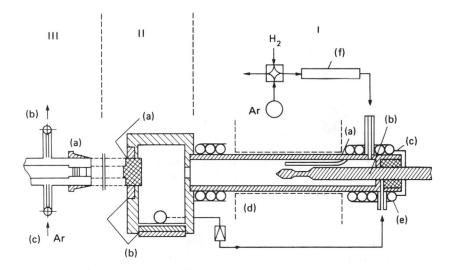

Fig. 15. Reactor for the high-temperature treatment of catalysts under atmospheric pressure connected to a spectrometer ES 200B:

I — reactor zone: *a* — reactor; *b* — specimen holder; *c* — Viton seals; *d* — removable furnace; *e* — cooled jacket; *f* — gas scrubbing; *II* — transfer unit: *a* — Teflon plug; *b* — rubber seals; *III* — spectrometer: *a* — adapter; *b* — outgassing; *c* — Ar

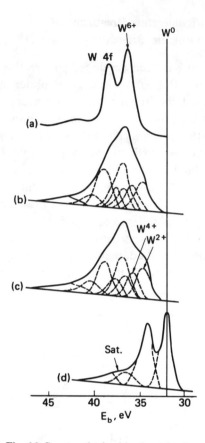

Fig. 16. Spectra obtained in the reduction of ammonium tungstate by H_2 in a microreactor ES 200B [195]:

a — initial $(NH_4)_{10}W_{12}O_{41}$; b — after treatment at 400 °C with H_2; c — after 550 °C, H_2; d — after 650 °C (coincides with the pure metal)

transfer unit. Next the specimen holder is introduced into the transfer chamber in a stream of argon, and the chamber is then outgassed. Usually the transfer time up to the obtaining of a high vacuum in the analysis chamber of the spectrometer is 20-30 min, which enables us to consider this method to be an express one.

The reliability of transfer was tested on specimens with an increased sensitivity to oxygen and moisture [195] such as W, Raney catalysts, Co, Fe, and reduced zeolites. Figure 16 shows the W $4f$ spectrum for some stages of obtaining metallic tungsten in a reactor by reducing $(NH_4)_{10}W_{12}O_{41}$ at 650 °C. The parameters of the W $4f$ spectrum (the broad low-energy peak belongs to plasmon) indicates that all of the tungsten has been reduced to the metallic state.

2.4.4   Transfer Glove Box

A special dry glove box was used to transfer specimens from catalytic reactors or arrangements in which catalysts were treated. The following manipulations can be performed in this box (Fig. 17): (a) releasing of specimens from sealed capsules or reactors; (b) grinding of specimens to a powdered state; and (c) pressing of specimens into a foil or grid screen with the aid of a special press installed in the box and fastened on the specimen holder. The glove box was first filled with purified argon. Elastic rubber containers were used to speed up the exchange of gases and maintain an excess pressure in the box. Next a capsule with a catalyst was inserted into the box in a counterflow of argon, and the box was sealed with the Teflon plug. Pressing was performed as follows. The substrate with the catalyst was placed between steel inserts. After this the glove box was placed in a hydraulic press in such a way

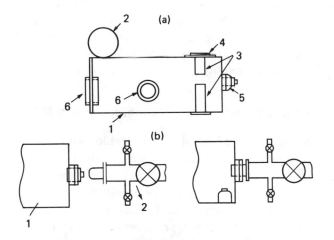

Fig. 17. Glove box for preparing catalyst specimens in an inert atmosphere (a) and how it is connected to a spectrometer (b):

(*a*): *1* — box; *2* — elastic rubber container for producing an excess pressure; *3* — inserts for pressing; *4* — annular gasket; *5* — Teflon stopper; *6* — glove box flanges; (*b*): *1* — box; *2* — transfer device

that the pressure would be transmitted to the specimen being pressed via the inserts. Motion of the upper insert was ensured by installing it on an annular gasket made from sheet vacuum rubber. This procedure prevented mechanical overloads from being applied to the body of the box when pressing a specimen.

A specimen was transferred into the spectrometer in the way described for a reactor. The time needed for preparing a specimen (including its pressing) and transferring it was 30-40 min. When the specimens were put on an adhesive tape, this time dropped to 15-20 min. A glove box can be used to transfer specimens sensitive to the air and moisture. When using various facilities for reduction and transfer (a glove box, reactor, preparation chamber), one obtains satisfactory coincidence of the degrees of reduction of the metals (Ni, Rh, Pd, Pt).

2.4.5   Reactor with Catalyst Sampling

A special microreactor (Fig. 18) was used to study the same catalyst specimen in various ways (also after a reaction). It is a thick-walled glass tube provided with pockets for a thermocouple and with a capsule that can be unsealed for taking a catalyst sample. A needle with a sleeve intended for taking samples of the reaction mixture is soldered into the reactor. A heater coil is wound directly onto the reactor walls, while the catalyst is placed on a porous glass disk. After the completion of treatment or a reaction, part of the catalyst is spilled into the capsule, which is sealed and placed into the dry glove box. After taking a sample for XPS, the other part of the specimen can be studied by various techniques such as EPR and NMR.

2.4.6   Preparation Chamber of Spectrometer XSAM 800

The spectrometer XSAM 800 is provided with a standard preparation chamber in which specimens are subjected to thermal and vacuum treatment at temperatures up to 600 °C, etching by Ar^+ ions, or the sputtering on of films by the thermal evaporation of metals from a tungsten filament. A gas admission system was fabricated for running experiments involving adsorption or treating specimens at low pressures (down to 10^{-4} torr). It included glass reservoirs with gases ($p = 500\text{-}700$ torr) separated from the "header" and the outgassing system by metal valves, and an outgassing and gas admission system made from stainless steel. A rotary and diffusion pumps were used for outgassing the preparation chamber. Treatment was conducted as follows: gas from a flask was admitted into the header up to a pressure of 10-20 torr, next a metal valve was closed, and the leak valve was used to produce

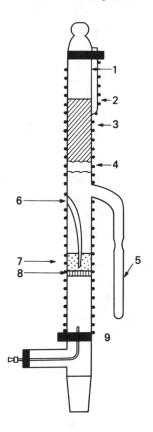

Fig. 18. Microreactor with facilities for taking catalyst samples during regeneration or a reactor:

1 — "pocket" for a regulating thermocouple; *2* — heater; *3* — forecontact; *4* — glass wool substrate; *5* — ampule for taking catalyst samples; *6* — pocket for a measuring thermocouple; *7* — catalyst; *8* — porous glass disk; *9* — needle for withdrawing reaction products

a pressure of 10^{-6}-10^{-5} torr in the preparation chamber. The rate of gas admission was regulated by a leak valve and also by the valve separating the diffusion pump from the chamber. The pressure in the chamber was measured with a Penning gauge. A quadrupole mass spectrometer operating in the mode of residual gas analysis was used for additional calibration of the pressure and controlling the purity of the gases.

2.4.7   System for Multifold Handling of Catalysts

For modelling the real conditions of catalyst treatment and also the proceeding of a catalytic reaction directly in a spectrometer, we developed a microreactor connected to the transfer chamber of a spectrometer XSAM 800 via vacuum seals (Fig. 19). This attachment has a number of additional possibilities and advantages in comparison with the reactor for the spectrometer ES 200B:

(a) a vacuum seal in the connection to the spectrometer together with a gate valve withstanding a pressure drop of 1 atm-10^{-6} torr makes it possible to clean the space of the specimen transfer unit more thoroughly from air, moisture, and other contaminants by repeated "rinsing" with purified argon and outgassing. This ensures a higher reliability of transferring specially processed specimens;

(b) the microreactor is outfitted with a manipulator for detaching the tip with the specimens from the transfer rod and for attaching it again after treatment. This ensures a high accuracy of maintaining and measuring the specimen treatment temperature within the range from 50 to 900 °C and the possibility of conducting many cycles of treatment and measurement with the same specimens;

(c) a special system of supplying gases with their thorough scrubbing allows treatment to be performed under atmospheric pressure with various gases and their mixtures at a preset ratio of the components and flow rate;

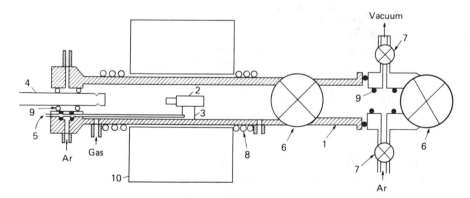

Fig. 19. System for multifold treatments of catalysts and their transportation to an XSAM 800 spectrometer:

1 — reactor; *2* — specimen holder; *3* — fixing device; *4* — rod; *5* — thermocouple; *6* — high-vacuum gate valve; *7* — fore-vacuum valve; *8* — cooling jacket; *9* — seals; *10* — removable furnace

(d) the reactor has an outlet for analyzing the reaction products with a chromatograph or mass spectrometer.

The stydying of specimens of MnO, W, and Mo with their high sensitivity to oxygen-containing contaminants proved the high reliability of the system and the good reproducibility of the experiments.

2.4.8   Measurement of Spectra in Different Stages of a Catalytic Reaction

Figure 20 schematically exemplifies procedures that can be used to study the surface of catalysts by EES and IS with the aid of the instrumentation described above. When studying the activity in a pulse microcatalytic ar-

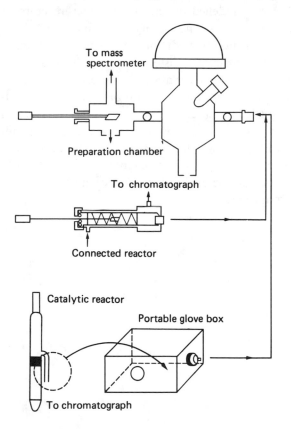

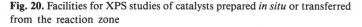

Fig. 20. Facilities for XPS studies of catalysts prepared *in situ* or transferred from the reaction zone

rangement, a specimen that is then directly studied by XPS is the catalyst. The specimen weighs about 10 mg. The spectra can be measured after introducing each pulse of the reaction mixture. The catalysts Pt/zeolite in the reaction of aromatization of lower alkanes were studied in this way [196]. When measuring the activity under flow (continuous) conditions the specimen is placed on a grid in direct proximity to the catalyst bed. It is then transferred in a sealed capsule into a glove box, is extracted, placed on a rod, and fed into the spectrometer. When a reactor is used with the taking of catalyst samples, the specimens after a reaction or special treatments are transferred into the glove box, where separate portions can be used for measurement by physicochemical methods.

The differential conditions of reactions such as hydrogenolysis and the hydrogenation of CO are modelled directly in the microreactors of the ES 200B and XSAM 800 instruments. The activity can be measured either directly on given specimens, or on specimens of the same kind under similar conditions. In Chaps. 3-5, the possibilities listed above are illustrated by specific examples of investigations in which a relation is established between the activity and the state of a surface, and the formation of a catalyst under the effect of the reaction medium is also studied. The above procedures when further developed will make it possible, in our opinion, to overcome the "alienation" of measurements by EES or IS, which require a high vacuum, from the conditions of catalysis because these procedures enable one to study not only the nature of the precursors, but also of the active centres themselves, and of some types of reactive intermediates.

3

The Use of Electron Emission and Ion Spectroscopies for Studying the Fundamental Characteristics of a Surface

All research associated with catalysis and performed with the use of EES and IS can be grouped by different aspects.

1. By the type of objects — the clean surfaces of single crystals and the more real polycrystalline and powdered catalysts.

2. By the type of tasks — fundamental studies of the basic properties of a surface and applied studies of individual characteristics of catalysts.

3. By techniques.

Frequently only studies performed on the clean surfaces of single crystals or films claim the role of fundamental studies, while investigations of more real systems are relegated to a secondary role. Here authors refer to the involved nature of interpreting spectral data, the insufficiently good characteristics of the objects, etc. One of the tasks of our monograph is to prove that the basic characteristics of a surface needed for the deep understanding of the mechanism of catalysis can also be determined on real objects. The task is not to oppose these two regions of research, but to attempt to compare their results and extract the maximum possible information when studying involved, but real systems. The latter becomes more and more possible in the light of the theoretical and methodological capabilities of modern physical methods described in Chaps. 1 and 2.

The present chapter analyzes the state of investigations of the most important properties of a solid surface for one of the most complicated classes — catalysts.

1. The chemical state of the components and the electron structure of catalysts.

2. The electron state of small clusters and surface atoms.

3. The composition of a surface and depth profiles of the components.

4. The local structure and morphology of a surface.

The level of solving these questions is different and depends substantially on the type of objects. For this reason, the possibilities of various techniques are illustrated by examples of studying both the well characterized model objects and complicated catalysts. Discussed in detail are the key moments in interpreting the data and extracting therefrom information on the electron structure of small clusters, the quantitative composition of surface layers with a thickness from a monolayer to several scores of atomic layers, local fragments of the structure of amorphous highly dispersed materials, etc.

3.1   Chemical State of Components and Electron Structure of Catalysts

By the chemical state of elements is meant the set of characteristics determining the chemical bond of a given element to its closest surroundings. These characteristics include the valence, effective charge, degree of ionicity or covalence of a bond, and the reaction with the close surroundings with the formation of stoichiometric compounds. In chemical literature, the state of an element is often described by its oxidation state, which also reflects the electron state of an atom in compounds or matrices of a single type. All these questions are studied broadly using XPS and high-resolution Auger spectroscopy with X-ray or electron excitation. The numerous examples given in monographs and reviews [7, 8, 22, 25-27] depict the effectiveness of determining the valence state of elements in various catalysts such as oxides, zeolites, and supported systems. In most cases, the valence state was identified empirically by comparing E_b or ΔE_b in catalysts and in reference compounds containing elements in known oxidation states, or by analyzing the oxidation states formed when these compounds are decomposed in the course of their reduction, oxidation, or ion bombardment. Since an element is often present in catalysts in several valence states, the procedures of peak synthesis, deconvolution, etc. are conducted for separating their spectra. Therefore, the accuracy of determining the degrees of reduction or the fraction of various states depends on the procedure used, the intensity of the lines, and so on. On supports even with a close surrounding of elements (e.g. oxygen), the lines may be shifted and broadened [27], which must also be taken into consideration when determining the valence states.

The data presented in Table 2 (see p. 21) illustrate the order of magnitude of the shifts observed when the oxidation states of the elements in catalysts change. To the very first approximation, a change in the oxidation state by unity should cause a shift of 1 eV. This also points to the linear relation between E_b (ΔE_b) and the oxidation state (n). Such relations were actually ob-

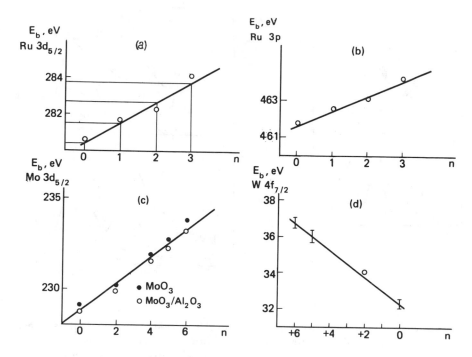

Fig. 21. Binding energy E_b of the core levels against the oxidation states of transition metals in catalysts:

a, b — Ru in zeolite RuNaY [27, 197]; c — Mo in MoO_3 and MoO_3/Al_2O_3 [198]; d — W in WO_3 and WO_3/Al_2O_3 [199]

served for the transition metals in some matrices of one type such as zeolites and oxides (Fig. 21). In other cases, the shifts are extremely nonuniform and depend greatly on the nature of the nearest surroundings [27, 34]. Consequently, the determination of the oxidation states of elements in catalysts requires thorough analysis of obtained data and verification of already available information.

A typical example is the determination of the states for the elements studied the most by XPS such as molybdenum, tungsten, and rhenium, for which the range of shifts in a transition from the highest states to the zero one is very great, namely, 5-7 eV. It was initially assumed that there is a linear relation between E_b and n [200, 201] as a result of analyzing the shifts in the spectra of reduced MoO_3 and calculating the clusters of MoO_6^{n-}. These data were then criticized by Patterson *et al.* [202] and Nikishenko *et al.* [203] on

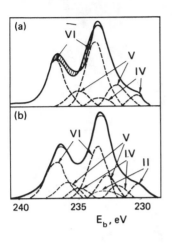

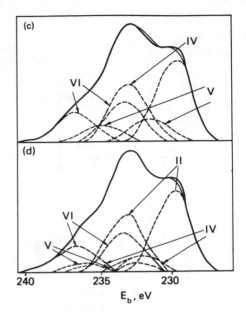

Fig. 22. Analysis (peak synthesis) of spectra of Mo $3d$ in reduced MoO_3 (Ar, 873 K) (a, b) and MoO_3/Al_2O_3 (H_2, 873 K) (c, d) with a different number of spin doublets [198]:

a, c — approximation by three doublets: Mo(VI), Mo(V) with E_b = 231.6-232.2 eV, and Mo(IV); b, d — approximation by four doublets: Mo(VI), Mo(V) with E_b = 232.4-232.7 eV, Mo(IV), and "Mo(II)" [198, 200]

the basis of the results obtained in the synthesis of the peaks of Mo(VI), Mo(V), and Mo(IV) in supported catalysts. Nikishenko *et al.* arrived at the conclusion that there is more likely a "band" of ambiguity rather than a linear relation between E_b of Mo $3d$ and n. Since two groups of data were obtained on various types of specimens, namely, unsupported and supported molybdenum oxides differing in the degrees of reduction and the parameters of the spectra, comparative studies were required to obtain more reliable conclusions on the states of Mo, especially Mo(V) [198]. Figure 22 presents experimental and synthesized spectra of reduced MoO_3 and MoO_3/Al_2O_3 obtained under the same conditions. By comparing the data of XPS and EPR relative to Mo(V)/ΣMo, and also using the laws observed in the shifts of the level W $4f$ for tungsten-containing oxides of the same type [204], Grünert *et al.* concluded that the Mo $3d_{5/2}$ peak with E_b = 232.4 − 232.7 eV belongs to the state Mo(V) and not the one with E_b = 231.8 − 232 eV, while the relation between E_b of Mo $3d$ and n_{Mo} = 6, 5, 4, 0 nevertheless has a linear nature (Fig. 21c).

The same conclusion follows for unsupported and supported tungsten catalysts [204], in which the states W(V), W(IV), and W(0) were found. Moreover, a W $4f_{7/2}$ peak was observed for tungsten oxide, which can be ascribed to W(II). However, Haber *et al.* [199] indicate that it should more probably be related to the pairs W(IV)-W(IV) localized in a WO_2 structure. When analyzing the Re $4f$ spin doublets of reduced Re catalysts on SiO_2 and Al_2O_3, the states Re(VI), Re(IV), Re(III), and Re(0) were singled out [205], although in this case the shifts when the oxidation state changes are not uniform. These examples reveal that even in complex catalysts up to three or four states can be separated quantitatively, although the difficulty of synthesizing the peaks of such systems does not enable one to obtain the parameters of the spectra of intermediate states with a high accuracy.

These difficulties can be eliminated partly by studying specimens with the use of synchrotron radiation that ensures a resolution of 0.2-0.3 eV. When analyzing one of the systems studied in the greatest detail, namely, $SiO_2/SiO_x/Si(111)$ by using a synchrotron source with $h\nu = 100 - 130$ eV [206], Braun and Kuhlenbeck succeeded in obtaining for the first time all four states of oxidized silicon: Si^+, Si^{2+}, Si^{3+}, and Si^{4+} (Fig. 23). The shift when the oxidation state changed by unity ($\Delta n = 1$) was 0.9 eV. This is a somewhat unexpected result because the value of E_{ER} changes substantially in the transition from silicon to its oxide. Braun and Kuhlenbeck succeeded not only in separating all the states of Si quantitatively, but also in determining how they are distributed when passing from the surface of an oxide film to Si(111).

There are quite a few elements for which transitions from one chemical state to another result in very small shifts, or are not attended at all by a change in E_b (see Table 2, p. 21). Examples of anomalous chemical shifts have also been observed. We can single out two basic reasons for the distinctions in the magnitudes and directions of the chemical shifts:

(a) a different effective charge (potential) on a given atom in the given electron configuration;

(b) differences in the relaxation energies for different electron configurations.

A linear relation between E_b and q in the form of $E_b = Aq + B$ is a quite good approximation if we compare series of compounds differing appreciably in their type of bond [1-3]. Within the limits of comparing simple calculations by the complete neglect of differential overlapping (CNDO) or electron populations after Mallican and thoroughly calibrated values of E_b, such relations were obtained for a number of elements and organometallic compounds. The values of the coefficients A and B for selected elements and electron levels

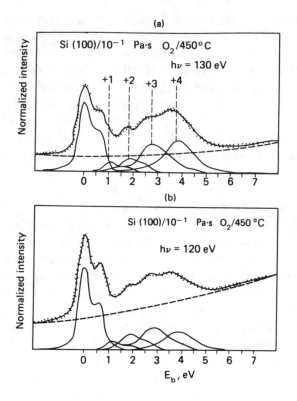

Fig. 23. Deconvolution of peaks of Si 2p obtained using a synchrotron source with $h\nu = 130$ eV (a) and 120 eV (b) for oxidized silicon [Si(100), O$_2$, 450 °C] [206]

are given in Table 5 [207]. A more accurate relation between E_b and q obtained for carbon-containing compounds also includes nonlinear terms:

$$E_b(C\ 1s) = 6.42q_C + 4.52q_C^2 + 285.8 \qquad (3.1)$$

But this relation remains monotonic for a broad number of compounds. In the opinion of Sundberg *et al.* [207], this also points to a monotonic relation between q and ΔE_{ER}. This conclusion is apparently correct only in some cases. Detailed measurements of the Auger parameter performed by Wagner *et al.* [177] reveal that ΔE_{ER} changes sharply even in a transition from a metal to its oxide. As we have already noted, the values of α' themselves can be a

Table 5. Coefficients A and B of the Equation $E_b = Aq + B$ Relating the Binding Energy of the Core Electrons E_b (eV) and the Atomic Charge q [207] for Selected Elements and Electron Levels

Element	Level	A	B	Element	Level	A	B
C	$1s$	11.5	284.8	Fe	$2p_{3/2}$	6.4	704.1
N	$1s$	7.00	401.4	Ni	$2p_{3/2}$	6.74	848.3
O	$1s$	4.23	534.1	Cu	$2p_{3/2}$	1.52	932.2
F	$1s$	4.28	688.8	Pd	$3d_{5/2}$	4.45	333.9
Si	$2p$	1.53	100.6	Pt	$4f_{7/2}$	3.17	71.1
P	$2p$	1.67	131.6	Cr	$2p_{3/2}$	1.98	576.1
S	$2p$	3.38	163.8	Mo	$3d_{5/2}$	4.49	228.9
Cl	$2p$	4.25	201.2	Sn	$3d_{5/2}$	1.81	485.8

qualitative criterion of the chemical state. To determine these parameters quantitatively in the absence of noticeable chemical shifts, it is necessary to analyze the shapes of the Auger transition lines. Sometimes, for instance when Cu^0, Cu^+, and Cu^{2+} are present, an analysis of the XPS and AES spectra allows one to readily uncover all the states of copper (see Fig. 2, p. 23). This analysis underlies numerous studies of copper-containing catalysts used in the synthesis of methanol and the water-gas shift reaction, for which the determination of the active form of copper is of a fundamental importance [208-210]. Without discussing the results of these works here, we shall indicate that at present it is possible to single out the states of Cu^+ and Cu^0 or Zn^{2+} and Zn^0 in mixed oxides quantitatively.

Considerable hopes in the determination of the charge and spin density of atoms were associated with analysis of the satellite structure [43-52]. But to date no quantitative description of this relation has been obtained, and analysis consists in finding an empirical relation between the intensity and energy of satellite shake-up and the state of the transition elements. The transitions $Cu^{2+} \rightarrow Cu^+$, Cu^0 and $Co^{2+} \rightarrow Co^{3+}$ are clearly identified, for example, by the vanishing of the relevant satellites.

Within the terms of an electrostatic model, an increase in E_b without a change in the oxidation state indicates a growth in the effective charge. The high ionicity of the bond of transition cations with the framework of zeolites according to E_b of the core levels has been confirmed by a number of authors

[211]. But of no less importance is the contribution of the Madelung interaction, which also leads to substantial shifts for cations of oxides or elements of zeolite framework [22, 197]. XPS has a low sensitivity to a change in the coordination state of elements, but when additional techniques are employed, information can be procured on the coordination of cations in a structure. For example, two components of the spectrum of Cu $2p_{3/2}$ observed previously by Minachev et al. [212] were ascribed by Narayana et al. [213] to two coordination states of Cu^{2+}. The latter was confirmed by data of EPR. The high-energy peak with E_b of about 936.2 eV was related to octahedrally coordinated cations of Cu^{2+} localized in large cavities of a type Y zeolite, while the low-energy peak (933.9 eV) was related to tetrahedrally coordinated ions of Cu^{2+} in the sodalite cages.

The set of techniques (XPS, ISS, SIMS, photoacoustic and Raman spectroscopy) employed in a series of works by Hercules et al. [214-217] made it possible to determine more reliably not only the chemical state of elements (Co, Mo, Ni, and W), but also their coordination and chemical interaction with the support. For example, to identify the Co^{2+} ions occupying tetrahedral and octahedral positions in the lattice of $\gamma\text{-}Al_2O_3$, the data used included the Co $2p_{3/2}$ binding energies, splitting between the main and satellite peaks, the Co/Al ratio obtained by XPS, ISS, and SIMS, the characteristic bands in photoacoustic spectra, and also the effect of screening the He^+ scattering from the differentially coordinated Co^{2+} ions.

Of importance is not only the qualitative, but also the quantitative determination of various chemical states and most frequently of the degrees of reducing transition metals to a zero-valence state. The difficulty of determining the true degree of reduction lies in the need to take account of the processes of sintering and redistribution of a supported phase in reduction. If these processes are disregarded, the degree of reduction is determined as

$$\alpha_{app} = \frac{I(Me^0)}{\Sigma\, I(Me^0) + I(Me^{n+})} \tag{3.2}$$

where $I(Me^0)$ and $I(Me^{n+})$ are the intensities of a peak of a metal and of the particles that were not reduced, respectively. The apparent degree of reduction may be understated because of aggregation of particles on the surface or be overstated when reduced metal atoms migrate from pores to the outermost surface. Assuming that the processes of aggregation and migration concern only the metal phase and that no transportation of ions over large distances occurs, the actual degree of reduction can be evaluated by the formula [218]:

$$\alpha_{act} = \frac{[I(Me^{n+})/I_s]' - [I(Me^{n+})/I_s]''}{[I(Me^{n+})/I_s]'} \tag{3.3}$$

where $I(Me^{n+})$ is the same as in Eq. (3.2), I_s is the intensity of the peak of one of the support elements, which is assumed to be constant, and the prime and double prime signify "before reduction" and "after reduction", respectively. Equation (3.3) is observed for many supported metal catalysts prepared by impregnation, although with substantial cation migration additional techniques have to be used to appraise the degree of reduction.

Information on the charge on the nonmetallic components of catalysts is significant for revealing the nature of the active centers. For oxides and zeolites, these components are various types of oxygen. Its effective charge can be determined by using the O $1s$ and KLL spectra. Considerable changes in the O $1s$ spectra of oxygen in a lattice were observed only with sharp changes in the covalence of a bond or in the acid-basic properties in oxides and zeolites [219, 220]. This is probably why such information cannot be used for a detailed description of the properties of oxygen. More definite information was obtained when analyzing KLL, KLV, and KVV Auger transitions of oxygen and the Auger parameter [221]. The splitting of ΔE (KVV-KLV) is a criterion of the charge on oxygen since the relaxation contributions for these two sublevels are identical (Fig. 24). A relation was also found between the ratio of the inten-

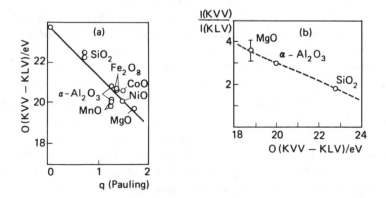

Fig. 24. Correlation between the energy splitting of O ($KVV - KLV$) peaks and the Pauling charge on oxygen in oxides (a) and between the ratio of the intensities of these peaks and their splitting (b) [222]

sities and the energy splitting of these transitions in which one or two valence levels participate, and a phenomenological model establishing the relation between the charge on oxygen and the parameters of the Auger spectra was proposed. To what extent this model is effective for appraising the Lewis basicity or acidity in oxides and aluminosilicates will apparently be shown by studies of specimens with known acid-base characteristics. The separation of the local density of the electron states from the Auger spectra could be the next step in appraising the charges on oxygen. But such a procedure can be performed with sufficient accuracy only for thin films with a view to the appreciable broadening and distortion of the spectra of insulators and the ambiguity of the position of the Fermi level.

The studying of the electron structure of a multicomponent catalyst by emission electron spectroscopy is an involved task in the general case. The valence bands of specimens containing *sp* elements, e.g. specimens of representative metal oxides and zeolites, have a low intensity and are blurred. And although Barr [23] indicates differences in the shape of the bands for various types of zeolites and oxides (X, A, mordenite, SiO_2, and Al_2O_3), and he attempts to appraise the degree of amorphization of the zeolites by the width of the valence band, the observed insignificant difference can also be ascribed to charging of the nonuniform surface of the powders. Sharper spectra can be obtained when the catalysts contain *d* elements. Figure 25 depicts spectra of the valence band of the catalyst Rh/TiO_2 that was subjected to reduction at various temperatures. The change in the shape of the bands near the Fermi level that had belonged to the *d* states of Rh and to the ions Ti^{3+} enables one to judge on the transfer of the electron density in the metal/semiconductor system. At a high temperature, a strong interaction between the metal and its support appears, which by these data results in an increased density of the *d* states on the metal. This conclusion was also confirmed by negative shifts in the Rh $3d$ spectra [222]. But a quantitative description of the electron structure of such a system (the effective charge, the degree of *spd* hybridization of the Rh-Ti bond) is hampered for reasons that will be discussed in Sec. 3.2 and Chap. 4.

As previously, the most detailed conclusions on the electron structure can be arrived at from the data of EES only for model systems, namely, alloys and stoichiometric compounds. For example, in the intermetallics Ni_3Ti, a distinct change in the density of the valence states (XPS) due to electron interaction of the components manifests itself [223]. Both experimental data and data obtained by calculations of clusters by the X_α-SW SCF method show the transfer of the electron density from Ti to Ni, as a result of which the

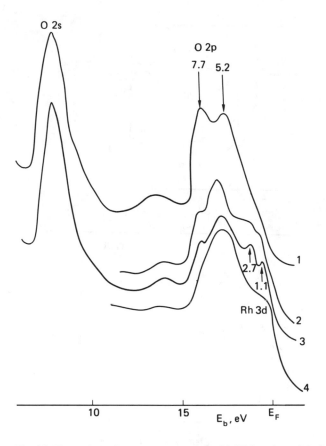

Fig. 25. X-ray photoelectron spectra of the Rh/TiO₂ valence band
[223]:

1 — TiO₂, 300 °C, H₂; *2* — 5% Rh/TiO₂, 300 °C, H₂; *3* — 5% Rh/TiO₂,
500 °C, H₂; *4* — 5% Rh/TiO₂, 300 °C, H₂ + Ar⁺, 1 keV, 30 min

antibonding Ni-Ti orbital localized near the Fermi level is half filled, while
the Ti acquires a charge (+1).

To illustrate the modern approach to studying the electron structure of
alloys and intermetallics, we shall present some data from a series of works
[98, 224, 225] in which there were investigated about 60 binary and ternary
alloys and intermetallics of Pd and Ni with 20 elements differing in their thermodynamic characteristics and electron structure. The spectra of the valence
band, core levels, and Auger transitions were measured with a high resolution
(X-ray monochromator) in a vacuum of 5×10^{-11} torr. The following conclu-

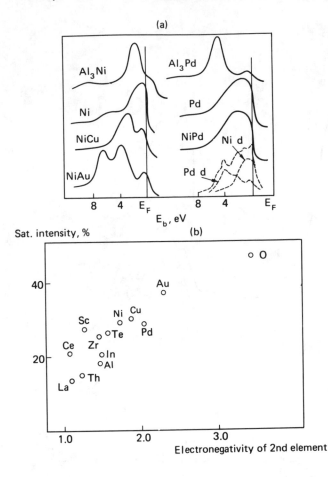

Fig. 26. Electron structure of some alloys of Ni and Pd [98, 225, 226]:

a — spectra of the valence band (the dashed line shows the theoretical total and partial density of the states in the NiPd alloy; *b* — intensity of the shake-up satellite in the Ni 2*p* spectrum against the electronegativity of the second component in Ni alloys

sions were made (Fig. 26): (1) in alloys with electropositive elements, the peaks of the *d* bands of Ni and Pd are shifted toward larger binding energies with the corresponding lowering of the density of the states at the Fermi level. This indicates filling of the *d* bands; (2) the *d* bands of Ni and Pd in such alloys are narrowed, which indicates strong Ni-Me and Pd-Me interaction; the interaction strength increases with a growth in the density of the states of the second metal in the region of the *d* states of Pd and Ni; (3) the shake-up satellite intensity in the Ni 2*p* spectrum correlates with the electronegativity of the second component; the satellite does not disappear completely in any of the alloys (including Ni-Cu), which confirms the retaining of the *d* vacancies in

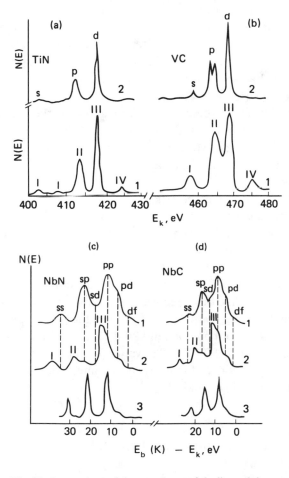

Fig. 27. Comparison of the true shape of the lines of the experimental Auger spectra, X-ray photoelectron spectra of the valence band, and the theoretical density of the electron states (DES) (in the relevant clusters) of some nitrides and carbides [99-101]:

a — TiN: 1 — Ti $L_3M_{2,3}V$; 2 — DES; b — VC: 1 — V $L_3M_{2,3}V$; 2 — DES; c — NbN: 1 — valence band; 2 — N KVV convolution; 3 — DES; d — NbC: 1 — valence band; 2 — C KVV convolution; 3 — DES

the valence band; (4) the filling of the *d* bands of Pd or Ni is due to hybridization with bands of the second component; (5) although in the formation of some intermetallics a substantial shift in E_b of the inner levels of Pd or Ni is observed, the transfer of the charge between the components of the alloys is insignificant; and (6) the determination of the true form of the Auger spectra of the Ni *LVV* transitions by the Cini-Sawatzky model [98] yielded additional information on the relation between the electron structure in alloys and the local Coulomb interaction of the holes in an Auger process. Notwithstanding the appreciable changes in the density of the states in the alloys, the potential of the Coulomb interaction changed insignificantly (by less than 1.4 eV). Only in alloys of Ni with Al, La, and Th did the spectrum acquire some details of a quasiatomic structure (see Fig. 4b, p. 35).

An important consequence of these studies for catalysis is the found relation between the electron structure of the alloys and the difference in the electronegativities of the components. Moreover, the data for the compounds of Pd and Ni with Al, Ti, Cr, Cu, Au, and also with La, Th, and Ce are of interest for metal oxide catalysts.

The procedures for determining the true shape of the type *KVV* Auger transitions are of general interest for a quantitative appraisal of the local density of states, at least in two-component catalytic systems such as nitrides, carbides, and oxides [97, 99-101]. A comparison of the true shape of the Auger spectra, the XPS spectra of the valence band, and the density of states calculated by the X_α-DV method in a single energy scale for a number of carbides and nitrides (Fig. 27) shows that the effective Coulomb interaction in these systems is low (especially in the spectrum of the metallic component), and the partial atomic charges on both components of a catalyst can be estimated directly from an experiment [97, 99-101]. The values of the charges are of a substantial interest as the initial data for studying the interaction of the local centers of catalysts with reacting molecules.

3.2   Electron State of Small Clusters and Surface Atoms

The spectra of the core levels and valence bands of small clusters differ greatly from those of bulk metals. The most concise relations between the spectral characteristics and the size of clusters have been obtained for simulated systems. Figure 28 presents a typical example of the change in the parameters of the core level spectra (XPS) and valence bands with a growth

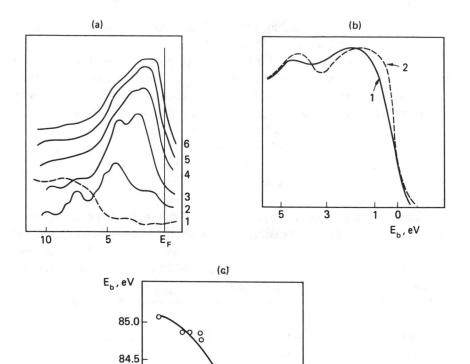

Fig. 28. Density of states in the valence band depending on the coating of SiO_2 with supported Pd [229] (a) and Pt [227] (b), and shift of Au $4f_{7/2}$ for Au clusters on amorphous carbon [229] (c):

a (Pd): 1 — SiO_2; 2 — $\theta = 3.9 \times 10^{13}$ atom/cm^2; 3 — 1.6×10^{14}; 4 — 4.0×10^{14}; 5 — 5.8×10^{15}; 6 — 1.8×10^{16} atom/cm^2 (bulk) b (Pt): 1 — $\theta = 3 \times 10^{15}$ atom/cm^2 (corresponds to $d_{av} = 1.8$ nm); 2 — 1.2×10^{16} atom/cm^2

in the coverage of the support by a metal (10^{14}-10^{16} atom/cm^2). The main features of the spectra with low coverages (small cluster sizes) are [226-246]:

    (1) an increase in the binding energy of the core levels;

    (2) an increase in FWHM of the core levels;

    (3) a reduction in the asymmetry of the core level lines;

    (4) a lower intensity of photoemission near the Fermi level, less spin-orbital separation, and a smaller valence band width; and

(5) an increase in the binding energy of the maximum of the d states in the valence band and a higher Fermi level.

It should be noted that even for the lowest coverages at which a dispersion close to an atomic one is presumed, the valence band is much wider than what is expected from calculations for an isolated cluster. Such changes in the spectra have been observed qualitatively for various metals, e.g. Pd, Ag, Au, Cu, and Pt. Amorphous carbon is used especially frequently as the support, although data are available for specially prepared SiO_2, Al_2O_3, TiO_2, alkali metal halides, Teflon, etc. and also for metal supports. In most cases, the specimens are prepared by evaporating the relevant metal onto the smooth surface of the support.

With an increase in the coverage, the spectra acquire the shape and position typical of bulk metals. The possible reasons for the observed changes in the spectra and, first of all, in the binding energies have been treated in a number of theoretical and experimental works [35, 37, 38, 226, 228-249] and have been discussed in detail in reviews [226, 228, 232].

The shifts and line broadening in the spectra of small clusters are associated with the following main factors:

(a) **initial state effects** — the intrinsic electron structure of metal clusters and charge transfer to the support;

(b) **final state effects** produced by a change in the relaxation energy, residual charge, etc. Let us analyze in greater detail the possible contribution of the two components to changes in the spectra.

We have already indicated in Sec. 1.1 that E_b is approximated by two terms [Eqs. (1.8), (1.9)], namely, E_j, the term of the energy of an orbital (the initial state) and E_{R_j}, the relaxation energy spent on screening a hole. As a result, E_{hole} and E_b diminish [36]. The quantity E_R, in turn, includes E_{IR} and E_{ER}. The energy of intra-atomic relaxation does not depend on the chemical and physical states of an atom and can therefore be omitted from our further reasoning. The quantity E_{ER} — the energy of extra-atomic relaxation, decreases in a transition from a metal to an isolated atom, and its value will depend on the relative ability of the support and bulk metal to screen a hole. The support is generally less effective, which leads to a positive shift relative to the metal. The incomplete screening of the hole potential in small clusters also explains line broadening [229]. On the basis of a simple model of a "particle in a box", evaluation of the dependence of the hole potential on the size yields $\Delta FWHM = 0.34\,eV$ for particles with $N_{el} = 500$, and $\Delta FWHM = 0.86\,eV$ for ones with $N_{el} = 20$.

A procedure based on determination of the Auger parameter has been proposed for the quantitative separation of the component ΔE_{ER}. The kinetic energy of an Auger electron in a j, k, l transition is calculated as follows [94-96]:

$$E_k(j, k, l, X) = E_b(j) - E_b(k) - E_b(l) - F(kl; X) + E_{R,s}(kl) \qquad (3.4)$$

where $F(kl; X)$ is the energy of interaction of two holes k and l in a multiplet with the final states X, and $E_{R,s}$ is the static energy of relaxation — the polarization energy:

$$E_{R,s}(kl) = E_{R,t}(kl) - E_{R,d}(k) - E_{R,d}(l) \qquad (3.5)$$

Here $E_{R,t}(kl)$ is the total energy of relaxation of two holes, $E_{R,d}$ is the dynamic energy of relaxation of one hole. If the latter have the same quantum numbers n and l, the total relaxation energy is $E_{R,t}(kl) = 4E_{R,d}(k)$, that is $E_{R,s}(kk) = 2E_{R,d}(kX)$. By Eqs. (1.9) and (3.4):

$$\Delta E_b(j) = -\Delta E(j) - \Delta E_{ER,d}(j) \qquad (3.6)$$

$$\Delta E_k(j, k, l, X) = -\Delta E(j) - \Delta E(k) - \Delta E(l) - \Delta F(kl; X)$$
$$+ \Delta E_{ER,s}(kl) \qquad (3.7)$$

If we consider a transition of the CVV type, then $\Delta F(kl; X)$ does not depend on the number of atoms in a cluster and can be disregarded. With identical final states, $k = l$, $E_{ER,s}(VV) = 2E_{ER,d}$, and

$$\Delta E_b = -\Delta E - \Delta E_{ER,d} \qquad (3.8)$$

$$\Delta E_k = \Delta E - 2\Delta E(\overline{V}) + 2\Delta E_{ER,d}(V) \qquad (3.9)$$

where $\Delta E(\overline{V})$ is the mean binding energy of the valence electrons participating in the Auger transitions. Provided that for the valence and core levels $\Delta E_{ER,d}(V) = \Delta E_{ER,d}(C)$, we have

$$\Delta E_k = -\Delta E - 2\Delta E_b(\overline{V}) + E_{ER,d}(V) \qquad (3.10)$$

In a very simple approximation (only core levels participate in an Auger transition), we arrive at Eqs. (1.29) and (1.30). The change in the modified Auger parameter is

$$\Delta\alpha' = \Delta E_b + \Delta E_k = 2\Delta E_{ER} \tag{3.11}$$

Thomas [247] has analyzed the influnce of the second order terms on ΔE_{ER}:

$$\Delta\alpha' = 2\Delta E_{ER} - \frac{4}{3}\Delta\left[\left(\frac{dk}{dN}\right)\left(\frac{dq}{dN}\right)\right] \tag{3.12}$$

where dk/dN is a negative quantity describing the change in the configuration of the valence orbitals due to the loss of a core electron; dq/dN describes the transfer of a screened charge to the valence orbitals, and $-\frac{4}{3}\Delta\left[\left(\frac{dk}{dN}\right)\left(\frac{dq}{dN}\right)\right]$ can vary from 0 to +1, i.e. the maximum value of the coefficient in Eq. (3.12) is 3 eV and characterizes the transition of a metal atom from the free to the condensed state. But in most cases this coefficient is not great for molecular systems (0.2-0.3 eV), and it grows for molecules with readily polarizable ligands. If we extend this trend to clusters of metals, we can expect overstated values of ΔE_{ER} to be obtained when no account is taken of the second-order term. Hence, the use of Eqs. (3.10) and (3.12), or in the simplest case of Eq. (3.11) when the relevant experimental data are available enables one to appraise ΔE_{ER}.

In a detailed investigation of model sputtered-on specimens by ESS techniques, Mason [226] appraised the relaxation shifts for five metals on amorphous carbon (Table 6). Their values can range from 30 to 100% of ΔE_b. But

Table 6. Relaxation Shifts in Spectra of Metal Clusters [226]

Shift (eV)	Metal				
	Ag	Au	Pd	Pt	Ni
Shift of core level (XPS)	0.41	1.00	1.17	0.75	0.9
Shift of Auger electron spectrum	1.2	1.68	2.18	1.96	0.9
$\Delta = 2\Delta E_{ER}$	0.8	0.68	1.01	1.21	0
ΔE_{ER}	0.4	0.34	0.50	0.61	0

Mason also indicates the rather low accuracy of determining E_k for broad and low-intensity Auger spectra, and that ΔE_{ER} may be overstated because the second-order terms [Eq. (3.12)] were disregarded.

A comparison of these data with the results of X-ray absorption spectroscopy, which were also used to assess ΔE_{ER}, allowed us to conclude that at least on inert supports (C, SiO_2), ΔE_b is due to the initial state effects. Mason ascribed the entire change in E_b to the size effect — the reduction in the true count of d electrons in the small clusters because of spd rehybridization. Calculations have shown [37] that the transition from the configuration $3d^8 4s^2$ (a nickel atom) to $3d^9 4s^1$ (metallic nickel) should result in a shift of 5.5 eV. The sensitivity of the core levels to the configuration of the valence electrons is due to the presence of strong Coulomb repulsion. Since the sp electrons are more liable to diffuse than the d electrons, their interaction with the inner electrons increases E_b. The photoelectron cross section of the d levels is much higher than that of the sp levels, therefore the deficiency of the d electrons is easily observed according to the lower density of states near the Fermi level (see Fig. 28).

When studying spectra of Pd sputtered onto Al_2O_3 (sapphire) and SiO_2, Kohiki [228] obtained a number of interesting and unexpected results once more proving the necessity of taking the initial state effects into account when interpreting the spectra of small particles. Attention was given to the relation between the polarizability of the support, which differs substantially for SiO_2, Al_2O_3, and C, and the contribution of $\Delta E_{ER,s}$ to ΔE_b. Silicon dioxide as the most polarized support makes the greatest contribution to the screening of a hole, which is expressed in the large value of ΔE_{ER} when going over from an atom to a metal (Pd) on SiO_2 (+1.1 eV) in comparison with Al_2O_3 (+0.65 eV). The spectra of Pd $3d_{5/2}$, Pd MVV, and the valence band were used to appraise all the components of Eq. (3.10) for various surface coverages. It was found that at low coverages ($0 < 1 \times 10^{15}$ atom/cm^2 on Al_2O_3 and $0 \leqslant 1 \times 10^{14}$ atom/cm^2 on SiO_2) the value of ΔE_{ER} does not change with a growth in the coverage, which is associated with the small contribution of the conduction electrons of the cluster itself. With a further increase in the surface coverage, the modified Auger parameter and ΔE_{ER} begin to grow, approaching the values typical of the bulk metal. Consequently, the changes in E_b at low coverages are due to the change in the mean energy of the valence electrons $\Delta E(V)$. Since the latter diminishes owing to spd rehybridization, the true chemical shift in this region is positive, namely, $+0.9$ eV for Al_2O_3 and $+0.8$ eV for SiO_2. Some data [228] can be interpreted not only from the viewpoint of the different contribution of ΔE_{ER}, but also from the viewpoint of the different cluster size. At lower surface coverages on SiO_2, the size of the Pd particles is smaller, which is reflected in the valence band spectrum typical of quasiatomic Pd. If we take equivalent coverages, e.g. 10^{14} atom/cm^2, and appraise ΔE_b and ΔE_{ER} by Eq. (3.11) as shown below:

Coverage	Substrate	ΔE_b Pd $3d_{5/2}$	ΔE_{ER}	ΔE_{chem}
1×10^{14}	SiO_2	1.1	1.05	+0.05
1×10^{14}	Al_2O_3	1.2	0.65	+0.55

we shall find that the true chemical shift on Al_2O_3 is positive, and it is close to zero on SiO_2. This difference is probably due to metal-support interaction instead of to intraatomic charge transfer.

Also considered were the tendencies of the shifting of levels depending on the mutual arrangement of the upper orbitals of the metal and support [226]. When the *spd* orbitals of the support interact with the valence orbitals of the metallic cluster, Coulomb repulsion occurs that changes the energy separation and raises or lowers the energy of the valence and core levels (Fig. 29). If the *p* level is localized above the valence band molecular orbital of Au $5d6s$, the shift is positive, otherwise it is negative. With a close arrangement of the levels of the support (in the given case a metal) and the cluster, the latter is included in the band structure of the support, which produces a negative shift.

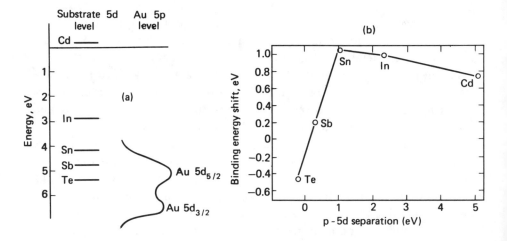

Fig. 29. Relative positions of the upper levels of *p* metals (substrate) and Au (cluster) in the valence band (a) and the dependence of the chemical shift of Au $4f_{7/2}$ on the splitting of the upper *p* levels and the $5d$ valence band of Au (b) [226]

It is proposed to consider the asymmetry of the core level peaks as still another criterion for appraising the contribution of the final and initial states. Incomplete screening of a hole should lead to a quasisymmetric shape of the core level peaks [235, 241]. This was observed for the clusters Pt/C and Pd/C [235]. The appearance of asymmetry of the Au 4f lines for small clusters was explained by the nonuniform distribution of the particles by size [231, 241, 250]. It should be noted that this serious problem is ignored unfoundedly in a number of publications devoted to supported films and clusters. The contribution of the initial states (transfer of the charge to the support) is indicated by some authors [231, 250]. The true chemical shift at low coverages of Pd on C was +0.3 eV, which corresponded to a charge of $q = +0.15$.

Wertheim *et al.* proposed [232] and then developed [233, 248, 251-253] a model differing from the other ones. It explained the positive shifts in the spectra of the core levels and the observed lowering of the Fermi level (Au) for deposited metal clusters. These changes are ascribed to a single effect — the presence of a unit positive charge on the cluster atoms after photoionization. It is considered here that the screening of a hole in the clusters themselves is the same as in bulk metals (ΔE_{ER} does not change), but since the clusters are deposited onto poorly conducting supports, the charge on the surface of the clusters is not neutralized instantaneously. The binding energy of a photoelectron will increase by the magnitude of the Coulomb interaction of the conduction electrons and the hole, i.e. by $e^2/2R$, where R is the radius of a spherical particle. Consequently, in this model, the shift is controlled exclusively by the size and shape of the metal particles. The contributions of the other effects (the size effect, metal-support interaction, insufficiently complete hole screening, shifting of the Fermi level as a result of a change in the ionization potential [253]) either compensate one another or are less than the charging effect.

The unit charge model, in the opinion of some authors [232, 233, 248], is supported by facts such as (a) the different direction of the shifts for the surface atoms of the bulk metal (negative) and the small applied clusters (positive); (b) the use of a simple ratio $P = 2e^2/\Delta E_b$, which gives dimensions coinciding in their order of magnitude with those found by electron microscopy, and (c) neutralization of the charge and no positive shift when the same cluster is deposited on the metal support (for instance, Au/Ni).

It should be noted that all these proofs can be treated critically. The model according to its definition itself can be applied only for clusters with metallic conductivity, although, judging from the theoretical [234, 237] and exprimental [226, 228, 237] data, the maximum shifts are observed for molecular clusters with quasiatomic valence levels. Negative shifts are observed for a number of supported metal systems [222]. As shown in Fig. 29, the direction of the shift when a metallic cluster is deposited onto a metal support is determined by the mutual arrangement of the orbitals of the two metals. Hence, the hypothesis on the necessity of taking the initial states into consideration when interpreting spectra again becomes competent.

Wertheim's model was criticized in greater detail by Bagus *et al.* [243, 249, 255], who on the basis of theoretical calculations of Si and Al clusters and measurements of Pt spectra on SiO_2 and Teflon proposed a model close in its concepts to Mason's one and describing the shifts within the limits of the size effect and metal cluster-support interaction. Judging from calculations, the first factor prevails in the region of small clusters containing surface atoms with a low coordination. For clusters of Al_{13} and a coordination number of five, the change in E_b should be 2.19 eV when E_{ER} decreases by only 0.3-0.4 eV. It should be noted that in this region Wertheim's model is in general illegitimate. But even in the region of its applicability, the shifts associated with the initial state should not be compensated by the shift associated with the charge [249]. For instance, Bagus considers that if the charge model is applied to Pt clusters with a size of 6-9 Å, then the maximum values of the "Coulomb" shift for them (with account taken of the substrate polarizability) will be 0.6-0.8 eV, whereas the total shift for clusters of even a larger size generally exceeds 1 eV [226, 228-247]. A discussion by two groups of authors [248, 249] revealed both the shortcomings of the two models (in neither of them the changes in E_{ER} were measured, and the size of the particles was characterized poorly) and the common viewpoint on the necessity of taking the metal-support interaction into account when interpreting the shifts in the spectra of small particles.

Consequently, the problem of interpreting the changes in the parameters of photoelectron spectra of small clusters has not been solved completely, neither theoretically nor experimentally. Although the general trends of the change in E_b, FWHM, and the valence band spectra can be reproduced quite well, the quantitative data diverge greatly. What has been generally adopted is that in most cases the observed changes can be ascribed to the small clusters themselves, and are not associated with the macroheterogeneity of charging, the polydispersed nature of the clusters, or different localization (adsorption on the surface or implantation into the bulk), although all these factors must be taken into account without any doubt. In particular, differences in the shifts for the same systems may be due to different cluster sizes with equivalent coverages. Unfortunately, only in a few investigations were the same objects studied by electron microscopy [237, 238]. Although a number of authors give preference to final state effects when interpreting the shifts in spectra of small particles, the above analysis shows that when the relaxation term is taken into consideration, the shift in most cases does not shrink to zero. The experimental values of ΔE_{ER} do not exceed 0.5-0.7 eV, whereas ΔE_b reaches +1.5-1.7 eV. This is why it is legitimate to consider the intra-atomic charge transfer and interaction with the support to be the sources of the shifts. This information is of direct interest for studying the electron structure of supported metal catalysts. In the present section, we have not treated the results of studying real catalysts, a discussion of which has been relegated to Chap. 4. We shall only note that both positive and negative shifts from 0.5 to 1.5 eV are typical of them.

Another matter of principle — finding of the cluster-metal interface — can be studied by analyzing the spectra of the valence bands whose sensitivity to the size of a cluster has been shown in many of the cited publications [226-246]. The results of these investigations are also considered in Chap. 4.

The appearance of photoelectron spectroscopy gave rise to hopes that the energy of surface atom levels could be determined. The employment of angular resolution procedures when measuring single crystals and oriented films with variation of the energy (synchrotron radiation) made it possible to differentiate quite accurately the signals from surface and bulk atoms of metals, e.g. Cu, Ag, Au [256-258], Ni [258], Ir [259], Pt [260], Re [261], Yb, Gd, Ta, and W [262] (Fig. 30). Since the coordination possibilities of the surface atoms, which contain less neighbors than the bulk ones, are not saturated and these atoms have a more greatly localized d band, their shift to lower binding energies was expected owing to an $s \rightarrow d$ transfer of the electron density. On the other hand, they partly overlap with the bulk valence band, and the nature of this interaction also tells on the magnitude and direction of a shift. The experimental values of the shifts range from -0.7 to $+0.6$ eV. The transition metals and the rare-earth metals with a half-filled d shell yield a positive shift, while for metals with $n_d > 5$, the shifts are always negative. Since the shifts are not large, special procedures should be used for determining them. The most detailed analysis of how to single out the peaks of surface atoms is given by Citrin *et al.* [256], who in addition to the shifts of the core levels also discovered a change in the valence band and its shift by -0.5 eV for surface atoms. To determine the components relating to the peaks of surface atoms, the lineshapes are analyzed with the aid of the Doniach-Sunjic function, which takes peak asymmetry into account (Fig. 31). Another criterion for singling

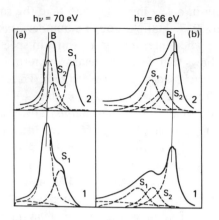

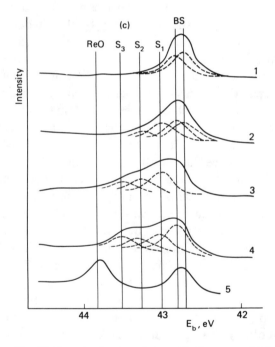

Fig. 30. Core levels of surface and bulk atoms of W (a), Ta (b), and Re (c) before and after adsorption [261, 262]:

a — W(111) $4f_{7/2}$, $h\nu$ = 70 eV: 1 — initial; 2 — adsorption of H_2; b — Ta $4f_{7/2}$, $h\nu$ = 66 eV; 1 — initial; 2 — adsorption of H_2; c — Re (0001) $4f_{7/2}$, $h\nu$ = 70 eV: 1 — initial; 2 — O_2, 1 L, 295 K; 3 — O_2, 3 L, 295 K; 4 — O_2, 6 L, 295 K; 5 — O_2, 300 L, 923 K; B = bulk atom level; S = surface atom level

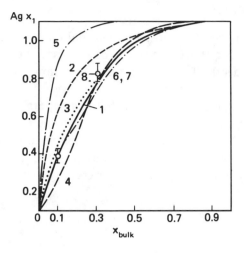

Fig. 31. Comparison of data on the composition of the first layer in alloys of Ag-Au(100) obtained by AES and evaluated according to different theoretical models [263]:

1 — Monte Carlo calculations; *2* — monolayer, ideal solution; *3* — monolayer, regular solution; *4* — monolayer, surface (ideal solution), bulk (regular solution); *5* — model of broken bonds; *6* — modified model of broken bonds; *7* — classical multilayer model; *8* (dots) — data of AES

out surface atoms is the adsorption of hydrogen and oxygen. Here their peaks experience appreciable shifts, whereas the core levels of bulk atoms do not shift. Citrin *et al.* [256] assume that the vacuum-solid interface is limited to the first atomic layer.

The data for a stepped surface of Ir(332) [262] or different structures of Au/Pt(111) [231] reveal that the shifts depend on the coordination number of the surface atom. For Ir, three components have been found in the Ir 4*f* spectrum that characterize the atoms in the bulk, on a smooth surface, and on the steps. The latter have the smallest binding energy. The observed shifts in the spectra of the surface atoms are explained exclusively by initial state effects, although incomplete screening of a hole should apparently also contribute to ΔE_b.

3.3   Surface Composition

Let us consider the three most fundamental questions associated with the quantitative analysis of the composition of a surface, the distribution of the components in the depth of a layer, and the dispersion of a supported phase by EES and IS:

(a) the theoretical and experimental approaches to an appraisal of surface segregation and a determination of the profiles of the concentrations in alloys;

(b) the employment of XPS intensities for appraising the distribution and dispersion of components deposited onto a porous support;

(c) studying the composition of the surface layers of powdered catalysts by XPS and ISS.

3.3.1   Surface Segregation in Alloys

The distribution of catalyst components in the bulk and on the surface is determined by a set of thermodynamic and kinetic processes that go on when the catalysts are formed. We can assert that in a general case the composition of a surface will not correspond to the average chemical composition of a system. But, except for simple binary systems, data on the composition of a surface can be procured only experimentally because many of the parameters needed for the calculation of complex heterogeneous systems are unknown. This is why the theoretical and experimental approaches to the determination of the composition of a surface can be compared only for bimetal alloys.

A variety of thermodynamic models have been developed for an equilibrium alloy. They enable one to predict the direction and degree of surface segregation [264-270]. Strictly, they may be applied for the clean surfaces of single crystals. Let us consider the fundamental principles underlying some models. For an ideal solid solution consisting of components A and B, the energy of formation of the alloy (the enthalpy of mixing equals zero) is

$$E_{AB} = \frac{1}{2}(E_{AA} + E_{BB}) \tag{3.13}$$

If the enthalpy of formation $\Delta H_f^0 > 0$,

$$E_{AB} < \frac{1}{2}(E_{AA} + E_{BB}) \tag{3.13a}$$

the alloys are endothermic ("clusterizing"), and depending on the temperature they form mono- or two-phase systems. A typical example is the alloy Cu-Ni for which what is known as the "cherry" model [271] has been proposed. It predicts a constant composition of the surface with segregation of copper on it in the region of low mutual solubility of the components. Another limiting case are exothermic ("ordering") alloys, for which $\Delta H_f^0 < 0$ and

$$E_{AB} \gg \frac{1}{2} (E_{AA} + E_{BB}) \tag{3.13b}$$

They include intermetallics of the type Pt-Sn and Pt-Pb. In these systems, one can expect a nonmonotonic change in the composition in depth because enriching of the surface in component A may lead to "pulling out" of component B into the second layer [271].

Most theoretical approaches to an analysis of surface segregation are based on the approximation of a regular solution [264-270]. Here a two-phase model is frequently considered that includes the bulk phase and the surface phase as a "monolayer". In this model, the concentrations of the components on the surface $c_{A,s}$ and $c_{B,s}$ depend on their concentrations in the bulk $c_{A,b}$ and $c_{B,b}$ and the energy (enthalpy) of segregation ΔH:

$$\frac{c_{A,s}}{c_{B,s}} = \frac{c_{A,b}}{c_{B,b}} \exp \left(-\frac{\Delta H}{RT} \right) \tag{3.14}$$

The segregation energy is generally written in the form of a functional dependence on several parameters determining the segregation process, namely, the atomization energy, the surface energies or surface tensions of the pure components, the mixing energy, etc. The expression for the segregation energy may include an additive term representing the contribution of the elastic deformation when the atomic radii of A and B fail to correspond. Also popular among researchers is Miedema's model [266], which treats the formation of an alloy as dissolution:

$$\frac{c_{A,s}}{c_{A,b}} = \exp \left[\frac{f\Delta H_{sol}(A, B) - g(I_A - I_B)V_A^{2/3}}{RT} \right] \tag{3.15}$$

where V_A is the molar volume of metal A, ΔH_{sol} is the heat of solution of A in B, I_A and I_B are the surface enthalpies, and f and g are constants. Equation (3.15) can be written in a different way as:

$$\frac{c_{A,s}}{c_{A,b}} = \exp(w, t) \tag{3.15a}$$

Here $t = 1000T$ and $w = w_1 + w_2 + w_3$; w_1 is the contribution of the heat of solution, w_2 is the contribution of the surface energy, and w_3 is the contribution of the difference in size ($w_3 > 0$). Notwithstanding the use of empirical and not completely physically substantiated parameters in the model, it predicts the direction of segregation quite well for many alloys [272].

Mention must also be made of the use of the Monte-Carlo numerical modelling method for appraising the surface segregation and ordering (local environment) first of all in exothermic alloys [273-275]. These calculations allow one to obtain c_A over the depth of a layer as a function of c_B, the temperature, the number of layers, and the thermodynamic parameters of an alloy.

Since the proposed theoretical models are of a semi-empirical nature and have a limited sphere of application (regular solid solutions), experimental data are needed as previously not only for a more precise determination of the surface composition, but also as the initial data for improving the theoretical approaches. Such results obtained by AES, XPS, and ISS techniques [263, 271, 276-278] confirm the presence of surface segregation even in pure equilibrium alloys annealed in deep vacuum.

Figure 31 illustrates the range of deviations of the composition of the surface from that of the bulk for one of the most studied alloys Ag-Au [263]. The surface composition was determined by AES (the data have been reduced to the first layer) and was calculated by various models. The best agreement with the experimental results was given by a modified model of regular solid solutions. This model predicts the enrichment of the first to third (fifth) initial layers of the surface and a monotonic transition from the surface to the bulk composition. But the considerations mentioned above reveal that the concentration profile for exothermic alloys should be nonmonotonic; this is confirmed by Monte Carlo calculations.

Let us see what the possibilities of modern EES and IS techniques are for obtaining information on the first atomic layer of a surface and the concentration profile. Although a number of authors [9, 27] employed XPS quite successfully for studying surface segregation, for homogeneous solid solutions its sensitivity to the enrichment of a surface is not high. This is why AES using low-energy electrons under 100 eV [6] was employed for a long time to appraise the composition of a surface and the concentration profile. But the early AES data were inaccurate, which was explained by (a) the difficulty of quantitative analysis, (b) averaging of the signal for several layers, and (c) ignoring of the effect of preferential sputtering.

After the appearance of theoretical publications devoted to the improvement of the quantitative analysis techniques [106, 107, 279, 280], the accuracy of the quantitative data of AES improved. We shall consider as an example the work of King and Donelley [263], who performed one of the most detailed studies of surface segregation and the structure of the surface of alloys by AES. The intensities of 14 Auger transitions of Ag and Au differing in their energy (from 69 to 2024 eV) were measured. This made it possible in conjunction with slow electron diffraction and scanning electron microscopy, and also Monte Carlo calculations, to determine the following characteristics of polycrystalline Ag-Au alloys:

(a) the composition of the first layer by the procedure of minimizing the difference between the calculated and measured ratios of the intensities of the Auger electrons with different energies [106, 280] (here the composition of the third layer was taken to be the same as the bulk composition). King and Donelley obtained 14 equations for the intensity ratio. The difference between the intensities was determined as follows:

$$Q(E)[R_{exp}(E) - R_{calc}(E)]^2 = \{R_{exp}(E) - r_{bk}[x_1 A_1(E)$$
$$+ x_2 A_2(E) + x_b \sum_{i=3}^{\infty} A_i(E)]\} \tag{3.16}$$

where $R(E)$ is the ratio of the intensities, r_{bk} is the back-scattering coefficient, x_i is the fraction of an element in layer i, A_i is the fraction of the signal emerging from the i-th level, and x_b is the fraction of the element in the bulk;

(b) the concentration profile in a layer depth;

(c) the crystallographic orientation of separate domains; and

(d) the concentration profile for the faces of different orientations.

In all cases, segregation of Ag was observed, while the Ag concentration in the second layer in some specimens was lower than in the bulk. For the alloy Au-Ag (10%), the concentrations of Ag in the first and second layers were:

Face	(100)	(101)	(111)
x_{Ag} (first layer)	0.29 ± 0.04	0.35 ± 0.04	0.15 ± 0.04
x_{Ag} (second layer)	0.08 ± 0.08	0.10 ± 0.08	0.11 ± 0.08

The data of AES and Monte Carlo calculations agree satisfactorily for all the faces of an alloy (Fig. 31). Close results for the composition of the first layer of Ag-Au alloys have been obtained by ISS [277]. This technique, notwithstanding some problems of quantitative analysis, is used successfully for analyzing the surface of other alloys, e.g. Ni-Cu [278].

The employment of ion etching for obtaining a concentration profile in alloys is made more complicated by ion-induced effects that were not always taken into consideration. At present, it is deemed the most reliable to use several techniques for appraising the composition of a surface, for instance, ISS, XPS, and SIMS.

Judging from XPS data obtained at a glancing angle of photoemission and from SIMS data, the surface of solid Ni-Co solutions with large contents of the second component becomes enriched in Ni [281]. The concentration profiles (SIMS) given in Fig. 32 for two specimens of Co-Ni reveal that the ratios of the secondary ions Co^+/Ni^+ does indeed reflect the composition of the surface, while the sputtering yields and the secondary ion yields for the two metals are close. Only a few authors have obtained reliable proofs of the oscillations of the composition in the depth of a layer expected from

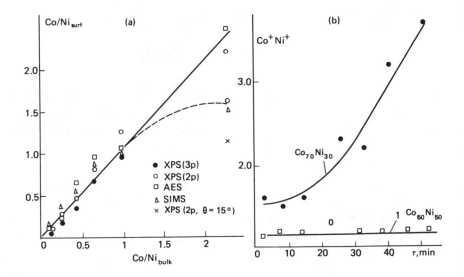

Fig. 32. Relation between the surface and bulk concentrations of Co-Ni alloys (a) and the SIMS concentration profiles for $Co_{50}Ni_{50}$ and $Co_{70}Ni_{30}$ specimens (b) [281];

Treatment: Ar^+, vacuum, 350 °C

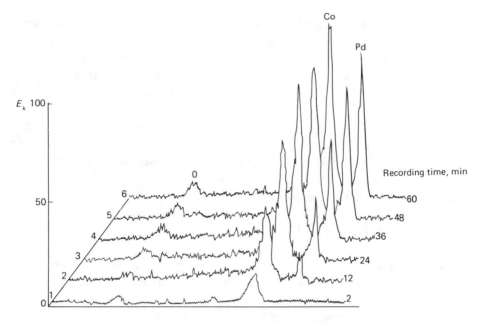

Fig. 33. Spectra of ion scattering (^{4}He$^+$) on the alloy Pd$_{30}$Co$_{70}$; Treatment: CO, 100 °C, 10^6 L

calculations for exothermic alloys and intermetallics. A special low-energy electron diffraction technique (quantitative LEED) made it possible to prove the nonmonotonic nature of the change in the composition in the first three layers of single crystals (111) of the alloys Pt$_{0.5}$Ni$_{0.5}$ (1) and Pt$_{0.78}$Ni$_{0.22}$ (2) [282]. The Pt concentration in them was:

	1st layer	2nd layer	3rd layer
1	88 ± 2	9 ± 5	65 ± 10
2	99 ± 1	30 ± 5	87 ± 10

Information on the composition of the second and deeper layers is also very important for catalysis, because in "corrosive" chemisorption a reaction may be attended by displacement or removal of the first layer, and also by penetration of a reagent into the deeper layers.

A change in the composition of the surface of a binary alloy caused by one or more reagents is called "segregation due to chemisorption" [271]. The most "corrosive" reagents include oxygen, CO, and NO. Even at low pressures and moderate temperatures, CO "pulls out" one of the components to the surface, e.g. Co in the alloy Pd-Co (Fig. 33). This is reflected first of all in the spectra of the scattered ions, which indicate that the first layer is mainly occupied by cobalt. When passing over to deeper layers (scattering of He$^+$), the concentration of Pd gradually increases, but according to data of both ISS and XPS, the surface layers of the alloy are also enriched in cobalt. Such data and their consequences for catalysis are considered in greater detail in Chaps. 4 and 5.

Feilman [283] has proposed an original way of appraising the surface segregation in dilute alloys from the difference between the energies of the levels of the surface and bulk atoms. Feilman's main assumption consists in that the surface energy in a metal Z increases when a bulk $(Z + 1)$ dopant exchanges places with a surface atom of Z

$$\Delta H_{seg} = E_b - E_s \tag{3.17}$$

where H_{seg} is the heat of segregation, E_b and E_s are the binding energies of the bulk and surface levels of the metal Z, respectively. The model is based on the approximation of "equivalent atoms". The dopant $(Z + 1)$ in the bulk is equivalent to a photoionized atom $(Z + 1)$. Fielman [283] estimated the heats of segregation in two ways: by Eqs. (3.15) and (3.17). In addition to obtaining good agreement between the two approaches, the change in the direction of segregation in adsorption was predicted. We can expect that when experimental data on the binding energy of the surface levels is accumulated, Feilman's model [283] will be quite popular for estimating the heats of surface segregation and predicting the surface enrichment.

3.3.2   Distribution and Dispersion of Components on Porous Supports

The expressions of the intensities [Eqs. (1.16)-(1.28)] considered above for homogeneous specimens are a particular case when dealing with catalysts. Most frequently, the surface layers of a catalyst are not homogeneous. This is due to (a) surface segregation of the components; (b) nonuniform distribution of a component over a grain; (c) changes in the dispersion of the supported phase and in the porosity of the support, etc. This is why an appraisal of the surface concentration or dispersion of the components in porous catalysts by XPS requires a comparison of the experimentally obtained intensities with those calculated on the basis of definite assumptions on how various factors affect the intensity. If we analyze the experimental ratios of the concentrations c_X/c_Y for supported catalysts, we can note three cases:

equivalence of their values for the surface and bulk ($R_s = c_X/c_Y$), overstated and understated ratios ($R_s > R_b$ or $R_s < R_b$). With the retaining of a constant surface concentration, the third case characterizes sintering of the supported particles to a size exceeding the value of λ.

The first attempts to relate the changes in the ratio of the intensities and the dispersion were undertaken for the catalysts Pt/SiO$_2$ and Na/aluminosilicate [284]. The Angevine-Delgass model developed in this publication consists in considering a semi-infinite support that carries cubic crystals of a supported component:

$$R = \frac{I_p}{I_{sup}} = \alpha d_p \qquad (3.18)$$

where

$$\alpha = R(\beta_i)R_{sup}\,\frac{\sigma_p}{\sigma_{sup}}\,\frac{1}{c_{sup}\lambda_{sup}} \qquad (3.19)$$

p and sup stand for the promotor (supported component) and the support, $R(\beta_i)$ is the ratio of the factors taking into account the asymmetry of the p and s lines, R_{sup} is the ratio of the analyzer transmission factors for the p and s lines, c_{sup} is the bulk concentration of the support component, $d_p = n_p/S_{BET}$ is the surface density of the supported component, n_p is the number of atoms of the supported component per gram, S_{BET} is the specific surface area of the support, and σ is the ionization cross section. As expected, a linear relation between R and the Pt content was observed for Pt/SiO$_2$ with a high dispersion (90-100%), but it failed to be observed with a low dispersion. A critical analysis of these data [182] showed that the description of a support by a semi-infinite layer is correct only at low values of S_{BET}, but not for SiO$_2$, where the thickness of the support "walls" is commensurable with λ. In this case, the expression obtained for a homogeneous solid solution holds (under conditions of monolayer distribution):

$$R = R(\beta_i)R_{sup}\,\frac{\lambda_p}{\lambda_{sup}}\,\frac{\sigma_p}{\sigma_{sup}}\,\frac{n_p}{n_{sup}} \qquad (3.20)$$

Equation (3.20) is similar to Eq. (1.28).

The most detailed mathematical model was proposed by Kerkhof [182]. In this model, the catalyst is assumed to be in the form of layers of a support (sheets) on which atoms or crystallites of the promotor are deposited. The overall expression for the intensities takes into consideration four types of attenuation of the escape of photoelectrons: (a) from the promoter in the promoter; (b) from the support in the support; (c) from the promotor in the support; and (d) from the support in the promoter. In the absence of support screening (low coverages and a high value of S_{BET}) and with monolayer distribution of the component, the expression becomes as follows:

$$\frac{I_p}{I_s} (\exp) \left(\frac{p}{s}\right)_{bulk} \frac{S(E_p)}{S(E_s)} \frac{\sigma_p}{\sigma_s} \frac{\beta_1}{2} \frac{(1 - e^{-\alpha_1})(1 + e^{-\beta_2})}{\alpha_1(1 - e^{-\beta_2})} \tag{3.21}$$

where $(p/s)_{bulk}$ is the ratio of the bulk concentrations of the promoter and the support, s are the transmission factors, α_1, β_1, and β_2 are dimensionless thicknesses of the layers of the pure promoter and support and of the support covered by the promoter:

$$\alpha_1 = c/\lambda_p \tag{3.22}$$

$$\beta_1 = t/\lambda_{ss} \tag{3.22a}$$

$$\beta_2 = t/\lambda_{ps} \tag{3.22b}$$

$$t = 2/S_{BET}\varrho \tag{3.23}$$

where c is the average size of the crystallites, and ϱ is the bulk density. With a monolayer coverage, α_1 is small, and

$$\frac{I_p}{I_s} = \left(\frac{p}{s}\right)_{bulk} \frac{S(E_p)}{S(E_s)} \frac{\beta_1}{2} \frac{(1 + e^{-\beta_2})}{(1 - e^{-\beta_2})} \tag{3.24}$$

Next the designation

$$Z = \frac{\beta_1}{2} \frac{1 + e^{-\beta_2}}{1 - e^{-\beta_2}} \tag{3.25}$$

is introduced. If $\Delta E < 400$ eV, then $\beta_1 \approx \beta_2 = \beta$, and

$$Z = \frac{\beta}{2} \frac{1 + e^{-\beta}}{1 - e^{-\beta}} \tag{3.25a}$$

$$\left(\frac{I_p}{I_s}\right)_{exp} = \left(\frac{p}{s}\right)_{bulk} \frac{\sigma_p}{\sigma_s} Z \tag{3.26}$$

Such an expression was obtained by Defossé when appraising the coating of aluminosilicates with adsorbed pyridine [285]. For the case when $t \leqslant \lambda$ and $\beta \approx 1$, the values of Z differ only slightly from unity, and there follows an important consequence:

$$\left(\frac{I_p}{I_s}\right)_{exp} = \left(\frac{p}{s}\right)_{bulk} \frac{\sigma_p}{\sigma_s} \tag{3.27}$$

i.e. the intensity is determined by the bulk concentration of the component distributed in a monolayer on the support. Equation (3.27) was used for high coatings of a supported component with account taken of support screening for NiO/Al_2O_3 [22]. A second important consequence following from the considered model is the possibility of appraising the dispersion from a comparison of the experimental intensity with what is expected for a monolayer:

$$\left(\frac{I_p}{I_s}\right)_{crystallite} \Big/ \left(\frac{I_p}{I_s}\right)_{monolayer} = \frac{1 - e^{-\alpha_1}}{\alpha_1} \tag{3.28}$$

The correctness of the model described by Defossé [285] is confirmed by an analysis of the data for specimens of Re_2O_7/Al_2O_3 for which a monolayer distribution of the supported phase was presumed, and also of Pt/SiO_2 measured by Angevine et al. [284]. They obtained a good agreement between the theoretical and experimental intensities (Re) and found a correlation between the dispersion of Pt and the parameter α_1.

According to Defossé [285], the model of a solid solution describes the cases when $t \leqslant \lambda$, while the Angevine-Delgass model describes the cases when $t \geqslant 3\lambda$, which corresponds to $S_{BET} > 300$ and $< 100 \, m^2/g$. Defossé [285] has proposed a general expression for the intensities that is correct within the entire interval of S_{BET}. In this model, a porous support is designed as a stack of various elements, namely, plates, spheres, or cubes. By changing the ratio between their size and mass, the value of S_{BET} can be monitored. The expression of the intensities

$$R = \alpha d_p q \tag{3.29}$$

includes the quantity q taking the porosity of the support into consideration. If the support is composed of plates, we have:

$$q = 1 + 2 \frac{e^{-t/\lambda_p}}{1 - e^{-t/\lambda_p}} \tag{3.30}$$

An example, in which the intensities of Mo/Al in catalysts MoO_3/Al_2O_3 with a variable value of S_{BET} was analyzed, has clearly shown that Eq. (3.29) describes the intensities throughout the entire interval of S_{BET} values.

Mukaida and Araya [286] have proposed a method for the more precise calculation of the surface composition of porous catalysts that takes the fraction of the voids in the support into account. The quantity $\varrho(1 - \varepsilon)$, where ε is the fraction of voids, is used instead of the density of the compact material. According to their data, the authenticity of the results obtained by this method for the catalysts Al_2O_3-SiO_2 and MgO-SiO_2 is higher.

Still another model enabling one to appraise the dispersion of a supported phase has been proposed by Fung [287], who introduced a complicated analytical expression relating the ratio of the intensities and the dispersion:

$$R_d = k \frac{W}{\varrho_p S} \frac{1}{d} \frac{\lambda_p(E)H_p}{\lambda_s(E')H_s} \psi[d, \lambda_p(E)] \qquad (3.31)$$

Here k is a constant taking into account the distribution of the supported particles on the surface of the support, S is the specific surface area of the support, W is the specific mass of the dispersed phase (g/g), ϱ_p is the density of the active phase, $\lambda_p(E)$ is the mean free escape depth of photoelectrons with the energy E (for the dispersed phase), $\lambda_s(E')$ is the same of photoelectrons with the energy E' (for the support)

$$\frac{H_p}{H_s} = R_p(\beta)R_s R\sigma \qquad (3.31a)$$

d is the particle size (diameter of a sphere, face of a cube, thickness of a disk), $\psi[d, \lambda_p(E)]$ is a function taking into account the shape and size of the particles, and also $\lambda_p(E)$. The drop in R_d produced by a growth in the particle size can be written as

$$N = \frac{R_d}{R_{d_0}} \frac{W_0}{W} \frac{S}{S_0} = \frac{\psi[d, \lambda_i(E)}{\psi_0[d_0, \lambda_i(E)]} \frac{d_0}{d} \qquad (3.32)$$

where N is the value of R_d normalized with respect to the mass of the dispersed phase (W_0) and the specific surface area (S_0) of a standard specimen with $d_0 = 1$ nm. For this specimen, we find k_0 from the relation between R_d and W_0. Next we assume that $k = k_0$, the value of N is found by Eq. (3.22) depending on the size and shape of the particles. Fung's model yielded good agreement of the size of Pt particles (XPS and TEM) for a specimen sputtered onto SiO_2 provided that the particles have a hemispherical shape.

A theoretical model [288] studied the possible influence of the geometry of individual catalyst particles (an oriented layer, sphere, hemisphere) oriented randomly relative to one another on the angular distribution of the emitted electrons and, consequently, on the intensity. It was found that for most systems of this kind, effects of anisotropy were not observed or balanced one another, and for any particle shape the intensity is described by a common expression and reflects the dispersion of the supported phase.

For binary oxide catalysts, the ratio of the intensities can be related to the ratio of the sizes of the particles of the individual components, i.e. of their specific surfaces. This allows us to approximately appraise the specific surface S_{BET} of a supported oxide from the relation

$$\frac{I_p}{I_s} = \frac{p}{s} \frac{\sigma_p}{\sigma_s} \frac{S_p}{S_s} \tag{3.33}$$

For low concentrations, we have $S_s \approx S_{BET}$, and

$$\frac{I_p}{I_s} = \frac{p}{s} \frac{\sigma_p}{\sigma_s} \frac{S_p}{S_{BET}} \tag{3.33a}$$

Such an estimate was performed for the series of catalysts Cr_2O_3/Al_2O_3 in one of our early studies [27]. A stricter model has been proposed [289] for coprecipitated RuO_2-Al_2O_3 specimens. It made possible an appraisal of the average radii (and, consequently, the surface) of the Al_2O_3 and RuO_2 grains.

We have treated different models, which from a comparison of the theoretical and experimental intensities in X-ray photoelectron spectra enable us to draw conclusions on the nature of distribution of a supported component (monolayer or crystallites) and appraise its dispersion or the ratio of the "active" surfaces of two components. Of importance is the fact that XPS allows one to determine the dispersion of an unreduced phase within the range of small particle sizes (1-3 nm) when X-ray phase analysis has a low sensitivity.

In the considered models, except for Fung's one [287], no account was taken of the possible nonuniform distribution of a supported phase over the external surface and inside the pores and its redistribution during various treatments. Moreover, when simplifying Fung's model [287], the factor accounting for the distribution of a supported component was also equated to unity. The question of the applicability of Kerkhof's model to very highly porous supports such as zeolite molecular sieves is also important because in these sieves the molecular uniform distribution of a supported component over the entire volume of a crystal is considered to be ideal instead of a monolayer. Actually, however, there may be deviations in the molecular distribution due to migration and segregation of the components and their agglomeration on the external surface.

Kaliguine et al. [290] applied Kerkhof's model to zeolite supports (the zeolite ZSM-5 with adsorbed ruthenocene). They attempted to take the surface segregation effect into consideration by assuming bidispersed distribution of the component (one part X_1 with a size of C_1 was assumed to be distributed inside the channels, and the other part X_2 in the form of microcrystallites C_2 on the external surface). Additional coefficients were introduced to take account of the probability of the photoelectrons passing through the layer on the external surface, namely,

$$\gamma_1 = \frac{C_2}{\lambda_{pp}} \tag{3.34}$$

and

$$\gamma_2 = \frac{C_2}{\lambda_{sp}} \tag{3.34a}$$

After simplification (including the condition of close values of E_k for the promoter and support), an expression was obtained similar to Kerkhof's equation:

$$\left(\frac{I_p}{I_s}\right)_{exp} = \frac{K_p}{K_s} \left[\frac{(1 - P_1/P_2^2)}{(1 - P_1)P_3} - 1 \right] \tag{3.35}$$

where

$$\frac{K_p}{K_s} = \frac{n_p}{n_s} \frac{\sigma_p}{\sigma_s} \frac{\lambda_{pp}}{\lambda_{ss}} \frac{D_p}{D_s} \tag{3.35a}$$

$P_1 = P_{ss} = P_{ps}$ is the probability of support and promoter photoelectrons passing through the support, $P_2 = P_{sp} = P_{pp}$ is the same for the promoter in channels, and $P_3 = P_{spi} = P_{ppi}$ is the same for the promoter on the external surface. The exponential multipliers are, accordingly, $\beta_1 = \beta_2 = t/\lambda_s$, $\alpha_1 = \alpha_2 = C_1/\lambda_p$, and $\gamma_1 = \gamma_2 = C_2/\lambda_p$. Equation (3.35) contains three variables C_1, C_2, and X_1 (in an implicit form) and, therefore, additional methods must be used for finding two of them. Kaliguine *et al.* [290] determined X_1 from the condition that all the Brönsted centers found in accordance with the adsorption of pyridine are exchanged for Ru ions. Although they obtained a quite satisfactory coincidence between the size of the crystals on the external surface found by XPS and electron microscopy, some of the assumptions and simplifications in this model seem to be quite arbitrary. Among them are the modelling of zeolite having pores of a molecular size by a stack of plates, the inadmissibility of using the concept of the specific surface area after Brunauer, Emmett, and Teller for a zeolite, and the inaccuracy in appraising the amount of metal in the channels only from the number of Brönsted centers. We cannot agree with the assumption that the parameters C_1, C_2, and X_1 in zeolites change absolutely arbitrarily and they cannot be estimated. Kaliguine *et al.* considered a nontypical case of surface segregation of an introduced component in zeolites, although when using the procedure of ion exchange it is generally distributed uniformly, and the model of a solid solution [27] may be applied to such a system. Conversely, when metals are reduced, their migration to the external surface is clearly registered by the increase in the intensities [291]. We shall treat matters associated with a quantitative analysis of metal-zeolite systems in Chap. 4.

3.3.3 Surface Segregation in Supported and Oxide Systems

The case $R_s > R_b$ corresponds to the enrichment of a surface layer with a thickness of $d \leqslant 6$ nm in one of the components. There are situations when the surface segregation makes the main contribution to the change in the intensities, namely, when $c \ll \lambda$, $c \gg \lambda$, $t \approx c$, etc. XPS can be used to obtain a nondestructive depth profile of powdered specimens: (a) by measuring peaks with different kinetic energies E_k [see Eqs. (1.16)-(1.28)]; (b) by varying the photoemission angle. In both cases, the quantity λ is varied. Other methods (e.g. varying of the radiation energy) require the employment of multiple-anode X-ray tubes or a synchrotron source. Finster *et al.* [292], proceeding from the ratio of the intensities of Ni $3p$ ($E_k = 1400$ eV) and Ni $2p$ ($E_k = 630$ eV) attempted to plot a depth profile using NiO/SiO_2 as an example. If an exponential relation between the concentration and the layer depth is adopted, the profile is determined as:

$$\Gamma = \frac{R_{exp}}{R_b} = \frac{Y_b}{X_b} \frac{X_0 n + X_b \lambda_X \sin\theta}{Y_0 n + Y_b \lambda_Y \sin\theta} \frac{n + \lambda_Y \sin\theta}{n + \lambda_X \sin\theta} \tag{3.36}$$

where X_b and Y_b are the bulk concentrations of components X and Y, X_0 and Y_0 are the surface concentrations, and n is a profile parameter. The concentration in a layer of thickness t for component X is

$$X_t = X_b + (X_0 - X_b)e^{-t/n} \tag{3.37}$$

and in the first layer is

$$X_1 = X_b + (X_0 - X_b)e^{-d/2n} \tag{3.38}$$

where d is the thickness of the first layer [292]. It follows from simple reasoning that $\Gamma_1 \approx 2\Gamma_{means}$. To plot a concentration profile, one must find n, which is difficult to do from two points. Moreover, the assumptions adopted by Finster *et al.* substantially lower the accuracy of the determinations, therefore we are dealing with a semiquantitative relation.

A decrease in the photoemission angle appreciably increases the surface sensitivity even for polycrystalline specimens [57]. But this effect is restricted by the size of the particles, and for roughly dispersed powders [$d > 1 \mu$m) no changes were observed in the surface sensitivity with wide variation of θ [22, 23]. Positive effects were observed for specimens with a smaller surface roughness, e.g. for intermetallic powders [104]. The other ways of determining the concentrations on the surface and in the near-surface layers of powdered catalysts are destructive or require the use of several techniques. The procedure of depth sputtering by Ar^+ ions is widely employed in surface analysis. The interpretation of the data for powders is quite ambiguous, however, because preferential sputtering of the components and ion-induced effects may distort the true picture. This procedure is nevertheless employed for determining the surface enrichment in oxide [293] and zeolite [294] catalysts. The reliability of the results can be improved when analyzing a removed (SIMS) or a remaining (XPS) surface layer.

The use of a combination of these techniques, and also of ISS has already yielded a number of interesting results. Hercules *et al.* [214-217, 295] explained the change in the slope of the plot of the scattering intensity of $^4He^+$ (I_{Me}/I_{Al}, where Me = Co, Mo, Ni, W, or Cu) against the bulk concentration of the metal by the transition of the surface coverage from a monolayer to an "islet" type (crystallites). A comparison of these data and the X-ray photoemission intensities depending on the calcination temperature pointed to migration of the cations either to the outer surface or into the support lattice. Brinen *et al.* [296] studied still more complicated systems — sulphidized alumina-cobalt-molybdenum catalysts by ISS. For more specific interpretation, the components of single scattering were singled out proceeding from the lineshapes of the individual phases of the sulfides and Al_2O_3. A model was proposed for conduction semiquantitative analysis of catalyst powders by ISS. The intensities of MoO_3/Al_2O_3 measured by ISS and XPS were used to plot profiles of the changes in the ratios Co/Mo, Co/Al, and Mo/Al with the help of the coefficients of elemental sensitivity. When analyzing the XPS intensities, nonuniform distribution of the components along the depth of a layer was assumed. The surface concentrations determined by ISS depend greatly on the preparation conditions. For instance, with the direct incorporation of Co and Mo into pseudoboehmite, no cobalt was detected on the surface at all [296].

When industrial catalysts are being used, it is important to know how the active components are distributed over a granule within the range of several micrometers. In this case, a microprobe, analytical electron microscopy and so on are used together with XPS. The distribution of Al in crystallites of the zeolite ZSM-5, for example, was studied in a synthesis process with the aid of XPS and X-ray microanalysis [297]. Even simple manipulations such as slicing tablets and measuring spectra before and after the grinding of a granule yield useful information on the nature of active component distribution in MoO_3/Al_2O_3 extrudates [298], mixed oxides V_2O_5-TiO_2, V_2O_5-SnO_2 [299], etc.

3.4   Structure of a Surface

When dealing with the EES and IS techniques, we indicated that although they are not structure-sensitive ones in the strict meaning of this term, an analysis of the features of spectra can provide information on the local structure and morphology of surface layers. If the morphology of a surface can be studied by local methods of analysis with the obtaining of an image in secondary or Auger electrons, secondary or scattered ions, then such structure elements as clusters, defects (their concentration, composition, and size) can be studied first of all by employing SIMS and high-resolution Auger spectroscopy. All these procedures can be applied to model objects such as various kinds of films, mono- and multilayer adsorbates on metals, alloys with partial segregation of their components and the presence of surface oxides or other compounds, and so on.

Notwithstanding the difficulty of interpreting SIMS data, the registration of the molecular and cluster ions and the change in the secondary ion emission coefficients in two- and multicomponent systems provide a notion both of the macroscopic (topography, work function) and local (cluster composition and size) structural features of a surface. In one of the first studies of this kind [300], both a change in the secondary ion emission of the components in the matrix and the appearance of mixed clusters of FeRu$^+$ typical only of alloys and not of a mechanical mixture were discovered on FeRu alloys. The presence of peaks of FeO$^+$, FeOH$^+$, RuO$^+$, and RuOH$^+$ indicated the partial oxidation of Fe$^+$ and the adsorption of H_2O on Ru, which was confirmed by XPS data. The detection of FeCl$^+$ and RuCl$^+$ peaks indicating the presence of Cl on the surface of the alloys was also significant.

Later SIMS was employed to study the features of the structure of bimetal catalysts prepared by sputtering films onto single crystals [301, 302]. XPS and thermodesorption were used as additional techniques. It was shown for Cu/Ru(001) specimens prepared at 540 K and 1080 K that with submonolayer coverages, the concentration of Cu is proportional to RuCu$^+$/Ru$^+$. The deposition of Cu at the two temperatures follows different mechanisms. An analysis of the ratio of the peak intensities for the dimers Ru$_2$, Cu$_2$, and clusters of RuCu yielded the following information: (a) on the relation between $\Delta\varphi$ and the secondary ion emission; (b) on the film structure, namely, the formation of three-dimensional clusters at 540 K and of very low two-dimensional clusters at high temperatures and low coverages; (c) on the real surface roughness; and (d) on the size of the Cu clusters. The latter was appraised approximately from the ratio RuCu$^+$/Cu$^+$ for two temperatures. The minimum Cu cluster consists of three to five atoms. We have already noted that the escape depth of molecular ions is smaller than that of atomic ones. Prigge and Baner [303] indicate that Me$_2^+$ clusters appear when a monolayer forms, whereas Garrison *et al.* [304] state that the appearance of such ions was observed during the building up of the second or third layer. Garrison *et al.* have also clearly registered steps on the curves showing the change in the emission of Cu$^+$, Cu$_2^+$, and Cu$_3^+$ when the first, second, and third layers were filled. The structure of a surface covered by an adsorbate can be studied by employing SIMS with angular resolution [305]. Peaks of the yields of C$^+$ and O$^-$ in the structure $c\,(2 \times 2)$ O/Cu(001) were observed in the directions $\langle 100 \rangle$ and $\langle 110 \rangle$, respectively. In accordance with theoretical calculations, this points to localizations of atomic oxygen between four copper atoms.

SIMS and XPS give a notion not only of the thickness, but also of the morphology of the oxide films formed when the polycrystalline alloys Ni-Co and Pd-Cu react with oxygen at elevated temperatures [281]. The oxide films are not continuous, but form islets, and from the ratio of the intensities of

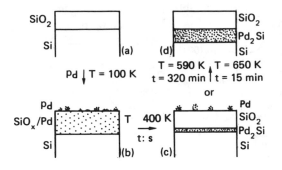

Fig. 34. Schematic representation of the local structure of the surface and near-surface layers for the system Si/SiO₂/Pd depending on the treatment [305]:

a — clean layer of SiO₂ on Si(111); *b* — sputtering of Pd at 100 K; *c* — annealing at 400 K; *d* — annealing at $T \geqslant 550$ K

NiCoO$^+$, Ni$_2$O$^+$, and Co$_2$O$^+$ it follows that a noticeable part of the surface is occupied by clusters of mixed oxides. In the oxidized Pd-Cu alloys, Cu$_2$O clusters on whose periphery atoms of metallic Pd are localized are the main fragments of the structure. The latter has been proved by the data of XPS, XAES, and SIMS.

A combination of the XPS, UPS, and HREELS techniques [305] made it possible to establish the surface and interface structure for more complicated systems such as Si/SiO₂/Pd (Fig. 34). It was found that even at 100 K palladium diffuses into the SiO₂ layer and occupies tetrahedral anion vacancies therein. With elevation of the temperature, owing to greater diffusion along the vacancies, the Pd atoms reach the SiO₂/Si interface and form the silicide Pd₂Si. The surface gradually becomes depleted in metallic Pd (clusters 1.6 nm in size) and at $T \geqslant 700$ K all the Pd passes into the oxide film. The diffusion kinetics depends on the energy needed to detach the Pd atoms and on the concentration of the oxide defects. The obtained data are of direct interest for studying the formation of real supported metal catalysts.

Recently, an interesting and promising technique was proposed for determining the local structure of a surface from the ultraviolet photoelectron spectra of physically adsorbed xenon (PAX) [82]. We have already mentioned that the shift in the $5p_{3/2-1/2}$ level of xenon can be used to find the local work function, which depends not only on the nature of the metal, but also on the real structure of the surface. Let us give examples of studies by the PAX technique. In one case, the system Ag/Ru(001) close to the one considered

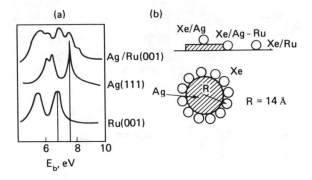

Fig. 35. Spectra of He I for the Xe $5p$ level of xenon adsorbed on pure Ru and Ag and on a thin film (0.2 monolayer) of Ag/Ru (a) and a model of Ag "islets" on the surface of Ru (b) [306]

above was studied [306]. Figure 35 presents the He I spectra of the Xe $5p$ level observed in the adsorption of Xe on clean metals and on a bimetal system. In the latter case, there are three adsorption centers: on Ag "islets", on RuAg centers, and on Ru. It is interesting to note that the number of islets coincides with the number of defects in Ru(001), which were also "titrated" with the aid of PAX. The average size of the supported metal islets determined by PAX is 2.8 nm. Since the work function of Ag and Ag/Ru coincides, it is concluded that Ag does not become alloyed with Ru. The surface structure of single crystal alloys consisting of $Pt_xNi_{1-x}(111)$ is of a different nature [307]. Although the surface becomes very rich in platinum, the latter is not in the form of islets, but forms a quasihomogeneous layer. The properties of this layer, however, differ greatly from the surface of clean Pt(111). This is indicated by the values of the macroscopic and local work functions (measured by a standard technique and by PAX, respectively), which not only coincide, but are also close to the work function for Ni(111). This unexpected result is interpreted as a consequence of the retaining of strong electron interaction of the metals in the alloy even with such a great surface segregation of the platinum.

Another model system of interest for catalysis, namely, Pt/Re(0001), has also been characterized by PAX [308]. Three types of centers were identified on it, namely, isolated Pt atoms, islets of Pt with a size of about 1.6 nm, and mixed Pt-Re centers, the latter forming the majority. Since in the given case the number of defects is smaller than the number of islets, a mechanism is assumed by which single Pt atoms are initially adsorbed on a surface, and they are centers of cluster nucleation.

The mechanism of growth of Cu and Co epitaxial films on Ni(100) has been studied by ARXPS [309]. The technique is very sensitive to the structure of the adsorbed layer and surface reconstruction. Such procedures have meanwhile not found any use in the studying of real catalysts. Among the reasons are the fundamental limitation in the use of ARXPS and the difficulty of interpreting the SIMS data. At the same time, SIMS can apparently extend our notion of the local structure of their surface. In particular, in cobalt-molybdenum-alumina catalysts [214, 215], the ratio of the intensities of the Al_3^+, CoO_2^+, and $CoAl^+$ clusters is related to the ratio of the clusters of cobalt oxide and aluminate. Such a procedure was used to describe the structure of the surface of supported Cu-Ru/SiO$_2$ catalysts [310].

An analysis of the shape of high-resolution Auger spectra is used to obtain information on the structure for various objects such as nonmetallic [311] and metallic [312] adsorption layers, and nonstoichiometric carbides, nitrides, and oxides [99-101, 313-315]. For instance, a complete analysis of the shape of N *KVV* Auger lines and a comparison of experimental data with calculations of clusters of iron nitrides [313] yielded a relation between the ratio of the intensities of individual components and the nitrogen content in the diffusion layers of iron nitride. Moreover, criteria were found for these amorphous materials that made it possible to appraise the ratio of the various highly dispersed nitride "phases". Data obtained by Göpel *et al.* [314] for single TiO$_2$ crystals treated in various ways (Fig. 36) are a brilliant example illustrating

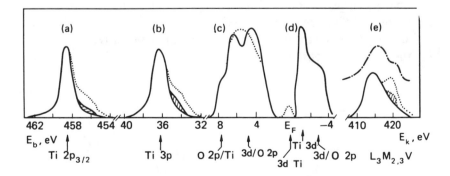

Fig. 36. XPS spectra of inner Ti 2p (a) and Ti 3p (b) levels, valence band of Ti (c), EELS spectra of conduction band (d), and Auger spectra of Ti L_3VV (e) for an ideal and defective surface of TiO$_2$ (100) [314]:

———— ideal stoichiometry, _ _ _ annealing at 1310 K, ... Ar$^+$, 10 min, 0.5 keV, _._._. Ti L_3VV spectrum for Rh/TiO$_2$ specimen reduced at 773 K [222]

the sensitivity of various EES techniques to the presence of point defects. The appearance of defects in annealing or ion bombardment of the oxide is reflected in a change in the shape of the X-ray photoelectron spectra of Ti $2p$, Ti $3p$, O $1s$, the valence band, the conduction band (EELS), but the Ti L_3VV spectrum describing an intra-atomic transition with the participation of only $3d$ states of Ti is the most sensitive. Even in the initial specimens, a slight asymmetry of the Ti L_3VV line is observed that vanishes only after prolonged annealing in oxygen. These features of Auger spectra can also be employed for analyzing real Me/TiO$_2$ catalysts. The Ti L_3VV spectrum obtained after the reduction of Rh/TiO$_2$ at 500 °C also points to the appearance of defects (Ti^{3+}), whereas the Ti $2p$ and Ti $3p$ spectra remain virtually without any changes at these concentrations (Fig. 36).

Attention is deserved by works in which attempts are made to establish the nature of the local environment and type of bond of atoms on a surface by using an ion bombardment technique [316, 317]. For example, the easiness of reducing Ti^{+4} ions in mixed SiO$_2$-TiO$_2$ oxides under small dozes of Ar$^+$ correlates with the nature of the interaction of the components and their bond to the supported metal atoms. As a whole, however, it should be noted that the possibilities of the surface analysis techniques in determining the local structure of catalysts are far from being exhausted. In this connection, we must mention the SAM technique, which has begun to be used for studying industrial catalysts [318, 319]. It has been employed to determine not only the composition of the surface of promoted Fe catalysts used in ammonia synthesis [318] that are greatly enriched in oxides of the promoters (Al, K, Ca), but also its morphology. The Al$_2$O$_3$ and CaO were found to be distributed in the form of separate islets, while the potassium covers the surface in a more uniform layer. When the catalysts are poisoned by sulfur (H$_2$S), it is deposited on the active portions of the surface.

4 The Study of Heterogeneous Catalysts by Electron Emission and Ion Spectroscopies

In Chapters 1-3, we dealt with the theoretical and methodological fundamentals of the EES and IS techniques and the approaches to studying the basic characteristics of the surface of solids by these techniques. In this chapter, we shall discuss the results obtained by EES and IS as applied to the main types of various heterogeneous catalysts, i.e. massive metals and alloys, oxides, other stoichiometric compounds, zeolites, and also to systems containing active components in a highly dispersed state. This classification is quite conditional because the fundamental problems associated with revealing the mechanism of action of real catalysts must be studied in close conjuction with model objects. Moreover, there is a whole class of catalysts that can be related to an "intermediate" type. There are compositions prepared from metals and oxides, zeolites and oxides, and also catalysts which in the course of preliminary treatments or service transform from one type of system to another. Nevertheless, the classification by systems seems to be expedient because for each class of catalysts we can single out the key problems associated with establishment of the surface properties, and their solution will help us reveal the mechanism of catalytic action of this class. From the viewpoint of practical application, such a classification enables one to summarize data on the mechanism of formation of an active surface for catalysts of specific processes.

Each section of this main chapter of our monograph has been built up as follows: first we formulate the general theoretical and applied problems associated with studying of the surface of a given class of catalysts, comment on the level of their solution by the beginning of the 1980's, and then discuss the new experimental data that enable one to advance in understanding the fundamental theoretical problems of catalysis, elucidating the mechanism of catalytic action, and in developing catalysts with improved catalytic properties.

It should be noted that many of the considered systems—metals, alloys, and semiconductors—are of interest not only in catalysis, but also in other fields of science and technology. We naturally took the relevant studies into account when considering the general problems, although preference was given first of all to investigations aimed at studying these systems in the capacity of catalysts.

4.1   Metals and Alloys

Metal-based catalysts occupy a leading place in catalysis as regards the scale of their employment in various processes. This circumstance probably explains the interest in them at all stages of catalysis science development. Many of the general theoretical concepts advanced at different stages of the development of the science of catalysis related primarily to metal systems. If we analyze the last stage beginning in the 1970's, then since structural and electron spectroscopy techniques were adapted the most to metal and alloy systems, they continued to remain at the center of attention of investigators both in the field of catalysis and in the field of surface physics. The determination of the mechanism of the catalytic action of metals and alloys depends on the level of understanding the role of local and collective structural and electronic properties of their surface in adsorption and catalytic interactions, i.e. this determination is closely related to the description of properties such as:

(a) the electron structure (degree of filling of the energy bands, transfer of the electron density between components);

(b) the surface electron states;

(c) the crystallographic and geometric structure; and

(d) the composition of the surface, the surface segregation, and ordering.

The fundamental concepts relating to a description of the bulk band structure of metals and alloys were proposed in the 1970's on the basis of data of photoelectron spectroscopy and theoretical calculations by methods of solid state physics [10, 31, 98, 224, 225, 320, 321]. One of the most important conclusions for catalysis is the one on the quite strong localization of the d states of active metals and on the retaining of a high degree of localization of these states in binary alloys. The phenomenon of surface segregation in alloys with various thermodynamic characteristics was established and confirmed by numerous investigations [265-276]. This explained the fundamental drawbacks of the theory of catalysis on metals and alloys that was based on simplified models of the band theory and took no account of surface segregation [322-326]. For example, the maxima of the d states of Ni and Cu in an alloy shift only slightly, and even with the addition of 60% of Cu sufficient for the complete filling of the d band of Ni, the density of the d states at the Fermi level is retained. On the other hand, in this alloy, surface segregation of the copper is observed, and a constant surface composition is retained in a large interval of bulk concentrations of copper [325, 326]. Only with a view to these data and to the structure of the adsorption and catalytic centers did it become possible to explain the laws of the change in the catalytic activity of this system in various reactions [325, 326]. The studies performed by EES techniques yield the general conclusion that the disturbance of the electron structure of metals in alloys is associated with their thermodynamic characteristics: for intermetallics they are maximal, and for alloys with close heats of formation of the components or stratifying within a broad interval of compositions, they are minimal. On the other hand, such a macroscopic approach did not explain the causes of the synergism of action of many supported bimetal systems containing an active and inactive components, or components that are virtually insoluble in each other [327-329].

We can also relate the determination of the role played by the surface structure of metals in catalysis to the achievements of the 1970's. This determination was based on data on the reactivity of molecules on various single crystal surfaces differing in the orientation and closeness of packing of the faces and the presence of various kinds of defects [1, 2, 330-332]. These sections consisting of separate structure elements (steps, terraces, kinks) and even atomic sites were classified by their reactivity with respect to various classes of reactions, and the concept of the "structural sensitivity" of reactions introduced in various terminology by Boudart [333] and Boreskov [334] was explained on an atomic-molecular level. Since the qualitative features of the reaction of molecules with metals are simulated by calculations in a cluster approximation, the hypothesis on the local nature of catalytic interactions was also supported for massive metals [326, 332, 335], although the simplified theory of these methods does not enable one to appraise the role of collective interactions for an entire surface. The electron structure of the most reactive regions of a surface has also not been revealed to the end. The main technique for studying the surface structure of single crystals is LEED, and these results are generally considered together with the results of studying the reactivity of molecules. With a view to these circumstances, and also to there being an enormous number of publications on these matters, we shall limit ourselves in the present section to a treatment of the surface structure of more complicated objects whose studying was begun only recently (single crystal alloys, bimetal films) by a set of techniques including EES and IS.

Surface segregation has been described qualitatively for many clean systems; less data have been obtained for systems including an alloy and a reaction medium. Moreover, as indicated in Chap. 3, in many cases the experimental data were either inaccurate or contradictory. We also find doubtful the simple transfer of results obtained on clean metals and alloys to real catalysts based on them and moreover to dispersed metals and alloys on supports. Information available at present makes it possible to compare the properties of the surface of clean metals and alloys, thin bimetal films, real metal and alloy catalysts of the Raney system type, and also catalysts produced by a novel technology—by special processing of alloys, intermetallics, and films.

In this connection, we shall attempt to discuss the following matters using objects studied the most as an example:

(a) the electron structure of a surface, the electron interaction of metals in alloys;

(b) the laws of the change in the chemical state of components and the surface composition of alloys in chemisorption and surface oxidation;

(c) modelling of the structures of alloys, metal oxide, and bimetal dispersed systems on clean and oxidized bimetal films;

(d) the formation mechanism and nature of surface centers in real catalysts based on metals and alloys.

4.1.1   Electron Structure of the Surface of Metals and Alloys

The matters associated with the electron structure, charge transfer, and surface segregation have been studied the most completely for single crystal specimens, thin films of one component deposited onto a single crystal of the other one, and polycrystalline foils. The characteristics of the surface electron structure determined by EES and IS include the surface electron states in the valence and conduction bands, the electron levels on the surface atoms, the local work function for individual sections of a surface, and the degree of charge transfer between the components in alloys. The specific features of the surface band structure of metals (Pt, Ni, Pd, Cu, Ag, Au, etc.) are indicated by data of UPS with angular resolution or with spin polarization, and of inverse photoemission [42, 87-89, 336-339]. Photoelectron spectroscopy monitors the presence of occupied surface levels below the Fermi level (1-2 eV), whereas IPS confirms the formation of vacant surface levels arranged immediately after the Fermi level. IPS measurements are especially effective for metals such as Cu, Ag, and Au having a wide *sp* hybridized "gap" in which the surface states (of the Shockley type) are localized [87]. The surface states observed

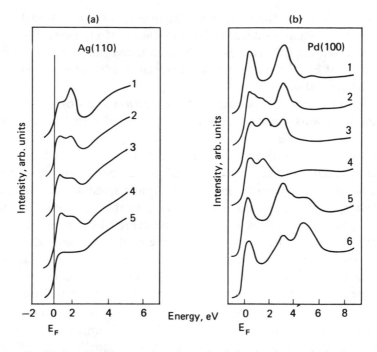

Fig. 37. Spectra of inverse photoelectron emission for clean and adsorbate-coated metal surfaces:

a — Ag(110) [87] ($h\nu$ = 9.7 eV): 1 — clean surface; 2 — 0.75 L O_2; 3 — 1.75 L O_2; 4 — 0.75 L H_2; 5 — 1.5 L H_2; b — Pd(100) [88]: 1 — clean surface; 2 — 1 L NO; 3 — saturation with NO; 4 — saturation with NO (θ = 30°); 5 — 5 L CO; 6 — saturation with CO

by photoelectron spectroscopy and IPS play an important role in adsorption and catalysis. Figure 37a shows the change in IPS spectra for Ag(110) identically oriented relative to the radiation source and detector [87]. One can see quite clearly that the surface states are the most reactive: in the adsorption of O_2 or H_2 even with an exposure of 1.5 L they are completely suppressed, whereas the "bulk" part of the valence band remains unchanged. The IPS spectrum for Pd(100) (Fig. 37b) has two peaks: one is directly at the Fermi level, while the other is 3 eV above the Fermi level. The first peak corresponds to the vacant d states of Pd [86]. When NO is adsorbed, a new structure appears with ΔE_F = 1.8 eV, while when CO is adsorbed, a structure appears with ΔE_F = 4.8 eV. Simultaneously, the emission of the surface states of Pd (especially with NO) is suppressed. The new bands in the spectra correspond to vacant 2π levels of the molecules. The overlap of the NO peaks and the d states of Pd points to intensive back donation of electrons $d \rightarrow 2\pi^*$. This is not observed for CO. The results obtained agree quite well with the different reactivity of Pd with respect to NO (dissociative adsorption) and CO (molecular adsorption).

The differences in the energy of the electron levels of surface atoms and in the local work function on various sections of a surface were mentioned in Chap. 3. These deviations are generally the most significant for coordination-unsaturated and the most reactive atoms. The inclusion of such data as initial information for calculating the interaction between molecules and metals will apparently enable us to find a quantitative relation between the charge state of the atoms of a metal and the reactivity. Up to now, the convergence of calculations and experimental results was verified by comparing the density of the valence states or the binding energies of the core levels averaged along the depth of a layer. At the same time, since the chemical shifts for the levels of the surface atoms do not exceed 0.1-0.2 eV in the majority of cases, such an approximation can be considered as sufficiently correct.

The further studying of the valence band structure in alloys containing elements with a different catalytic activity is of substantial interest. Within the confines of simple correlations between the reactivity and difference between the electronegativities of the components, the appraisal of the true charge transfer between the components is important. Table 7 presents the values of the chemical shifts for a number of alloys and intermetallics of Ni and Pd with electronegative and electropositive elements [225]. There is noted a qualitative tendency of growth of the shift with an increase in the electronegativity of the second component and of its fraction in the alloy. For instance, in the series $Al_{0.1}Pd_{0.9}$, AlPd, and Al_3Pd, the shift in the Pd $3d$ spectrum increases from 0.3 to 2.5 eV. But such large shifts are explained not by charge transfer between the components, but by changes in the density of states in the valence band of Pd or Ni and by the final state effects. When sp elements are added, the d subband of Ni or Pd is filled, but the d states of these metals are retained in the subband of a dissolved component even when saturated solutions form. The contribution of the final states was taken into account by using the model of "equivalent cores" [340], according to which the shift is determined by thermodynamic characteristics such as the difference between the heats of solution of an element with the core Z and Z + 1 in solvent A. For the alloy Al-Pd (element Z), $\Delta H = -84$ kJ/mol, while for Al-Ag (Z + 1), $\Delta H = -10$ kJ/mol. The large difference in ΔH does indeed cause substantial chemical shifts. Although this reasoning is of a qualitative

Table 7. Shifts and Satellite Structure of Inner Level Spectra in Alloys and Intermetallics of Pd and Ni with Electropositive and Electronegative Elements [225]

Compound	ΔE_b, eV	Compound	ΔE_b, eV	ΔE_{sat}, eV	r_{sat}, %[b]
Pd	0[a]	Ni	0[c]	5.8	29
$Al_{0.1}Pd_{0.9}$	0.3	$AlNi_3$	0	6.55	21
AlPd	1.9	AlNi	0.2	7.20	19
Al_3Pd	2.5	Al_3Ni_2	0.75	8.00	12
$TiPd_3$	0.6	Al_3Ni	1.05	—	10
VPd_3	0.4	TiNi	0.4	7.20	26
VPd_2	0.9	CuNi	− 0.2	6.00	30
VPd	0.9	PdNi	− 0.05	6.20	28
V_3Pd	1.3	AuNi	− 0.5	5.40	30
NiPd	0.5	$LaNi_5$	− 0.05	6.00	23
$NiPd_3$	0.9	LaNi	0.20	—	—
$TaPd_3$	0.8	La_7Ni_3	0.05	—	—
$TaPd_2$	1.3	La_3Ni	0	—	—
TaPd	1.3	$CeNi_5$	− 0.05	6.40	25
YPd_3	0.4	$CeNi_2$	0.35	6.85	23
$LaPd_3$	0.6	CeNi	0.30	6.90	20
$LaPd_2$	0.9	Ce_7Ni_3	0.3	7.00	19
LaPd	1.2				
La_7Pd_3	1.4				
$CePd_3$	0.7				
Ce_3Pd_5	1.1				
CePd	1.0				
Ce_7Pd_3	1.4				
CePdAl	1.4				

[a] Level $3d_{5/2}$, E_b = 335.2 eV.
[b] Relative intensity of satellite, %.
[c] Level $2p_{3/2}$, E_b = 852.6 eV.

nature, it is evident that the shifts of the inner levels in alloying reflect not only charge transfer between the components.

Let us now analyze the data for bimetal systems that are the most interesting for catalysis according to three main aspects: (a) the electron interaction of the components; (b) surface segregation (including that produced by chemisorption); and (c) the change in the chemical state of the components and the composition of the surface during mild oxidation.

4.1.2   Platinum Alloys

These systems are of the greatest interest as model catalysts of hydrocarbon transformations, primarily of the reforming process. Re, Sn, Pb, Ir, etc. are used as the second component in platinum reforming catalysts. Godfrey and Somorjai [341] modelled a Pt-Re catalyst by growing a film of Re on Pt(111) or of Pt on Re (0001). Annealing at 1150 K leads to intensive mutual diffusion of the two metals. Here the total shift of the Pt $4f_{7/2}$ spectrum was $+0.9$ eV, and of the Re $4f_{7/2}$ spectrum, $+0.4$ eV. It is interesting to note that initially in the formation of an epitaxial layer of Pt on Re(0001), i.e. in the absence of appreciable interaction between the metals, the shift for Pt was already $+0.4$ eV. Treatment of the bimetal system in O_2 (800 K) results in partial oxidation of the Re to ReO_x ($2 < x < 4$), the Pt catalyzing the process of decomposition of the Re oxide under very mild conditions (673 K, vacuum). The oxidation of the Pt-Re alloy has been studied in greater detail by SIMS and AES [342]. The reaction of the massive alloy $Pt_{60}Re_{40}$ with O_2 at 300-900 K and a low O_2 pressure (1.3×10^{-6} torr) caused strong surface segregation of Re. In clean alloys in accordance with Miedema's model [Eq. (3.15)], platinum should segregate. But when chemisorption occurs, it is necessary to take into consideration the heats of formation of the relevant surface compounds. In the model of Tomanek et $al.$ [343], the surface state of such a system can be appraised by the equation:

$$\frac{c_{A,s}}{c_{B,s}} = \frac{c_{A,b}}{c_{B,b}} \exp \frac{Q_{ch.seg}}{RT} \tag{4.1}$$

where

$$Q_{ch.seg} = Q_{seg} + (E_A - E_B)\,\theta \tag{4.1a}$$

Q_{seg} is the heat of segregation in the absence of an adsorbate, $Q_{ch.seg}$ is the heat of segregation due to chemisorption, E_A and E_B are the energies (heats) of chemisorption of an adsorbate on the pure components (or the heats of formation of the relevant surface compounds) A and B, respectively, and θ is the fractional coverage. In the given case at $\theta \geqslant 0.15$, the value of $Q_{ch.seg} \gg 0$, i.e. the segregation of Re is expected. Unlike a previous investigation [341], no bulk oxidation of the components was observed under these conditions (data of XPS), which made it possible to ignore the changes in the sputtering yield coefficients (Y^+) when appraising the surface composition.

In a detailed study performed by XPS [344], an attempt was made to establish the electron structure of the alloys Pt-Pb and Pt-Sn obtained by the consecutive sputtering on of both components. The shape of the Pt $5d$ valence band and the Pt $4f$ core levels was close for both systems whose alloying was established in particular by the absence of a relation between these characteristics and the sequence of sputtering on the components. The basic distinctions of the Pt-Me alloys from pure Pt are the narrowing of the valence band and a decrease in the asymmetry parameter of the Pt $4f$ line (from 0.34 to 0.21). Since the asymmetry is associated with electron-hole excitations in the conduction band, its decrease is explained by the shifting of the $5d$ band relative to the Fermi level. The distinctions between Pt-Pb and Pt-Sn consisted in the greater segregation of lead on a surface in comparison with tin, which agrees with the lowest heat of sublimation of Pb, and in the presence of a shift of $+0.6\,eV$ in the Sn $3d_{5/2}$ spectrum. This served as the grounds for a quite paradoxical conclusion on the stronger electron interaction of Pt and Pb in comparison with Pt and Sn. It is based on the interpretation of the positive shift as a result of the decrease in E_{ER}, and on the absence of such a shift (Pb) as a larger compensation of E_{ER} at the expense of the initial negative shift of the Pb level.

Investigations performed by XPS [345], ISS [346], and SIMS [347] delved into the reaction of these alloys with oxygen. Cheung [345] observed the oxidation of both Sn (with the formation of SnO or SnO_2) and Pb (PbO_x with $x < 1$) at room temperature and an oxygen pressure of 200 kPa. With a growth in the platinum concentration, the fraction of the oxidized states diminished, and electron interaction between the components was retained (a shift in the Sn $3d_{5/2}$ spectrum). The suppression of the ability of the tin and lead to oxidize is associated with the decrease in the density of the sp states in the d subband of Pt. At lower exposures (10 Pa·s), but at higher temperatures (600-700 K) no oxidation of the intermetallic Pt_3Pb was observed [347], while the oxygen depending on the temperature was either on the surface of the alloy or penetrated into the near-surface layers. The surface of the intermetallic was enriched in lead, and its segregation increased in the presence of oxygen. Finally, for the alloy $PtSn_3$ obtained electrochemically and oxidized at 200 °C (100 L), the suppression of the intensity of the Pt ISS peak has in general given rise to the assumption that the chemisorption of oxygen predominates only on Pt, but not on Sn.

Consequently, the data obtained by different techniques for Pt-Me alloys prepared in different ways vary appreciably, especially in the appraisal of the degree of oxidation of the second component. At the same time, common qualitative trends important for catalysis can be singled out, namely, the presence of electron interaction between the components, the surface segregation of the second component that grows in the presence of oxygen, and the retardation of its oxidation when the concentration of Pt grows (or the easier decomposition of the surface oxide).

The studying of the alloys Pt-Pd, Pt-Ru, and Pt-Ir within a broad interval of compositions made it possible to establish a noticeable mutual influence of both components on their electron and redox properties [348]. In accordance with simple thermodynamic models, surface segregation of Pd is observed in the initial Pt-Pd alloys, while surface segregation of the second component is observed in the other two alloys. When definite threshold concentrations of the second component are reached (13-20%), both metals are oxidized; within the interval from 200 to 600 °C at 1 atm various platinum oxides form: PtO, α-PtO_2 and β-PtO_2. At lower concentrations of the second metal, oxidation neither of the dopant nor of the main metal is observed. The unusual phenomenon of the oxidation of Pt in alloys under conditions when pure Pt does not become oxidized was explained by the flowing of the oxygen by a spillover mechanism from the initially oxidizing second metal to Pt. Hilaire *et al.* [348] explained another unexpected effect—the presence of threshold concentrations of the second metal below which no oxidation process occurs at all—by what is called a percolation transition. This mechanism presumes that oxidation is possible only when a definite number of Me-Me bonds are present. In their opinion, both mechanisms are rather speculative. It is not clear, for instance, why oxygen cannot dissociate directly on Pt. More detailed information is needed on the degree of modification of the electron properties of the components in these alloys.

The ability of the alloys Pt_xNi_{1-x} (111) to adsorb hydrogen and hydrocarbons, and also to catalyze hydrogenation reactions of hydrocarbons differs noticeably from this ability for the pure metals [307]. In particular, the bond with the reactants is weakened, although, as noted in Chap. 3, the upper surface layer of these alloys is greatly enriched in Pt, while beginning from $x = 0.5$ it contains practically only Pt. This behavior is explained by the electron interaction between the components in the first and second layers, the latter being enriched in nickel [307]. This interaction manifests itself in lowering of E_b of the Ni level and increasing of E_b of the Pt level in XPS [349]. Qualitatively, the conclusion on the charge transfer from Pt to Ni is also consistent with the magnitude of the local work function [307], which is lower than on pure Pt(111). Among the other investigations of platinum alloys, we shall note that of van Langeveld *et al.* [350] in which the surface segregation of Cu was observed in the very dilute alloy $Pt_{0.98}Cu_{0.02}$, and the investigation of the intermetallic Pt_3Ti [351] whose mild oxidation resulted in segregation on the surface of TiO_x.

4.1.3 Other Solid Solutions and Intermetallics

A broad range of solid solutions and intermetallics containing Pd, Ru, Ni, Fe, Co, and Cu are of interest as promising catalysts of reactions of partial oxidation and selective hydrogenation of hydrocarbons, methanation, the synthesis of hydrocarbons from CO and H_2, the synthesis of methanol, etc. Since in these reactions the intermediate or product is an oxidizing agent, it is important to study the initial stages of oxidation of these alloys in addition to clean surfaces.

Co-Ni alloys are a continuous series of solid solutions close to ideal ones whose surface by Miedema's model (Table 8) should be enriched in Ni. Judging from data of XPS, AES, and SIMS, the composition of the surface layers of Co-Ni is close to the bulk one except for a specimen with a high cobalt content, i.e. $Co_{60}Ni_{40}$ [281]. Enrichment of the latter specimen is especially noticeable when measuring the photoemission at a glancing angle and from SIMS data (see Fig. 32) [281]. The enthalpy of surface segregation determined by Eq. (3.14) is 20 kJ/mol within the interval from 623 to 823 K.

Table 8. Theoretical [Calculated by Eq. (3.15)] and Experimental Appraisals of the Direction of Surface Segregation in Selected Binary Alloys

Solid solution	ΔH_{sol}, kJ/mol	W_1	W_2	W_3	W	Experimental data
Co in Ni	− 1	0	− 0.6	0	− 0.6	No segregation
Ni in Co	0	0	0.6	0	0.6	Segregation
Ru in Co	− 3	0	− 3.2	0.6	− 2.6	Co segregates
Cu in Pd	− 33	− 0.9	1.5	0.4	1.0	Cu segregates
Pd in Cu	− 43	− 1.2	− 1.8	0.4	− 2.6	Pd segregates

Note. ΔH_{sol} is the heat of solution of A in B, W_1 is the contribution of the heat of solution (the formation of an alloy), W_2 is the contribution of the surface energy, W_3 is the contribution of the difference between the molar volumes of A and B, and $W = W_1 + W_2 + W_3$.

Treatment of the alloy $Co_{50}Ni_{50}$ in H_2 was attended by diffusion of the nickel to the surface (the Ni/Co ratio changed from 1.0 to 1.3). This was possible because of the formation of the hydride of Ni. But the most substantial changes occur under the effect of oxygen (Fig. 38). Reaction with oxygen at 250 °C and $p = 5 \times 10^{-6}$ torr causes oxidation of the components and a

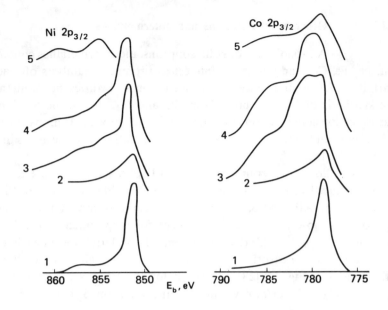

Fig. 38. Change in the state of the components of the alloy $Co_{50}Ni_{50}$ in oxidation depending on the temperature and pressure [278, 352]:

1 — initial state; *2* — 293 K, 5 torr; *3* — 523 K, 5×10^{-6} torr; *4* — 523 K, 5×10^{-6} torr ($\theta = 15°$); *5* — 673 K, 5 torr

change in their ratio on the surface. Oxidation is identified by the shift of the Me $2p_{3/2}$ lines, and also by the appearance of Ni_2O^+, Co_2O^+ peaks (Fig. 39), but even in the first measurements the intensity of CoO^+ or NiO^+ peaks is not high. The ratio Co^+/Ni^+ in specimens of $Co_{10}Ni_{90}$, $Co_{50}Ni_{50}$, and $Co_{70}Ni_{30}$ grows 2, 2.6, and 3 times, respectively. Moreover, the enrichment of the surface in Co is indicated by the ratio $Co\ 2p_{3/2}/Ni\ 2p_{3/2}$. A still greater surface segregation of cobalt occurs when the oxygen pressure is increased to 1-5 torr. It is noticeable already at room temperature, while at 400 °C the ratio Co/Ni ($2p$) in the alloy Co_{50}/Ni_{50} increases five times [352]. The difference in the oxygen pressure affects the degree of oxidation of the metals and the thickness of the oxide film. At a low pressure and 250 °C, 85% of the Co is oxidized to Co^{2+}, but an appreciable part of the Ni is present in the form of Ni^0. At a high pressure, the cobalt oxidizes partly at room temperature, at 250 °C it is oxidized completely, and at 400 °C only oxides of Co and Ni are present on the surface. In both modes at 250 °C, CoO forms, while at 400 °C (5 torr of O_2) Co_3O_4 appears. With a growth in the concentration of

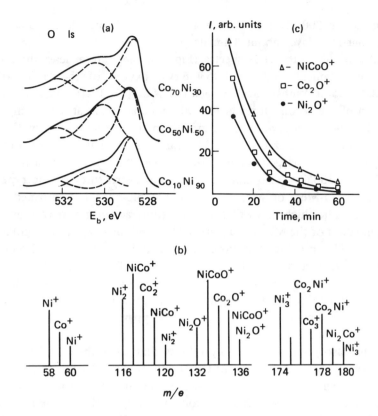

Fig. 39. Chemisorption of oxygen and the formation of an oxide film on Co-Ni alloys [281]:

a — forms of oxygen; b — clusters $(Me_nO)^+$ in a mass spectrum of the alloy $Co_{10}Ni_{90}$; c — change in the intensity of the cluster ion peaks against the Ar^+ sputtering time

Ni in an alloy, its oxidation becomes more difficult; pure Ni is oxidized under the same conditions worse than alloys. This result differs from that obtained by Hilaire *et al.* [348], where the oxidation of the second component promoted the formation of Pt oxides by the oxygen spillover mechanism. It is quite probable that in this case, as in Pd-Cu alloys [353], the oxygen is adsorbed dissociatively on the Ni.

The Co/Ni ratio obtained from the $3p$ spectra is lower than in the bulk, which points to the near-surface layers becoming depleted in Co; the gradient of its concentration over the layer grows with elevation of the temperature from 200 to 300 °C, and only at 400 °C (2 h, 5 torr) does the composition level out in a layer about 3 nm deep.

According to the O $1s$ spectra (Fig. 39), there are at least three forms of oxygen: atomically adsorbed (530.6 eV), oxygen of the lattice of oxides (528.9 eV), and "subsurface" oxygen (583.4 eV [27, 354]). The relative concentration of the second form at $\theta = 90°$ is 60%, while at a glancing angle of $\theta = 15°$ there is a simultaneous growth in the concentration of O_{ads}^- and in the fraction of oxidized metals. This may signify that the oxygen dissociates only on Ni°, the oxygen being on the surface ("top") centers, or that the peak with $E_b = 530.6$ eV also characterizes bridge structures of the type Me_1-O-Me_2, Me_2-O-Me_2, or Me_1-O-Me_1, whose presence is confirmed by SIMS (Fig. 39). The thicknesses of the oxide films according to various estimates (attenuation of the Me^0 signal, the SIMS depth profile, or an angular relation [55, 57, 113]), provided that the coverage is uniform, should be 2-6 nm (523 K, 5×10^{-6} torr) and over 10 nm (673 K, 5 torr). Since the Ni signal in all cases is registered in the surface layer (e.g. in the first SIMS measurements), the oxide film of Co has an islet nature. The presence of mixed clusters of the $CoNiO^+$ type also indicates that the islets apparently have small linear dimensions (<1 nm [302]).

The adsorption of NO also produces changes in the chemical state of the components and segregation in Co-Ni alloys. The NO dissociates already at room temperature. This is witnessed by the appearance in the spectrum of N $1s$ of a line with $E_b = 399.6$ eV corresponding to the nitride [9], and in the spectrum of Co $2p_{3/2}$ of a line with $E_b \approx 779$ eV. Elevation of the adsorption temperature to 250 °C caused complete vanishing of metallic cobalt, which mainly transforms into an oxide; nickel is in three states: metallic, oxide, and, possibly, nitride. Like oxygen, NO "pulls out" cobalt to the surface—the Co/Ni ratio grows from 1.5 to 1.9 times on the alloy $Co_{50}Ni_{50}$ (250 °C), and 2.8 times on the alloy $Co_{20}Ni_{80}$ (400 °C). The degree of cobalt oxidation at 250 °C grows in the series CO < NO $\leqslant$ O_2 [281, 352]. In all cases, chemically induced segregation occurs.

The **Pd-Cu alloys** form a continuous series of solid solutions, while at low temperature they have two ordered phases—PdCu and $PdCu_3$. An appraisal of the surface segregation requires that two factors be taken into account, namely, the difference in the enthalpies of sublimation, which for Cu-Pd is -38.6 kJ/mol, and the difference between the atomic radii $d_{Pd} - d_{Cu} = 0.09$ Å. It follows from Miedema's equation (Table 8) that the surface of the alloys should become enriched in Cu, but it must be noted that the contribution of the size difference in Miedema's theory is taken into consideration very

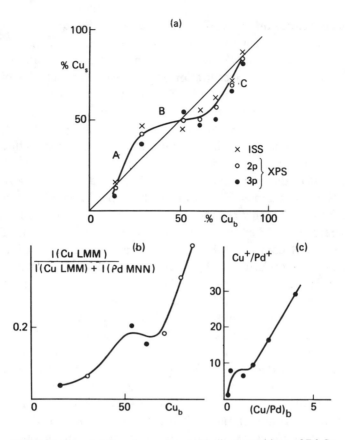

Fig. 40. Relation between the surface and bulk compositions of Pd-Cu alloys from data of XPS and ISS (a), AES (b), and SIMS (c). Preliminary treatments: Ar^+ + annealing, vacuum, 623 K

approximately. We have studied specimens of the compositions $Pd_{85}Cu_{15}$, $Pd_{70}Cu_{30}$, $Pd_{47}Cu_{53}$, $Pd_{29}Cu_{71}$, $Pd_{20}Cu_{80}$, and $Pd_{13}Cu_{87}$, which by X-ray diffraction data are solid solutions. The depth of the layer being analyzed was varied by employing different methods: ISS, SIMS, AES, and XPS (from a monolayer to 2-3 nm). The relations between the surface composition and the bulk one obtained for the initial specimens (Ar^+, 300 °C, vacuum) have a close form (Fig. 40): a linear section for compositions of 10-30 at.% Cu (A), a close to constant composition of the surface in the region of 40-60% (B), and a second linear section for compositions of 60-80% (C). XPS data show that at concentrations of Cu up to 40 at.%, the surface is enriched in copper, and at Cu concentrations above 60%, in palladium. The atomic ratios obtained from ISS data with a view to the elemental sensitivities $\gamma_{Cu}/\gamma_{Pd} = 0.66$ [137] also point to the enrichment of the surface in Cu in region A and to a certain depletion thereof at low concentrations of Pd (B). Although it is difficult to interpret the data of SIMS and AES without the introduction of correction factors, the deviation from a linear relation in region B can be seen quite clearly. Since this region coincides with the region of existence of the intermetallic PdCu, it can be assumed that the constancy of the composition is due to ordering of the surface layer. The regions of constant composition on the very surface (ISS) and in the near-surface layer (XPS) are displaced relative to each other. A "surface intermetallic" forms in the upper layers already at 35% of Cu in the bulk, and then when the Cu content reaches 55%, the surface again begins to become enriched in copper. In the deeper layers (2-3 nm), the beginning of region B coincides with a volume ratio of Cu/Pd = 1 and up to 60-65% of Cu the diffusion of the latter from the bulk into this layer is insignificant. Hence, the replenishing of the monolayer with copper (ISS) leads to a certain reduction in Cu/Pd in the near-surface layer.

Since the literature contains information that the surface composition of these alloys is close to the bulk one [107], and also that the surface becomes enriched in copper in the region of low Pd concentrations (9 and 25%) [353], the possible reasons underlying the discrepancy in the experimental data should be considered. The preferential sputtering of one of the components in ion bombardment may be one of these reasons. Indeed, during prolonged bombardment (four hours), the surface becomes depleted in copper (the alloy $Cu_{50}Pd_{50}$), but in our case these changes are smaller than the observed deviations from the bulk when the composition changed. Sampath Kumar and Hegde [353] indicate that appreciable segregation of Cu is observed in CuPd alloys at $T > 350\ ^\circ C$. Indeed, annealing at 500 $^\circ C$ also resulted in the surface becoming richer in copper. The quantitative differences may also be due to the absence of true equilibrium.

The information on the concentration profiles of elements in the upper layers is important, especially for alloys of the Pd-Cu type, for which a non-monotonic distribution in the layer depth can be expected. Figure 41 shows the changes in the intensities (ISS) on several alloys depending on the duration of He$^+$ sputtering. Its increase from 200 to 1000 s causes the intensity of both ISS peaks to grow, which is most likely due to cleaning of the surface, and also to the change in the ratio of the intensities of Cu and Pd. This ratio drops for a specimen with a low Cu content, while in other cases a blurred maximum is observed. We have already noted that bombardment by Ar$^+$ ions

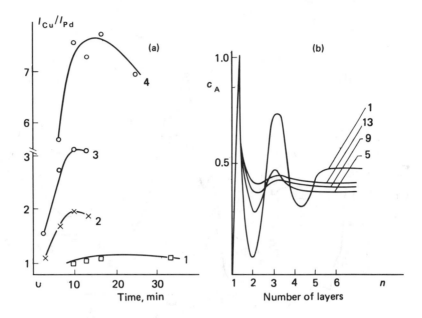

Fig. 41. Composition of the upper surface layers of Pd-Cu alloys by ISS data:

a — change in the ratio of the intensities of the sputtering peaks I_{Cu}/I_{Pd} against the duration of bombardment by He^+ ions (2 keV); 1 — $Pd_{38}Cu_{62}$; 2 — $Pd_{29}Cu_{71}$; 3 — $Pd_{20}Cu_{80}$; 4 — $Pd_{13}Cu_{87}$; b — theoretical depth profile for ordered alloys described by the interaction potential $V_1 = \dfrac{2}{z}\dfrac{1}{\Delta H_{AB}} - \dfrac{2}{\Delta H_A} - \dfrac{2}{\Delta H_B}$ [263, 275]

insignificantly lowers Cu/Pd. Since the rate of sputtering by He^+ ions is lower by an order of magnitude [113], the growth of this ratio is not associated in our case with preferential sputtering, but reflects the distribution of the components in the upper layers. Consequently, in specimens whose surface is enriched in Pd, the concentration of Cu in the second layer is higher. The blurred nature of the oscillation is apparently connected with the insufficiently high layer resolution achieved in this experiment. The decrease in Cu/Pd ISS in prolonged experiments may be a result of the predominating removal of Cu.

The influence of preliminary treatments on the electron state of metals and the surface composition has been studied in greater detail for the alloy $Pd_{47}Cu_{53}$. In an initial specimen, E_b of Cu $2p_{3/2}$ is lower by 0.5-0.6 eV, while E_b of Pd $3d_{5/2}$ is higher by 0.3 eV than in the corresponding spectra of the pure metals. These shifts are typical of these alloys [31] and possibly point to the transfer of the electron density from Pd to Cu. The greatly overlapping

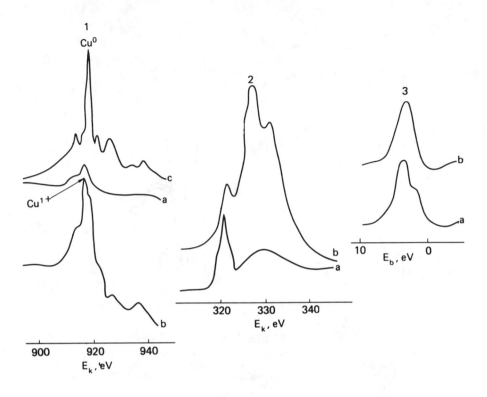

Fig. 42. Changes in the state of the components with various treatments of the alloy $Pd_{47}Cu_{53}$:
1 — Cu L_3VV spectra: a — O_2, 623 K, 0.1 torr; b — a + Ar^+, 5 min; c — b + Ar^+, 2 h + 673 K, vacuum; *2* — Pd MVV spectra: a — O_2, 623 K, 0.1 torr; b — a + Ar^+, 2 h + 573 K, vacuum; *3* — XPS spectra of valence band; a — O_2, 623 K, 0.1 torr; b — a + Ar^+, 2 h + 573 K, vacuum. The top curves coincide with the spectra of a pure alloy

bands of the *d* states of the metals (Fig. 42) are the main contributors to the XPS or UPS valence band spectra. The surface composition of this specimen differs insignificantly from the bulk one. Sharp changes in the composition and valence state of the components are observed during oxidation (673 K, O_2, 0.1 torr). All the techniques register a growth in the surface concentration of copper: XPS and SIMS—from 10 to 15 times, AES—30 times, and ISS— two or three times. Moreover, the Cu $2p_{3/2}$ level of Cu^0 shifts to higher binding energies; judging from the Cu L_3VV spectra, part of the copper transfers into the state Cu^+. The Pd MVV spectra also change noticeably: when the alloy is oxidized, the narrow peak with $E_k \approx 320.8$ eV has the highest intensity, whereas in the spectrum of the initial specimen the broad peak with

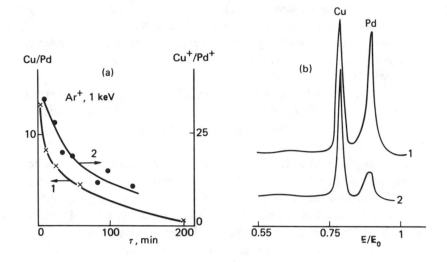

Fig. 43. Surface segregation of Cu in the oxidation of the alloy $Cu_{53}Pd_{47}$:

a — depth profile, XPS (1) and SIMS (2); b — change in the intensity of the scattering peaks of $^4He^+$ on Pd and Cu: I — Ar^+ + 623 K, vacuum; 2 — O_2, 623 K, 0.1 torr

$E_k \approx 326.6$ eV predominates. The Pd MVV signal excited by X-ray radiation was much weaker than the same spectrum obtained by electron excitation. For instance, the ratio of the intensities of Cu L_3VV and Pd MVV in a specimen of the initial alloy was 50 times lower in AES than in XAES. Oxidation produces changes in the valence band spectra: the peak of the local d states of copper stands out very clearly (Fig. 42).

Hence, in the oxidation of the alloy Pd-Cu, surface segregation of the copper in the form of Cu_2O is observed. Information on the depth profiles of the components and the thickness of the oxide film has been obtained by comparing the data received by different techniques. Figure 43 shows the changes in the ratio Cu/Pd (XPS) and Cu$^+$/Pd$^+$ (SIMS) depending on the duration of ion etching of the initial specimen and two specimens of $Cu_{53}Pd_{47}$ oxidized under identical conditions. The initial composition is reached after sputtering a layer of 10-15 nm. The changes in the Cu L_3VV and Cu $2p_{3/2}$ spectra here point to the removal of Cu_2O (no reduction of copper(I) oxide after ion bombardment under close conditions was observed [93]) and to the restoration of the unreconstructed structure of the alloy. The Cu/Pd ratios obtained by SIMS and XPS agree qualitatively with one another. The large changes in Cu/Pd that follow from AES are possibly due to the higher surface sensitivity of this technique. However, the change in the Cu/Pd ratio in oxidation by XAES data is considerably lower and is close to that found by XPS. The great weakening of the Pd MVV signal (AES) may be caused by the greater contribution of the multiply scattered electrons due to a change in the surface roughness [90, 106].

The ISS data for the oxidized alloy differ substantially from the results obtained by three other techniques (Fig. 43b). The Cu/Pd ratio after oxidation only doubles. If the curve in Fig. 40 is employed as a calibrating one, then by the ISS data, the Cu concentration in an oxidized alloy is 75-80%, whereas by XPS data it is 93-95%. The result should be the opposite one when a continuous Cu_2O film forms. But if the film forms in islets, the contribution of the Pd atoms on the surface to the ISS signal will be higher than to the XPS one. The oxidation of Cu and the surface segregation of Cu_2O are consistent with the difference in the enthalpies of formation of Cu_2O (-320 kJ/mol) and PdO (-80 kJ/mol), the latter oxide being unstable at 400 °C. Close results were obtained by Sampath Kumar and Hegde [353] in the oxidation of PdCu (500 K and 10^6 L of O_2). But these authors gave no data on the degree of enrichment of the surface and the forms of oxygen adsorption. An analysis of the O $1s$ spectrum obtained in our case indicates three types of oxygen on the surface of an oxidized alloy: lattice (530.5 eV, 31%) and two forms of chemisorbed oxygen (531.9 eV, 32%, and 533.1 eV, 37%). Unlike NiCo alloys, the concentration of the "near-surface" oxygen is higher, which is typical of Cu, while the high binding energy of the second O $1s$ peak possibly indicates the molecular nature of adsorption. Although the XPS data reveal no oxidation of Pd, the presence of clusters of $PdCuO^+$ (SIMS) in addition to $PdCu^+$ and the change in the structure of the Pd $M_{4,5}VV$ spectrum may be due to the formation of mixed clusters of the type Pd-O-Cu, Pd_2-O-Cu, etc. Again, as for CoNi, we can assume small linear dimensions of the Cu_2O islets having in view the considerable intensity of $PdCu^+$ in the mass spectra of the oxidized alloy.

Possible mechanisms of oxidizing Cu-containing alloys have been suggested by Sampath Kumar and Hegde [353, 355] and by van Pruissen *et al.* [356]. They are based on an analysis of XPS and AES data obtained for Cu-Pd, Cu-Sn, Cu-Ge, Cu-Fe, and Cu-Mn alloys annealed in vacuum after their reaction with oxygen (up to 10^6 L, 300-600 K), and also after oxygen desorption. In alloys of copper with palladium or with manganese, the predominating oxidation of the copper was observed, whereas in the Cu-Ge alloy, the oxides Cu_2O and GeO form. Subsequent annealing of the oxidized alloys in vacuum is attended by a number of topochemical reactions leading to the reduction of the copper and the formation of an oxide of the second component, e.g.:

$$2Cu_2O + Sn \rightarrow 4Cu + SnO_2 \qquad (1)$$
$$Cu_2O + Mn \rightarrow 2Cu + MnO \qquad (2)$$
$$Cu_2O + GeO \rightarrow 2Cu + GeO_2 \qquad (3)$$

No oxides remain on the surface at all for the alloy Cu-Pd:

$$Cu_2O + Pd \rightarrow 2Cu + PdO \qquad (1)$$
$$PdO \rightarrow Pd + O_{ads} \qquad (2)$$
$$O_{ads} + O_{ads} \rightarrow O_2 \text{ (gas)} \qquad (3)$$

The transition of oxygen from the copper to the second component is presumed to be facilitated by spillover. Such a mechanism assumes a close arrangement of the components, i.e. is consistent with our observations of the formation in such systems of oxide islets on whose periphery there are atoms of the first component in the metallic state.

The formation of such structures and the mutual influence of the components of an alloy on their ability of being oxidized are supported by data obtained for the alloys Ni-Ru [357], Co-Ru [358], and Co-Pd [359]. The surface of the solution of Ni (4.9%) in Ru(111) after annealing in a vacuum is enriched in Ru, although Miedema's or Abraham's [360] models predict the surface segregation of nickel. The enthalpy of segregation of Ru is very low—4 kJ/mol. In Co-Ru alloys (1.5; 5 and 21%), Co segregates on the surface, which is consistent with theory (see Table 8). These differences may be caused by the prehistory of the specimens and the annealing procedure. Reaction with oxygen produces close changes in the chemical states of the components of both types of alloys. The valence band features reflect the formation of complicated oxide structures (Fig. 44), while the spectra of the inner levels witness the predominating oxidation of the unnoble metal. As for Ni-Co or Pd-Cu, several forms of oxygen are on the surface, but unlike the former alloys, such a pronounced segregation of Ni or Co was not observed in ruthenium-containing specimens. Like platinum, ruthenium promotes the decomposition of the oxide of the second component (NiO). The dissociation temperature of NiO lowers to 700 K as a result of an $O^{2-} \rightarrow O^-$ transition by the scheme [357]:

$$Ni\text{-}Ru + O_2 \xrightarrow{300\ K} Ni\text{-}O + O^-_{ads} \qquad (1)$$

$$Ni\text{-}O + O^-_{ads} \xrightarrow{700\ K} NiO + \frac{1}{2}O \qquad (2)$$

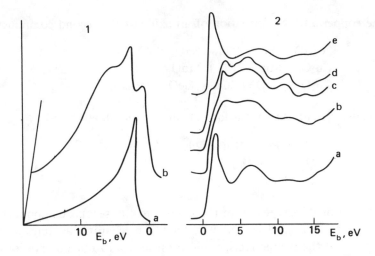

Fig. 44. Valence bands of partly oxidized alloys:

1 — Ru-Co [358]: *a* — pure alloy $Ru_{21}Co_{79}$; *b* — O_2, 373 K, 10^5 L; *2* — Ru-Ni [357]: *a* — pure alloy $Ru_{4.9}Ni_{95.1}$; *b* — O_2, 300 K, 10 L; *c* — O_2, 300 K, 10^3 L; *d* — *b* + 450 K, vacuum; *e* — *d* + 700 K, vacuum

$$NiO + Ru \xrightarrow{700\ K} Ni\text{-}Ru + O_{ads}^- \qquad (3)$$

$$O_{ads}^- + O_{ads}^- \xrightarrow{700\ K} O_2\ (gas) \qquad (4)$$

Employment of low-energy electron diffraction enables one to directly watch the change in the surface structure of alloys during oxidation. The adsorption of oxygen on the alloys $Pt\text{-}Co_{20}(001)$ and (111) at room temperature does not cause the appearance of superstructures; raising of the temperature to 700 K leads to the formation of nuclei of the oxide CoO with a monatomic depth, while an increase in the exposure is attended by their lateral growth and coalescence with the formation of a continuous oxide layer [359] (Fig. 45). The CoO nuclei decorate the platinum surface with the formation of various metastable structures, while the appearance of a compact monolayer of CoO results in the complete suppression of oxygen adsorption. With an increase in the exposure, the structure of the surface layer gradually transforms into Co_3O_4; after annealing in vacuum, the structure of the initial alloy is completely restored. A glance at Fig. 45 reveals that the models of partly oxidized alloys proposed on the basis of a set of data obtained by nonstructural EES and

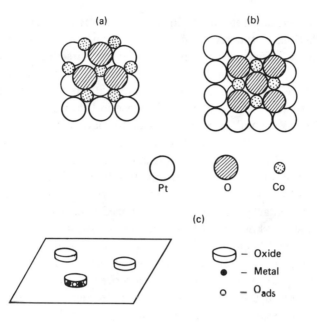

Fig. 45. Models of surface structure of partly oxidized alloys:
a and *b* — by data of SED: Pt-Co [360]; *a* — quasihexagonal layer; *b* —
phase c (2×2); *c* — by data of XPS and SIMS: Co-Ni, Ru-Co, and Pd-Co
[281, 358]

IS techniques and the structural technique of low-energy electron diffraction are qualitatively similar, although LEED naturally yields more detailed information. This can explain the increased interest in studying the single crystal surfaces of alloys, but they are generally limited by dilute solutions.

In addition to the results considered above, interesting information has been obtained for the alloy Fe-4% Sn(200), (110), and (111) [361]. A comparison of LEED, ELS, and XPS data uncovered a relation between the state of the Sn segregated on the surface and the surface structure. When going over from the ordered structure $c(2 \times 2)$ to a disordered one, a slight but sharp shift in the Sn $3d_{5/2}$ spectrum occurs on the face (100). In the absence of a phase transition [the face (110)], the shift increased monotonically with a growth in the coverage of the surface by tin. The set of data points to the formation of an ordered "FeSn" structure on the surface.

An analysis of data on the surface segregation and oxidation of alloys that are solid solutions or intermetallics thus leads to some general conclusions.

The surface segregation determined in various ways is consistent qualitatively with the predictions of theoretical models, although sometimes there were not only quantitative differences, but also the reverse direction of segregation. These differences, on the one hand, may be due to inaccuracy of the thermodynamic models that take into account especially poorly the contribution of elastic deformation caused by the difference in the size of the component atoms. On the other hand, they may be due to inaccuracy of the techniques of determining the surface composition, and also to the absence of true equilibrium in the alloys. The questions associated with the possibility of a change in the composition because of ion stimulating effects when analyzing the upper layers by SIMS and ISS still remain an object of investigations [118, 362-364]. In particular, the employment of mixed flows of argon and helium made it possible to show that the surface composition of the alloy Au-Cu being analyzed by ISS does not depend on the density of the current of primary Ar^+ ions used in sputtering (0.5-90 $\mu A/cm^2$). This signifies that the redistribution of the atoms in a surface layer proceeds more rapidly than sputtering. The relation between the preferential sputtering of one of the components and surface segregation was indicated by Kelly [118]. The finding of a quantitative relation between these parameters will make it possible to interpret more reliably the data of SIMS obtained not only in the static mode, but also in an intermediate one. It should be noted that the joint use of three or four techniques for appraising segregation, as was shown using the example of analyzing the alloys Co-Ni, Pd-Cu, Pt-Ni, and others [307, 365], improved the authenticity of the data on the composition of the surface proper. In the last case, the surface segregation is described quite well by a model taking into consideration elastic deformations [366].

The existence of composition fluctuations both in semi-infinite crystals and in thin films of ordered solid solutions or intermetallics, as well as the possibility of the formation of ordered surface phases when they are absent in the bulk have been predicted by theoretical Monte Carlo calculations [273-275, 367], calculations by the cluster variation method [368], etc. These calculations enable one to obtain a relation between the local environment on the surface of an alloy and its composition, face orientation, thermodynamic parameters, temperature, etc. (Table 9) [274, 275, 367]. These theoretical results have a direct bearing not only on the prediction of the composition of an active ensemble, but also on the finding of factors allowing one to regulate its size and composition.

For example, for the alloys Pt_3Pb, the degree of enrichment in lead and the statistical weight of di- and polyatomic ensembles of Pt are determined

Table 9. Surface Composition ($n_{Pt,s}$), Fraction of Mixed Bonds P_{Pt-Pb}, and Fraction of Ensembles Pt_n on Various Faces of Pt_3Pb [367]

Surface type	$n_{Pt,s}$	P_{Pt-Pb}	Fraction Pt_n (0-6) in first coordination sphere						
			0	1	2	3	4	5	6
(001)	0.20	1.00	1.0	0.0	0.0	0.0	0.0	—	—
(111)	0.47	0.70	0.05	0.27	0.46	0.22	0.0	0.0	0.0

by the orientation of the face [367] and the temperature of treatment of the alloys. To date, little experimental evidence is available confirming or refuting the correctness of these calculations. We have already mentioned the uncovering of a nonmonotonic nature in composition changes in alloys such as Pt-Ni, Pt-Sn, and Pd-Cu [263, 276, 281, 369] and the possibility of surface ordering in Fe-Sn, and Pd-Cu [361, 370-372]. Quite good agreement between the data of AES, TDS of adsorbed CO, and Monte Carlo calculations have been obtained by Kok *et al.* [370, 371]. We hope that this sphere of interest exceedingly important for catalysis will be further developed.

The electron interaction of the components in alloys has meanwhile been studied only qualitatively. The shifts of the inner levels and the change in the shape of the valence band show that such interaction manifests itself primarily in solid solutions and intermetallics that have acquired order. Although quantitative appraisals of the charge on components are extremely difficult without obtaining additional data or calculations, the chemical shifts or local work function are sufficiently reliable criteria of the changes taking place on a surface in the course of chemisorption or oxidation. The behavior of the components of oxidized alloys can also indicate the degree of electron interaction between components. The surface segregation during oxidation is described only qualitatively by Tomanek's model [343]. An appreciable mutual influence of the components on the ability of being oxidized, the growth of the oxides, their stability, activation, and oxygen spillover has been detected in many cases. Having in view that these processes are kinetically controlled, the oxidation-reduction conditions will also affect the final composition and structure of the surface layers of alloys. One of the most typical cases observed under conditions of oxidation at low pressures is the formation of two- or three-dimensional islets of oxides decorated by a metal, or of more intricate clusters.

As a result, bonds similar to metallic ones remain on a surface in addition to ionic Me—O bonds close in their nature to the bonds in bulk oxides. Quasimetallic bonds also form on a surface. They are due to the presence of clusters of the type O—Me—Me—O or two-dimensional oxides, and also of "intermediate" oxides that are the precursors of stable bulk compounds. Such a sequence of the oxidation state was observed, for instance, for the intermetallic Ni_3Al [373] or the system Fe/Ni(100) [374]. In the former case, under rather rigorous conditions (300-500 °C), but at a low oxygen pressure, there was observed an intermediate oxide "Al_2O_3" characterized by a binding energy of Al $2p$ of 73.6 eV and by an extremely high value of E_b for O $1s$ i.e. 534 eV. Within the limits of an electrostatic model, the bond in such an oxide is more covalent than in γ-Al_2O_3, or the aluminium is in an unusual divalent state. In the system Fe/Ni(100), which initially is a thin film of Fe (two monolayers) on a single crystal of Ni, the following transformation of the surface structure attended by shifts of the Fe $2p_{3/2}$ level occurs in oxidation: structure   O $c(2 \times 2)$ (0.3 eV) → monolayer   of   FeO   (1.2 eV) → three-dimensional islets of FeO (2.5 eV) → Fe_3O_4 (3.7 eV). The presence of the structures shown in Fig. 45 is also indicated by SIMS data for the oxidized intermetallic Pd_3Sn [375], the ratio between the oxide and quasimetallic clusters being determined by the ratio $SnO_2^-/SnPdO^+$ in SIMS.

It is important that even with considerable segregation of an oxidized component and the formation of a three-dimensional oxide, the unoxidized component, for example Pd in the alloy Pd-Cu, remains on the very surface and, consequently, can participate in catalysis. Since the nature of the oxidation of alloys depends on the treatment conditions, the obtained data cannot be applied directly to real bimetal catalysts that are reduced and oxidized at atmospheric or an increased pressure. But we can expect the general trends in the behavior of alloys to also hold under real conditions.

4.1.4   Bimetal Thin Films and Catalysts Based on Them

The modelling of alloys by thin films and their studying by EES, IS, LEED, TDS, etc. is a special field of research [376]. The advantages and restrictions associated with the use of thin films for studying the properties of multicomponent catalysts have been analyzed by Frick and Jacobi [377]. Moreover, thin films have special structural and electron properties that may substantially modify adsorption and catalytic properties in comparison with bulk objects. The depth of the probe when employing techniques such as XPS or AES in this case is naturally limited by the film thickness, which improves the surface sensitivity of the relevant techniques. The examples of the unusual electron properties of thin sputtered films given in Chap. 3 can be supplemented by data of UPS and LEED for Pd/Al(111). According to these data, when a monolayer is deposited, a unique quasiatomic state of Pd is realized that is difficult to produce even when using specially developed procedures for preparing highly dispersed metallic systems on supports.

The electron, structural, and catalytic properties of such systems can be modified in three main ways:

(a) by depositing an inactive metal onto an active one;

(b) by depositing an active metal onto an inactive one;

(c) by depositing an active metal onto an active one (differing with respect to selectivity).

The first type includes the systems Cu/Ru(0001) and Au/Ru(0001) studied in detail in a series of investigations by Vickerman and his co-workers [301, 302, 378, 379]. Both systems were found to be structurally similar: the deposition of Cu or Au at low temperatures leads to the appearance of metal islets, and at high temperatures, to the formation of epitaxial layers. By theoretical calculations, the electron properties of the first Cu layer on Ru should differ greatly from Cu(111) owing to Cu-Ru interaction and to the formation of Cu $3d$-Cu $4d$ surface states [380]. Indeed, the evaporation of Cu does not simply lead to inhibition of the reactivity of Ru relative to CO or hydrocarbons, but leads to appreciable modification of the properties of the Cu/Ru system. The number of centers of the low-temperature adsorption of CO or of the centers of pi-adsorption of ethylene increases [301, 378, 379, 381]. According to Berlowitz *et al.* [382], the activity of this system in the dehydrogenation of cyclohexane grows by an order of magnitude when going over from submonolayer to monolayer copper coverage. By SIMS data, mixed Ru-Cu centers are also observed for coverages exceeding a monolayer [301, 302]. Consequently, the action of the copper does not consist only in the cluster effect. Unlike copper, gold mainly behaves like a blocking agent and reduces the statistical fraction of multiatomic ensembles needed even for the dissociation of hydrogen and hydrogenation reactions [381].

The second method—the modification of nickel with tungsten [Ni/W(110)], changes the activity in the methanation and hydrogenolysis of ethane because of the weaker reactivity of the surface centers with respect to CO and H_2 [382]. The coating of Ru(0001) with a second active component, e.g. Ni (less than 10 monolayers) also changes the electron and catalytic properties. When a monolayer is deposited, $\Delta\varphi$ is 0.4 eV, while with higher coverages, the work function grows by another 0.26 eV, which equals the difference between the work functions on pure Ru(0001) and Ni(111). These changes result in a stronger bonding of the CO to the nickel monolayer than to the pure metals.

The system Pt/W(100) [383] can exemplify how deposition can be used to model various types of metal catalysts such as thin films, microcrystals, and alloys, and also to vary the contribution of the ensemble and ligand effects to the interaction with reacting molecules, e.g. with CO. At low coverages, Pt forms a stressed hexagonal structure on W, and therefore, although the dissociation of CO on such a surface is suppressed owing to the great dilution of the W centers, the bonding of CO on the Pt centers differs greatly from what is observed for bulk Pt(111). With a higher content and annealing at $T > 600$ K, the Pt multilayers agglomerate into microcrystallites, and part of the W localized adjacent to Pt becomes available for the adsorption of CO. The influence of Pt manifests itself again and leads to an increase in the fraction of molecularly bound CO in comparison with the dissociative fraction. At 1200 K, the surface alloy Pt-W forms, and chemisorption of CO proceeds only on the W centers (TDS data). But the latter are greatly modified owing to electron interaction with Pt. XPS and TDS data reveal a charge transfer from Pt to W (a shift of 0.4 eV). The ligand effect increases the d-pi backdonation in the W—CO bond and raises the strength of the chemisorption bond of the molecular CO.

Sometimes the advantages associated with the specific features of thin bimetal films are realized for the production of new highly active catalysts. Of appreciable interest is the studying of bimetal systems Cu/RE (rare-earth elements) that exhibit an extraordinary activity in methanol synthesis in comparison with the industrial $CuO/ZnO/Al_2O_3$ catalysts or with the systems Cu/REO$_x$ prepared by coprecipitation [384, 385]. Since the activity depends very greatly on the conditions of the oxidizing treatment of type Cu-Ce or Cu-Nd alloys, thin films of Cu-RE (Nd) alloys deposited on a support of Cu(100) were studied for optimization of the catalytic properties [386]. The electron interaction of Nd with Cu is confirmed by shifts in the Cu $2p_{3/2}$ $(+0.25$ eV) and Nd $3d_{5/2}$ $(-0.4$ eV) spectra. According to data of XPS, ARUPS, AES, LEED, and measurements of $\Delta\varphi$, the kinetics of the reaction of oxygen with the film includes the rapid stage of oxygen diffusion under the surface of the Nd layer, a reaction with Cu, and diffusion into the bulk of the film with the formation of oxide aggregates. Of importance for catalysis is the reaction at the Cu/REO$_x$ interface discovered by Nix et $al.$ [385, 386]. Under definite oxidation conditions, an "intermediate" oxide forms instead of Nd_2O_3. The system Cu-RE has been studied in [387, 388].

4.1.5   Catalysts Based on Alloys, Intermetallics, and Their Hydrides

Raney Skeleton Catalysts. Traditionally, the Raney catalysts are considered to include the Raney nickel system or skeleton catalysts based on multicomponent Ni-Al-Me alloys, although at present leaching with an alkali or acid is employed in the preparation of catalysts with a high specific surface area based on alloys of many metals (Cu, Pt, Pd, Cr, etc.). The interest in these systems is due to their widespread employment in reactions of hydrogenation, hydrogenolysis, and the anode oxidation of fuel in electrochemical generators. Data on the structure and morphology of Raney catalysts [389] show that they consist of microcrystals of nickel (50-100 Å) coated with a film of Al oxide and hydroxide. Characteristics such as the dispersion of Ni^0 and the specific surface area depend on the leaching conditions, and also on the nature and concentration of the alloying additions (Ti, Mo, Fe, Mn, etc.). The role of these factors in the formation of an active surface on two-, three-, and multicomponent Raney nickel catalysts have been studied by XPS, SIMS, XRD, Auger microprobe, etc. [390-396].

In the initial phases of $NiAl_3$, Ni-Al (a mixture of $NiAl_3$, Ni_2Al_3, and Al), and Ni_2Al_3, strong surface segregation of the Al oxide film is observed, while the nickel remains in the metal state [393]. After leaching, the ratio $(Al/Ni)_s$ decreases by an order of magnitude, both components being present in the surface layer in three states, namely, as a metal, oxide, and hydroxide. More detailed data on the structure of the surface of these catalysts have been obtained by SIMS [394]. Clusters of $NiAl^+$, $Al_2O_2^+$, and $NiAlH^+$ were registered in addition to the Al^+ and Ni^+ ions. The initial Ni^+/Al^+ ratios diminished in the series Ni-Al $>$ $NiAl_3$ $>$ Ni_2Al_3; the profiles of their changes depending on the duration of sputtering for three specimens also differ (Fig. 46). For $NiAl_3$, a monotonic decrease in Ni/Al is observed, while for Ni-Al the sharp drop in this ratio during the initial period is attended by a further gradual increase in its value.

The most unusual profile has been obtained for Ni_2Al_3. At the initial instant (6-7 min), Ni^+/Al^+ grows from six to seven times, next gradually diminishes, and emerges onto a plateau, its final state being 2.5 times higher than the initial one. The studying of these specimens by XPS before and after the SIMS experiments made it possible to appraise the extent of ion-induced changes in the surface and conclude that in all the specimens the Ni^0 particles are covered by a film of aluminium oxide whose thickness is maximal for Ni_2Al_3 (>20 Å). This is consistent with the fact that it is the most difficult to leach this intermetallic. The layer structure of a two-component catalyst

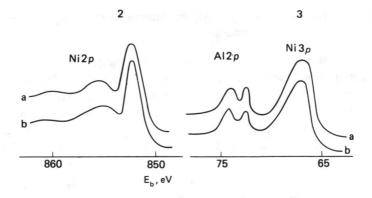

Fig. 46. Changes in the Ni^+/Al^+ ratios against the duration of ion sputtering of Raney Ni catalyst specimens (1) and XPS spectra of the Ni $2p_{3/2}$ (2) and Ni $3p$ + Al $2p$ regions (3) [394]:

1 — SIMS data (Ar$^+$, E_p = 1 keV): *a* — Ni-Al (1:1); *b* — NiAl$_3$; *c* — Ni$_2$Al$_3$; *2* — Ni $2p_{3/2}$ spectra: *a* — Ni$_2$Al$_3$; *b* — NiAl$_3$; *3* — Al $2p$ + Ni $3p$ in Ni$_2$Al$_3$ before (a) and after (b) the SIMS experiment

can be represented schematically as Al$_2$O$_3$, Al(OH)$_3$, Ni^{2+}/Ni^0/Ni$_x$Al$_y$. In more complex catalysts, these structure elements are retained, but the achievement of conditions under which surface segregation of oxide additives occurs is important. Under the conditions of autoclave high-temperature leaching, additives of Fe or Mo (a three-component system) or Mo, Fe, Cu, Cr, and Ti (a six-component system) segregate to a larger extent on the surface near the Ni^0 crystallites [393-395]. This ensures a high hydrogenating activity and the stable action of these systems in the production of hydrogen peroxide from substituted anthraquinones. By TDS data, the segregation of the additives not only stabilizes the dispersed phase of the nickel, but also increases the concentration in the activated hydrogen system because of its spillover from the nickel onto the oxides [395].

Leaching of the intermetallic phases of Pt-Al and Pd-Al, according to XPS data [396], causes destruction of the initial alloys and the formation of an intricate metal-oxide structure. In the initial compounds rich in Al of the type PdAl$_3$ or PtAl$_{2-4}$, the Me$_1$—Me$_2$ bond is preserved notwithstanding the very great enrichment of the surface in Al oxides. This is witnessed by the positive shifts in the Pt and Pd spectra, respectively. After leaching, E_b approaches the values for pure Pt and Pd, while the degree of enrichment of the surface in Al diminishes somewhat. In general, the structure of these systems is close to that of the Raney nickel catalysts.

Catalysts Based on Intermetallics and Their Hydrides. A comparatively new class of catalysts for reactions of hydrocarbon transformations, methanation, and Fisher-Tropsch synthesis is based on intermetallics of Group VIII or VI metals and what are called the hydride-forming elements of Group IV or VI, the rare-earth metals, or the actinides (Ti, Zr, Hf, V, Nb, La, Th). These compounds can absorb large amounts of hydrogen and form bulk hydrides. An important factor determining the activity and stability of these systems is their oxidizing activation [397]. As a result of the subsequent reduction in H_2 or in hydrocarbons, structures form that generally have a higher activity in comparison with similar metal-supported catalysts [398]. The formation of the surface of these systems, namely, the chemical state of the components, the composition and structure of the surface have been studied by XPS [117, 399-407]. It has been found that in the oxidizing treatment of specimens of Zr-Ni-H or Zr-Ni-Cu-H, the active components segregate on the surface, which initially was covered to a considerable extent by a film of Zr oxides [399]. The subsequent reduction of catalysts of the $NiO-ZrO_{2-x}$ type in H_2 results in the formation of the system Ni^0-ZrO_{2-x}, which has a higher activity in hydrogenation or hydroisomerization reactions than Ni/ZrO_2 [398, 399]. The use of ARXPS and ion etching revealed the structure of Ni-Zr by layers and led to the assumption on the mechanism of the surface segregation of Ni [117]. The top layers of oxidized catalysts (200 °C, air) are oxides of Ni and Zr, the NiO islets being directly on the surface, while ZrO_2 is localized in the second layer. Next in the surface layer 10-40 Å thick, the Ni/Zr ratio is minimal, while enrichment in Ni is again observed in deeper layers. Finally, at a depth of 1000-1600 Å, a bulk stroichiometric composition is reached. This unexpected result is explained by diffusion of Ni through the thick film of ZrO_{2-x} by a mechanism of the exchange of places with the oxygen of the oxide film.

The ability of the active component to diffuse in this way depends on the nature of the second component in the intermetallic [400, 401]. The oxidizing treatment of Ni_xSi_y does not lead to enrichment of the surface in nickel, whereas in the intermetallics Ni_xTh_y an appreciable migration of nickel to the catalyst surface is observed. The activity in methanation correlated with the surface concentration of Ni in both systems.

Definite trends were established in the change in the chemical states of Group VI elements (Cr, Mo, W) and Zr in the relevant intermetallics and hydrides depending on the treatment conditions and the amount of hydrogen in a hydride (Table 10) [403-407]. As for the Ni-Zr and Co-Zr systems [402], the surface of untreated catalysts consisting of a Group VI metal, Zr, and

Table 10. Binding Energies of Electrons (eV) of the Components in the Hydrides of the Intermetallics Depending on the Treatment Conditions [404]

Treatment conditions	Me-O$_{stoichiom.}$		Me-O$_{nonstoichiom.}$		Me^0-Me$^\delta$ (− H)	
	Zn 3$d_{5/2}$	Cr 2p	Zr 3$d_{5/2}$	Cr 2p	Zr 3$d_{5/2}$	Cr 2p
Initial ZrCr$_2$H$_{3.6}$[a]	182.3	576.2	—[b]	—	179.0	573.9
Vacuum, 773 K, 1.5 h	183.1	576.8	180.6	575.8	178.9	574.0
Ar$^+$, 45 min + vac., 773 K, 2 h	183.1	576.2	180.7	—	178.9	574.2
Air, 1073 K, 5 h	182.2	576.2	—	—	—	—
		Mo 3$d_{5/2}$		Mo 3$d_{5/2}$		Mo 3$d_{5/2}$
Initial ZrMo$_2$H$_{0.8}$	182.4	232.9	—	230.1	178.9	228.2
Vacuum, 773 K, 2 h	182.8	233.3	181.4	—	179.6	228.4
Ar$^+$, 45 min + vac., 773 K, 2 h	183.2	233.5	181.4	—	180.0	228.6
Air, 673 K, 2 h	182.2	233.2	—	—	—	—
		W 4$f_{7/2}$		W 4$f_{7/2}$		W 4$f_{7/2}$
Initial ZrW$_2$H$_{1.8}$	182.6	35.4	—	—	179.0	31.0
Vacuum, 773 K, 2 h	183.7	36.0	181.8	—	180.0	31.6
Ar$^+$, 45 min + vac., 773 K, 2 h	183.9	36.0	182.0	—	180.1	32.0
Air, 673 K, 2 h	182.6	36.1	—	—	—	—

[a] The data relating to ZrCr$_2$H$_{3.6}$ were obtained by Khan Ashraf *et al.* [403].
[b] Not found.

H is enriched in Zr, the surface layer consisting of a complex mixture of oxides and metals: Cr(0, III), Mo(VI, IV, 0), W(VI, 0), and Zr(IV, 0). The fraction of the surface oxides grows in the series Zr-Cr-(H) > Zr-Mo-(H) > Zr-W-(H). Thermovacuum treatment at 773 K causes the surface segregation of Group VI metals, the metal/oxide ratio in hydrides increasing quite noticeably. Oxidizing treatment at 673-703 K causes oxidation of the Group VI metal first of all. It is interesting that chromium is oxidized only to its trivalent state. Unlike Ni-Zr or Ni-Cu-Zr, oxidation does not produce appreciable surface segregation of the active metal (except for Mo). Still another feature of these systems is the very easy reduction of the surface oxides formed in oxidation, or of the initial oxide films (in the latter case only in hydrides).

The self-reduction of the metals (Cr, Mo, W, Co, Ni) in the relevant hydrides of the intermetallics is explained by the evolution of activated hydrogen from their lattice [407]. For instance, at 350 °C in vacuum, a reduction of 90% is achieved only for Co in Co-Zr-H; treatment of the relevant intermetallic with molecular hydrogen yields only a reduction of 40%. The high reactivity and mobility of hydrogen of hydrides was confirmed in experiments with mechanical mixtures of intermetallics or even of oxides of Cr and Zr formed in the rigorous oxidizing treatment of Cr-Zr with hydrides of other metals, e.g. $ZrNi_{0.16}Co_{0.84}H_{1.4}$ [407]. A high degree of reduction of the oxides under these conditions was achieved because of the spillover of the hydrogen evolving from the hydride.

Hence, catalysts based on intermetallic hydrides are complex systems in which various types of interactions such as Me_1-Me_2-O and Me_1-Me_2-H manifest themselves. For example, in the oxidation-reduction of systems having the structure Zr-ZrO_{2-x}-ZrO_2-Me of Group VI-H, reversible shifts were dis-

Table 11. Chemical State of Group VIB Elements and the Surface Composition of the Intermetallic Hydrides Me (Group VI)-Zr-H [404]

Specimen	Treatment conditions	Metal	Relative content of chemical states, %					Me/Zr
			VI	IV	III	II	0	
$ZrCr_2H_{3.6}$	Initial	Cr	—	—	75	—	25	0.88
	623 K, vacuum, 1 h	Cr	—	—	57	13	30	1.21
	773 K, vacuum, 1 h	Cr	—	—	25	25	50	1.22
	1023 K, air, 15 min	Cr	—	—	95	—	5	1.02
	1023 K, air, 5 h	Cr	—	—	100	—	—	0.88
$ZrMo_2H_{0.8}$	Initial	Mo	54	18	—	—	28	1.1
	623 K, vacuum, 2 h	Mo	10	7	—	—	83	0.8
	773 K, vacuum, 2 h	Mo	9	5	—	—	86	1.0
	673 K, air, 2 h	Mo	100	—	—	—	—	1.4
$ZrW_2H_{1.8}$	Initial	W	35	—	—	—	65	0.3
	623 K, vacuum, 3 h	W	23	—	—	—	77	0.5
	773 K, vacuum, 3 h	W	20	—	—	—	80	0.5
	703 K, air, 4 h	W	82	—	—	—	18	0.3

covered in the Zr $3d_{5/2}$ spectra belonging to ZrO_2 and Zr; the value of E_b increased with a growth in the deficiency of the surface in oxygen, i.e. in reduction [404, 405] (Table 11). Although the cause of this unexpected effect has not been established completely, it may quite possibly reflect the changes in the electron interaction of various dispersed phases that occur in redox treatments and affect catalytic activity.

Amorphous Alloys. These materials have a number of unique physicochemical characteristics and exhibit unusual catalytic properties. Their studying in the capacity of catalysts is also of a theoretical interest in connection with the possibility of appraising how short-range and long-range order arrangement on a surface affect the nature and energetics of catalytic reactions. But meanwhile the reasons why the catalytic properties of amorphous and crystalline alloys differ have been determined far from completely. The difficulty consists in that materials produced on the basis of strongly modified amorphous alloys are employed in catalysis. For instance, to increase their active surface area, the amorphous alloys Ni-Zr and Cu-Zr were treated prior to catalysis in hydrofluoric acid [406, 407]. And although some authors [406-408] indicate correlation between the electron deficiency of Ni or Cu revealed by XPS and the catalytic activity in hydrocarbon hydrogenation reactions, this can hardly be explained only by the nature of the precursor. The reason is that the nickel in such systems is greatly oxidized, while such a shift is also observed in the traditional Raney nickel catalyst. The conclusion that a more highly dispersed phase of the metal forms on the basis of amorphous alloys [406] seems to be more convincing. Bertolini *et al.* [407] and Tomanek *et al.* [408], by thoroughly cleaning the surface of Ni-Zr alloys and controlling their structure, attempted to avoid the complications associated with the slight oxidation of the surface of amorphous materials. The alloys were shown to exhibit a greater activity in the hydrogenation of CO [407] and a very high selectivity in the hydrogenation of 1,3-butadiene to butenes [408]. The nonadditive effect of Ni and Zr in an alloy with respect to CO (the CO dissociates more readily on the alloy than on the pure components) has been explained by the ligand effect [409]. By XPS data [409], the band of d states of Ni in an alloy shifts to higher binding energies, so that the d states of Zr contribute more to the density of the states near the Fermi level. Further studies will show to what extent this effect is related to the amorphization of alloys.

Other Catalysts. In addition to oxidizing activation, treatments such as nitridation (ammonia synthesis), carbidization (CO hydrogenation), and sulfidizing (hydrofining, hydrocracking) are employed to produce active catalysts based on alloys. The choice of the optimal activation conditions requires information on the electron structure and composition of the surface of such systems in comparison with their activity. The increased activity of the nitrided alloy FeTi in ammonia synthesis is explained according to XPS data by the inclusion of fine Fe particles in the FeN matrix [410]. Here virtually complete breaking of the intermetallic bonds occurs, and no appreciable electron interaction between the Fe and TiN is revealed. The increase in the activity is explained exclusively by the ensemble effect. The same conclusion follows for fused Mn-Fe catalysts, which exhibit a high selectivity in the formation of alkenes in the Fisher-Tropsch synthesis close to that for coprecipitated oxide systems [411]. SIMS and XPS data [411] show that notwithstanding the differences in the nature of the precursors, the surface layers of two types of catalysts activated in CO are close. They include the oxide MnO and various Fe-containing phases, namely, Fe^0, Fe_xC_y, and Fe_3O_4. The surface layers hundreds of nanometres thick are depleted of iron, which explains the increased selectivity of these systems in the formation of 1-alkenes with a medium chain length [411].

When using XPS to study the Ag-Pt catalyst of methanol oxidation to formaldehyde, the main attention was given to the state of the adsorbed oxygen [412]. Both metals in the alloy Ag-9.7% Pt were shown to retain their individual properties with respect to oxygen. The observed changes in the catalytic activity and reaction kinetics in the presence of Pt were ascribed to the promoting effect of Pt in the reaction of formaldehyde decomposition. Another catalyst of partial oxidation based on an Ag-Zn alloy was also studied by XPS [413]. The catalyst surface was found to be very rich in zinc, but it does not affect the kinetic parameters of processes of epoxidation or the complete oxidation of ethylene.

4.2   Oxide Catalysts

Oxide systems are widely used in catalysis primarily as catalysts of partial or complete oxidation. Moreover, they are active forms, promoters, or supports of most real catalysts. The interest in these systems is due to the significance of such fundamental problems, already noted for alloys, as the role of the local and collective properties of the surface and bulk of solids in the proceeding of a catalytic reaction. For this purpose, information is needed on the electron and structural characteristics of the surface of oxides and on the ways of forming them. Since many oxide catalysts contain transition elements, one of the factors determining catalytic activity is the valence and coordination state of these elements. The fundamental problems of catalysis on oxides have been discussed in monographs by Boreskov [414] and by Krylov and Kiselev [415]. They draw attention to the necessity of procuring more detailed quantitative data on the properties of the surface of such systems.

A number of these properties are studied by EES and IS:

(a) the electron structure of stoichiometric and defective oxides;

(b) the relation between the state of the oxygen and the acid-base properties;

(c) the chemical state of the transition elements depending on the conditions of formation of the oxide systems;

(d) the state of the components and the surface composition in binary and multicomponent oxide systems and catalysts based on them.

4.2.1   Electron Structure of Oxides. State of Oxygen

The problems associated with the kinetics and mechanism of the oxidation of metals and semiconductors, with the formation of oxide films, their composition, and structure are of interest for investigators in the field of catalysis, corrosion, thin coatings, and materials for microelectronics. We have dealt with part of them involving the oxidation of single crystals and films of alloys and of interest for modelling metal oxide systems in the preceding section. Here we shall discuss data on the bulk band structure, the formation of defects and their energy states in the forbidden band, and shall see how this information can be used to form a surface with a controllable set of defects. The thorough studying of these matters, in our opinion, could shed new light on the theory of catalysis on semiconductors [416].

The electron structure and formation of defects contributing to the states in the forbidden band have been studied in detail for oxides with the structure of rutile: TiO_2 [416-418], SnO_2 [418-420], RuO_2 [421, 422], α-Fe_2O_3 [423], etc. LEED and EES data on oxide single crystals have been generalized in a review by Kurtz and Henrich [424]. Figure 36 (see p. 131) presented spectral data obtained by various techniques of electron emission spectroscopy that made it possible to appraise semiquantitatively the formation of defects in the structure of TiO_2 [314]. A comparison of the results for this oxide and the pseudoisomorphous oxide SnO_2 [417] reveals that the oxygen vacancies formed in the ion bombardment of TiO_2 cause new states near the top of the forbidden band, whereas for SnO_2 the levels produced by defects are near the bottom of the forbidden band. These states in SnO_2 do not exhibit donor properties of interest for catalysis or adsorption. Bombardment by electrons

(E_k = 160 eV) [418] is a more selective way of creating a definite set of new energy levels in the forbidden band of an oxide. The gradual removal of oxygen follows the mechanism of interatomic Auger decay of the upper metal electron-hole states. It is attended by the reduction of $SnO_2 \overset{e}{\to} SnO \overset{e}{\to} Sn$. Data of EELS show that n-type states with E = 1.4 eV due to oxygen vacancies appear in the forbidden band whose width is 3.6 eV. The number of such states is N = 1.6 × 10^{21} cm^{-3}, which yields the average formula $SnO_{1.94}$. It should be noted that the concentration of defects is even higher than when chemical additives such as Sb are introduced. A shortcoming of the method, however, is that the oxide decomposes partly with the formation of Sn^0.

A comparison of LEED and XPS data (Ru $3d$, O $1s$) for different surfaces of single RuO_2 crystals {(110) and (100)} showed [421] how the initial structure affects surface reconstruction associated with the loss of oxygen and the formation of metallic Ru. It was also shown that equal amounts of RuO_3 are present on both RuO_2 faces, which apparently stabilizes the surface structure. Structures with a different number of defects appear depending on the conditions of preparation and treatment of nickel oxide [425]. The O $1s$ (E_b = 531.4 eV) and Ni $2p_{3/2}$ (E_b = 856.1 eV) spectra were relegated to the highly defective structure of the oxide. The intensity of these peaks increased for a single crystal of NiO(100) heated in vacuum and in O_2, especially with glancing photoemission angles. This was a witness of localization of O^- and Ni^{3+} at the surface/vacuum interface.

Not only the detection of defects in oxides, but also the determination of their energy characteristics and their division into surface and bulk defects are important for catalysis. Photoelectron spectroscopy using synchrotron radiation, AES, and LEED have been employed to study these questions using specimens of $SrTiO_3(001)$ as an example [426]. Quite a number of defects form depending on the temperature and duration of annealing in vacuum or in oxygen, and also on the conditions of ion bombardment (Fig. 47). The low-energy peak in the Ti $3p$ spectrum was related to surface Ti^{4+} atoms that do not cause the formation of a noticeable number of defects in the forbidden band. The difference between the levels of the surface and bulk Ti^{4+} ions of the defects is 2 eV, which is greater than what was observed for potassium in $KNbO_3$ (1.4 eV) [427]. The still lower energy peaks in the Ti $3p$ spectrum are associated with the new states in the forbidden band having a pronounced d nature. One of them relates to a surface associate of the type of Ti^{3+}-O, and the other relates to bulk ions Ti^{2+}. It is most important that Roberts and Smart [426] succeeded in determining the position of the levels produced by defects in the bulk and surface band structures of the oxide. Their contribu-

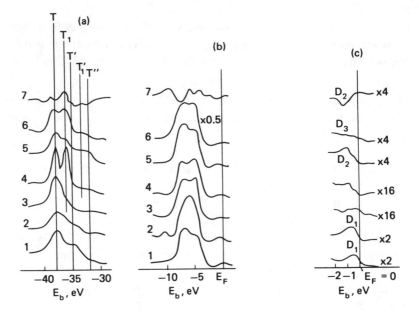

Fig. 47. Photoemission spectra ($h\nu = 100$ eV) of an annealed surface of single crystals of SrTiO$_3$ (001) treated with Ar$^+$ or oxygen [426]:

a — Ti $3p$ spectra; b — valence band; c — gap; 1 — vacuum, 1100 K, 60 min; 2 — Ar$^+$, 0.5 keV, 15 min; 3 — vacuum, 1100 K + 900 K; 4 — vacuum, 1100 K + O$_2$, 10^3 L; 5 — multifold annealing in vacuum, 900-1100 K; 6 — 5 + O$_2$, 10^3 L; 7 — 6 + 5; T, T_1, T', T_1', and T'' — bulk and surface atoms of Ti^{4+}, Ti^{3+}, Ti^{2+}; D_1, D_2, and D_3 — states in the gap produced by defects

tion to adsorption and catalysis can be appraised when studying how adsorbates interact with surfaces prepared in various ways. We can cite as an example the work done by Rocker and Göpel [428], who studied the adsorption of CO and H$_2$ on stoichiometric and defective surfaces of TiO$_2$(110).

Attention was drawn to the unusual low values of E_b for the levels of metals in surface oxides in the preceding section. This is explained by the difference in the coordination states of the metals, the deficiency of oxygen, and also by the difference in E_{ER}. There were cases of stabilization of unusual metal oxide structures, e.g. of α-Zr-O [429]. Such a structure exists at oxygen concentrations up to 25%. The state Zr $4d$ does not change here, but only the Zr $4p$ and O $2p$ bands broaden. The structure α-Zr-O formed owing to the high solubility of oxygen in zirconium is more stable than the oxide ZrO$_2$. In particular, in the ion bombardment of α-Zr-O no predominating removal of oxygen occurs, whereas the surface of ZrO$_2$ becomes greatly depleted of oxygen.

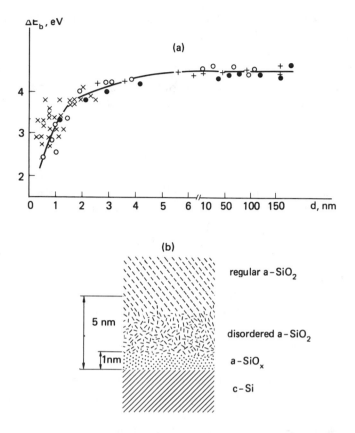

Fig. 48. Dependence of the position of the Si $2p$ level belonging to silicon oxide on the thickness of the oxide film [relative to E_b of Si $2p$ (Si^0) equal to 99.4 eV] (*a*) and the presumed structure of the surface and the SiO$_2$/SiO$_x$/Si interface (*b*) [429, 430]

The structure of thin SiO$_2$ films, and also the SiO$_2$/SiO$_x$/Si interface is being studied intensively in connection with the development of materials for microelectronics [206, 429, 430]. The laws of the formation of various intermediate states of Si were established (see Fig. 23, p. 92). Also found were the shifts associated with the change in E_{ER} in thin films [430] (Fig. 48), and in the bond angles and lengths in Si—O—Si tetrahedra. Near the Si/SiO$_2$ interface in the bulk ($>40\,\text{Å}$), the Si $2p$ peaks with E_b equal to 103.02, 102.55, and 102.05 eV correspond to the three diffraction maxima with angles of 144, 180, and 120° [431]. Negative shifts in the Si $2p$ spectra were observed at the interface (10-40 Å) [431, 432]. A layered model of this systems was proposed on the basis of the Si $2p$ shifts. It included regular α-SiO$_2$, distorted α-SiO$_2$, defective α-SiO$_x$, and c-Si. This model was confirmed by the measurement of 50 different specimens of single Si(111), (100) crystals [433]. In a review by Barr [54] devoted to studying the state of silicon in various materials including catalysts, not only Si $2p$ shifts, but also the peaks of the energy losses are analyzed. Barr considers that these data can be used to distinguish Si in various oxide materials and identify structural blocks such as Si—O—Si or Si—O—Al in crystalline aluminosilicates. It should be noted that the changes in the Si $2p$ spectra in the last-named systems are insignificant and depend on the conditions of specimen treatment [27]. Considerable clarity in the solution of this problem can be introduced by comparative studies of Si-containing oxides by several techniques, e.g. MASNMR on ^{29}Si nuclei, XPS, and SIMS. Nevertheless, data for oxide films are of substantial interest when studying the electron structure of the relevant bulk oxide systems.

The position of the Fermi level in the forbidden band is a critical parameter in the theory of catalysis on semiconductors [416]. It was considered that the serious discrepancies between the predictions of this theory and real catalytic properties are associated with the major dependence of the position of the Fermi level on the presence of microimpurities in the oxides and with the inaccuracy of its determination [415]. The presence of a static charge on the surface of insulators and semiconductors does not permit this characteristic to be determined with the aid of the standard XPS technique. But recently attempts have been made to determine the Fermi level with the aid of a specially developed "bias-reference" XPS technique [434-437]. Its essence [434] consists in that a noble metal, e.g. gold, is deposited on a specimen insulated from its metal holder. Next the charge of the insulator is compensated by means of an electron emitter with an equal bias potential on the specimen holder and on the emitter. The observed apparent shifts of the inner levels are corrected relative to the shift in the levels of the metal that is a marker of charging. The main condition is that the equalizing of the Fermi levels of the conductor and insulator is achieved at definite (low) bias potentials.

Table 12 presents the values of the Fermi level obtained for selected single crystal and polycrystalline oxide materials of interest as catalysts.

The bias-reference XPS technique was also employed for appraising the true surface charge of a surface, which by Lewis's theory determined its acid-base properties [436, 437]. Correlation was found between the position of the Fermi level of an oxide and the point zero charge (PZC) determined by the adsorption of water [436]. The value of E_F for γ-Al$_2$O$_3$ equal to 3.6 eV was

Table 12. Position of the Fermi Level, Other Electrophysical Characteristics, and the Point Zero Charge (PZC) for Selected Oxide Materials [436]

Specimen	E_F, eV	PZC	Width of forbidden band, eV	Work function, eV
SiO$_2$	5.4	2.0	8.6	5.0-6.3
AlPO$_4$	4.7	4.0	—	—
Varizite	4.5	4.0	—	—
Mullite	3.8	8.5	—	—
α-Al$_2$O$_3$	3.6	8.8	9.0	5.6-6.3
γ-Al$_2$O$_3$	3.5	8.8	—	—
Gibbsite	3.2	8.6	—	—
Boehmite	3.1	8.9	—	—
Mg-Al (spinel)	3.5	8.8	—	—
MgO	1.7	12.0	7.2	4.0-4.5

adopted as the datum level. An analytical expression was obtained relating the PZC and the Fermi level:

$$PZC = -2.9E_F + 16.8 \qquad (4.2)$$

For Lewis acids of the SiO$_2$ type, the highest Fermi level is observed and, accordingly, the lowest PZC. Conversely, for Lewis bases of the MgO type, the PZC equals 12 at a value of the Fermi level of 1.5 eV. By this correlation, a neutral surface should have a Fermi level of 4 eV at PZC = 7. Further approbation of the bias-reference XPS technique on different insulator specimens will show how accurate it is in determining these very important characteristics of a surface.

Another approach to an appraisal of the acid-base properties is the determination of the energy state of oxygen in oxides by the O $1s$ spectra, and also by the O KVV Auger transitions. But in addition to examples of systems for which correlation between the chemical shift and the charge on oxygen was observed, and from a more general viewpoint between the chemical shift and the acid-base properties [219, 220, 438], such correlation has not been found quite often [434]. For example, in zeolites with a substantial change in the acidity (a change in the degree of decationization or dehydroxylation), E_b for O $1s$ remains approximately constant [27], which is possibly associated with the compensating contribution of the initial and final states for the various types of oxygen found by other techniques.

Stoch [438] and Zazhigalov *et al.* [439] attempted to identify various types of oxygen in oxides by XPS. Four types of oxygen were earmarked:

1. "Lattice" oxygen O^{2-} with $E_b = 531.5$ eV at $r_{cation} < 0.6$ Å, and 530-531 eV at $0.6 < r < 1.07$, although within each group of oxides no precise correlation between E and r was observed. It was also found that for the tetrahedral state, E_b for O $1s$ is higher than for the octahedral state. No clear relation was discovered between the degree of oxidation of a surface and E_b. What was established the most reliably was an increase in E_b with a decrease in the coordination number and an increase in the covalence of the oxides.

The other types of oxygen, namely, O_2^-, O_2^{2-}, and O^- were modelled by various treatments.

2. The peroxide ion-radical O_2^- was obtained by treating KCl in an oxygen plasma; here a peak with $E_b = 532.5$ eV was ascribed to it, whereas the peak observed when heating K_2O_4 (531 eV) was ascribed to K_2O.

3. The ions O^{2-} were obtained by annealing SrO_2 in vacuum at 450-500 °C; here the value of 528.8 eV was ascribed to SrO, 531 eV to SrO_2, and 530.3 eV to $Sr(OH)_2$.

4. The ion O^- was obtained by vacuum treatment of Fe_2O_3. This ion is characterized by an O $1s$ peak with $E_b = 531$ eV and a level in the forbidden band of 1.5 eV. Alternatively, the O $1s$ peak can be ascribed to OH groups, but the degree of dehydroxylation of the surface was not controlled. Stoch [438] considers that the following shifts should be observed in a transition from O^{2-} to the other oxygen types:

Ion	O^{2-}	O_2^{2-}	O^-	O_2^-
E	0	2.2	2.0	1.5
e/atom	2	1	1	0.5

This classification, in our opinion, requires further verification and elaboration, which will be possible with thorough comparison of E_b for O $1s$ in oxides and in the chemisorption of oxygen, and also with account taken of the change in E_{ER} by determining the Auger parameter.

Reverting to the problem of acidity determination from XPS data, we can note that it is expedient for systems in which the acidity varies with the retaining of short-range order, e.g. in specimens of V-P-O [439], for which correlation was observed between the acidity measured by the chemisorption of ammonia and E_b for O $1s$.

Additional information on the degree of hydration or hydroxylation of the surface of oxides has been obtained by ISS (Fig. 49) [440, 441]. A low-energy peak whose intensity grew with increasing hydroxylation of the surface has

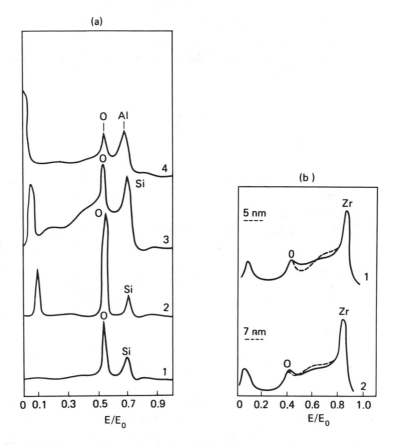

Fig. 49. Influence of the surface hydroxylation of compact and porous specimens on the intensity of the low-energy scattering peak in ISS spectra (*a*) [440] and influence of the film thickness on the shape of the main scattering peaks [441] (*b*):

a — ISS spectra of compact and porous oxides: *1* — silicate glass split in vacuum of 8×10^{-8} Pa; *2* — the same split in air; *3* — silica gel dried in air; *4* — Al(OH)$_3$; *b* — ISS spectra of ZrO$_2$ film on SiO$_2$: *1* — solid curve — 50 nm; dashed curve — 5 nm; *2* — solid curve — 50 nm; dashed curve — 7 nm. The low-energy peak belongs to scattered H$^+$ or OH$^-$ ions

been detected in the spectra of the scattered ions of non-porous Al and Si oxides. An identical peak has also been found in porous powders of silica gel and gibbsite [Al(OH)$_3$]. Its origin is associated with the scattering of H$^+$ ions and also, possibly, of fragments of the type of Si$^+$, SiO$^+$, and O$^+$, whose secondary electron emission coefficients grow in hydroxylation. A similar peak, and also the "tails" of peaks produced by inelastic scattering were used as a test for determining the structure of thin oxide films of ZrO_2 [441]. ISS also revealed an increase in Zr/O with a growth in E_p from 0.5 to 1.5 eV. This was explained by the greater yield of O^+ because of interaction of the lattice with a larger number of primary scattered ions.

4.2.2   Transition Metal Oxides

Two main tasks can be singled out when studying transition metal oxides by EES and IS:

(a) determination of how various treatments affect the chemical state of transition elements and the stoichiometry of a surface in specimens close in their composition to real catalysts;

(b) the identification of the chemical states in the precursor bulk oxides required for studying how active phases form in supported catalysts.

Many publications have been devoted to the determination of the chemical states in various oxides [27, 34, 198-205] (see Table 2, p. 21). We have considered the fundamental possibilities and limitations of XPS and XAES in solving this problem in Chaps. 1 and 3 (see Table 2 and Figs. 21 and 22). Here we shall deal in somewhat greater detail with identification of the chemical states of elements in oxides of variable valence that are of the greatest interest for catalysis, and also with works in which nontraditional solutions of these problems have been proposed. The latter include the studying of oxides by SIMS and FABMS, sometimes with the use of heavy ions such as Kr [442-444]. Figure 50 depicts individual sections of mass spectra procured by sputtering Sn oxides by Kr atoms [444]. The oxides SnO and SnO_2 cannot be distinguished by XPS data because the change in E_b associated with the increase in the charge in SnO_2 is compensated by the lowering of E_b associated with the transition from a tetrahedral to an octahedral environment. The same compensation of the shifts was predicted theoretically and was observed experimentally for the oxides PbO and PbO_2 [445]. Clusters of $Sn_xO_yH_z$ (x = 1-3, y = 0-4, and z = 0 or 1) were registered in the positive SIMS mode. The yields of Sn_2O^+ exceeded Sn^+ in both oxides. Clusters of the type SnO_x^- (x = 1 or 2 for SnO and x = 1-3 for SnO_2) were observed in the negative SIMS mode. Since the Sn—O bond in SnO_2 is stronger than in SnO, the yields of Sn^+ and SnO_x^- are higher. Moreover, the increase in the coordination of tin in SnO_2 causes clusters of SnO_3^- to appear. These oxides can also be distinguished using the yields of O^+. In this case, the ratio of the yields, R_{O^+} (SnO_2/SnO), was 1.8 ± 0.4. The group of oxides of the type ZnO, CdO, and Cu_2O is readily identified according to the shape and structure of the relevant

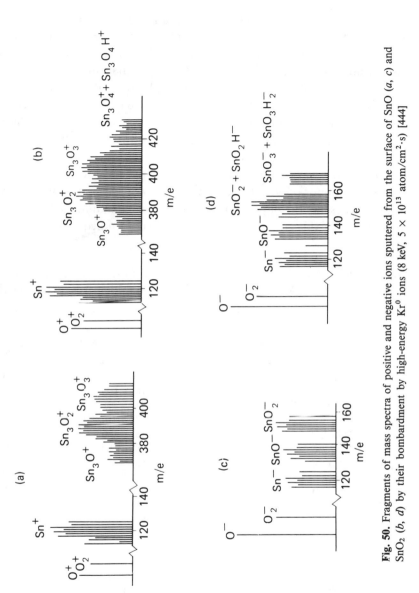

Fig. 50. Fragments of mass spectra of positive and negative ions sputtered from the surface of SnO (a, c) and SnO$_2$ (b, d) by their bombardment by high-energy Kr^0 ions (8 keV, 5×10^{13} atom/cm$^2\cdot$s) [444]

Auger spectra (see Figs. 2 and 4, pp. 23 and 35), which differ greatly from the corresponding spectra of the metals or higher oxides (CuO). The Cu L_3VV spectra were also used to determine the stability of copper oxides with respect to ion bombardment or the action of X-ray radiation [446]. In accordance with the enthalpies of formation [$\Delta H(Cu_2O) = -168$ kJ/mol and $\Delta H(CuO) = -155$ kJ/mol], CuO is easily reduced to Cu_2O, whereas the lower oxide is sufficiently stable even under prolonged bombardment. Chudinov *et al.* [447] proposed a special procedure for distinguishing the three states of copper by photoelectron spectra.

To identify various surface phases on supports, the spectra of different oxide compounds of Ni [34, 448] and Co [34, 449] were measured many times. Table 13 presents the most reproducible values of E_b obtained for these compounds. The oxides of Re, Cr, Mo, and W were studied as prototypes of the

Table 13. Binding Energies E_b of the Inner Levels Characterizing Various Valence and Coordination States of Transition Elements in Oxide Systems

Element	Compound	Treatment conditions	E_b, eV (formal oxidation states)	References
Co	CoO	—	780.3 (II)	[449]
	Co(OH)$_2$	—	781.1 (II)	"
	Co$_3$O$_4$	—	779.5 (II, III)	"
			780.6 (II, III)	[104]
	CoAl$_2$O$_4$	—	781.8 (II)	"
	CoMoO$_4$	—	781.2 (II)	"
Ni	NiO	—	854.5 (II)	[448]
	Ni$_2$O$_3$	—	855.8 (II, III)	"
	Ni(OH)$_2$	—	856.5 (II)	"
			855.8 (II)	[34]
	NiSiO$_3$	—	856.5 (II)	[448]
			856.7 (II)	[34]
	NiAl$_2$O$_4$	—	856.3 (II)	[448]
Cr	Cr$_2$O$_3$	—	576.2 (III)	[450]
			576.6 (III)	[34]
	CrO$_3$	—	579.1 (VI)	"
			580.3 (VI)	[450]
Mo	MoO$_3$	—	233.8 (VI)	[198]
		873 K, vac., 1 h	233.4 (VI)	[203]
			232.7 (V)	[198]
			231.8 (IV)	"
			230.2 ("II")	[200]

Table 13 (*concluded*)

Element	Compound	Treatment conditions	E_b, eV (formal oxidation states)	References
W	WO$_3$	—	36.6 (VI)	[204]
		923 K, H$_2$	36.8 (VI)	,,
			35.8 (V)	,,
			34.9 (IV)	,,
			34.0 ("II")	[454]
			32.2 (0)	,,
	Al$_2$(WO$_4$)$_3$	—	36.7 (VI)	[204]
		973 K, H$_2$	37.1 (VI)	,,
			36.0 (V)	,,
			34.8 (IV)	,,
			32.2 (0)	,,
Re	Re$_2$O$_7$	373 K, vacuum	46.5 (VII)	[205, 450]
			44.3 (VI)	,,
	NH$_4$ReO$_4$	—	46.0 (VII)	,,
			46.4 (VII)	[316]
			46.2 (VII)	[456]
	Ba(ReO$_4$)$_2$	—	46.2 (VII)	[205, 450]
	Ba$_3$(ReO$_5$)$_2$	—	46.6 (VII)	[450]
	ReO$_3$	—	44.3 (VI)	[205]
			44.5 (VI)	,,
	NH$_4$ReO$_4$	Ar$^+$	45.2 (VI)	[316]
	ReO$_2$	—	44.2 (VI)	[205]
			43.2 (IV)	,,
			42.5 (IV)	[456]
	NH$_4$ReO$_4$	Ar$^+$	42.1 (III)	[316]
	Re$_3$Cl$_9$	—	43.2 (III)	[205]
	Re (metal)	—	40.6 (0)	,,
			40.7 (0)	[456]
	NH$_4$ReO$_4$	Ar$^+$	40.1 (0)	[316]

relevant supported catalysts. It should be noted that there is often a film of a higher oxide on the surface of oxides with an intermediate valence, or disproportionation occurs:

$$2\mathrm{Me}^{n+}\mathrm{O}^{n-} \rightarrow \mathrm{Me}^{(n-1)+}\mathrm{O}^{n-} + \mathrm{Me}^{(n+1)+}\mathrm{O}^{n-1}$$

consequently the most authentic values of E_b can be obtained when the oxides are prepared *in situ*, or when the changes in their spectra are analyzed with various treatments [204, 451-453]. In the series of Re oxides, Re_2O_7 and ReO_3 have been identified the most authentically (Table 13); the data are not consistent for lower states of Re. The Re $4f_{7/2}$ peak with $E_b = 42.1$ eV observed in the ion bombardment of NH_4ReO_4 has been related to Re(III) [316], whereas Minachev et al. [450] and Broclawik et al. [454] ascribe this very low value of E_b even to Re(IV). A detailed analysis of the states of Re in individual compounds and on supports [205] makes it possible to ascribe the spectrum with $E_b = 43.0$-43.2 eV to Re(IV).

An analysis of the states of Mo(VI, V, and IV) has been conducted in detail in Chap. 3 (see Fig. 22, p. 90). In addition to these states, a state has been found in the reduced oxides of Mo and W that can formally be related to Me(II). But since such spectra were observed only for individual oxides, the interpretation of Haber et al. [199] and Chappel et al. [452] seems more plausible. They presume that such spectra belong to Me(IV) ions forming pairs of the type Me-O-Me on the edges separating the octahedra in the structure of the relevant oxides.

Let us consider in greater detail the reduction of one of the prototypes of active phases of hydrodesulphurization catalysts—oxide compounds of W [199]. Table 13 presents characteristics of the spectra of initial and reduced WO_3 and $Al_2(WO_4)_3$. Within the interval from 673 to 973 K, the reduced states W(V, IV, "II", and 0) form. It is interesting that here E_b of W(VI) grows and accordingly ΔE [W(0)-W(VI)] also grows. We also observed such a phenomenon for ZrO_2-Zr_2O_{2-x}-Zr-Me (Group VI) [406]. This is possibly connected with the specific electron interaction of W and O in the defective structure of the oxide, which also tells on E_b of O $1s$. The reduction of WO_3 to W^0 terminates at 973 K, whereas in $Al_2(WO_4)_3$ a considerable fraction of the states W(VI) and W(IV) remains. It is assumed that these ions stabilize in the amorphous structure of Al^{3+} and O^{2-}, as in the supported system [204]. The position of O $1s$ in the reduced aluminium tungstate is close to that in Al_2O_3 (531.1-531.6 eV). The W/O ratio in the reduced WO_3 gradually diminished; for $Al_2(WO_4)_3$ in the low-temperature region, the W/O and Al/O ratios also drop because of the removal of oxygen, while the drop in W/Al points to a structural rearrangement of this compound. The great decrease in W/O at 973 K may be associated with agglomeration of the metal phase registered by XRD [204].

ISS has been employed to determine the structure of the first and deeper layers of $V_6O_{13}(001)$ [457]: the first layer was found to be deficient in vanadyl ions, so that vanadium ions from the second and deeper layers are active in the exchange with the oxygen of the gas phase. A potassium addition segregates on the surface of the V_6O_{13}, the maximum K/V ratio being 0.5.

This result is very important for revealing the promoting effect of potassium. ISS data show that it is localized directly on vanadyl ions in the first layer [450].

The oxides β-MnO$_2$ and α-Mn$_2$O$_3$, whose XPS spectra differ only slightly, can be identified by ISS [132]. The phase transition of β-MnO$_2$ to α-Mn$_2$O$_3$ is attended by a growth in the Mn/O ratio of 40%.

4.2.3   Binary and Multicomponent Oxide Systems

There are regions of homogeneity (solid solutions, chemical compounds) and of heterogeneity in two-component oxide catalysts, hence the composition of the surface and the chemical state of the components may differ greatly from those in the bulk. Such catalysts can be analyzed by XPS with a sufficiently high accuracy [Eqs. (3.18)-(3.28)]. With the existing differences in the dispersion of individual phases, one can use a model of a supported catalyst, although it is often quite sufficient to limit oneself to the approximation of a solid solution [Eqs. (1.18) and (3.20)]. As for alloys, surface segregation should be expected in solid solutions of binary oxides. Within the limits of a monolayer model, segregation is determined by Eq. (3.14). When using the values of the surface tensions, the composition of a surface is related to these parameters:

$$\frac{X_{s,1}}{X_{s,2}} = \frac{X_{b,1}}{X_{b,2}} \exp \frac{-(\sigma_2 - \sigma_1)\, a}{RT} \tag{4.3}$$

where σ_1 and σ_2 are the surface tensions of the pure components, and a is the area occupied by an atom or molecule, which in this model is assumed to be the same for all the components. The surface tension in oxides can be estimated very approximately by the formula

$$\sigma = \frac{Ye^2}{4\pi^2 x_0} \tag{4.4}$$

where Y is Young's modulus, x_0 is the interatomic distance, and e is the radius of action of the interatomic forces [458].

Using these expressions, Menon and Prasad Rao [458] showed that segregation in the solid solution ZnO-MgO affects not over two layers, in other words, as for alloys, the sensitivity of XPS must be increased to study segregation, or SIMS and ISS must be employed. In later publications, a number of systems were studied that have a region of mutual solubility, namely, CuO-ZnO [459,

460], Fe-Mn-O [461-463], TiO_2-SiO_2 [464], and NiO-MgO [465]. In calcined CuO-ZnO catalysts, there have been detected both single-phase systems that are solid solutions of one component in the matrix of the other one and multiphase systems. At 553-773 K, the compositions of the surface and bulk are close, while at higher temperatures the Cu/Zn ratio diminishes. This is explained by the additional solution of CuO in ZnO. In the multiphase systems in all cases, Zn segregates on the surface. A complicated change in the surface composition was observed for Fe-Mn-O systems—selective catalysts of the synthesis of alkenes from CO and H_2. Copperthwaite *et al.* [461] indicate that in the oxide of Mn with additions of Fe (5%), the surface becomes enriched in Fe, especially after treatment in a CO atmosphere. Here, judging from the C $1s$ and Fe $2p$ spectra, part of the iron forms a mobile carbide phase. AES data for Fe/Mn oxide systems reduced in H_2 point to segregation of Mn on the surface [461, 462], whereas after treatment in CO/H_2, the surface composition was determined by the conditions of preliminary reduction.

A detailed study of Fe-Mn oxide systems produced by the coprecipitation of hydroxide-carbonate compounds reveals that the surface composition depends both on the Fe/Mn ratio and on the treatment conditions [463]. In calcined specimens, the iron and manganese are in the trivalent state, and the surface composition is close to the bulk one (Fig. 51). The reduction of the surface of oxides with a high Fe content is attended by great enrichment of the surface in Mn, whereas with a high Mn content, conversely, great enrichment of the surface in Fe is observed, i.e. the dissolved component diffuses to the surface. With subsequent treatment in a CO-H_2 mixture, the changes in the surface composition are insignificant if the specimen was preliminarily reduced at 623 K. The ability of iron to be reduced to the metal state grows with an increase in its content, and in specimens of $Fe_{90}Mn_{10}$ the degree of reduction (α) of iron to Fe^0 at 723 K reaches 60%; the manganese is present in the form of MnO. In catalysts treated with the reaction mixture, the carbide phase is present in addition to these forms of the metals, although by XRD data, a solid solution remains in the bulk. Hence, the composition of the surface of Fe-Mn systems is determined by numerous factors, one of the most important of them being the atmosphere in which the catalysts are treated.

In TiO_2-SiO_2 specimens (5-15% of TiO_2) prepared by the coprecipitation of alkoxy derivatives of Ti and Si, the surface became deficient in Ti^{4+}, although the opposite occurs in titanium glass (SiO_2-15% of TiO_2) [464]. According to appraisals by Kerkhof's model, in the first case microcrystallites of TiO_2 particles 1-2 nm in size form; a part of the Ti^{4+} may also be encapsulated in a matrix of SiO_2.

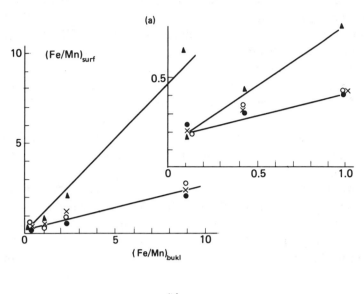

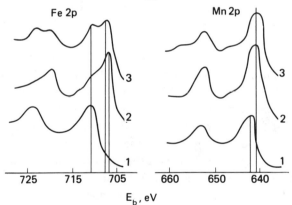

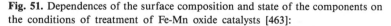

Fig. 51. Dependences of the surface composition and state of the components on the conditions of treatment of Fe-Mn oxide catalysts [463]:

a — relation between surface and bulk compositions: ▲ — air 673 K; o — H_2, 673 K; x — H_2, 723 K; • — H_2, 623 K + CO/H_2, 553 K; b — changes in the spectra of the $2p$ levels of Fe and Mn: 1 — air, 673 K; 2 — H_2, 623 K; 3 — CO/H_2, 553 K

The state of the cations and oxygen and the surface composition of spinels have been studied by XPS [27, 466]. A thin surface layer (about 10 Å) of a spinel $Cu_xCo_{3-x}O_4$ ($0 \leqslant x < 1$) mainly contains Co(III) ions in whose Co $2p$ spectra a weak satellite peak is observed at a distance of 9.7 eV. Such a spectrum structure was also found in $ZnCo_2O_4$ [467] and was ascribed to Co(III) in octahedral positions. The density of the Co cations on the surface of the Co-Cu spinel is lower than in the bulk, while the ratios $O/\Sigma(Co + Cu)$ and Cu/Co exceed the relevant bulk values. The growth in the Cu content was attended by a decrease in the O/Me ratio and by an increase in the surface OH groups. Ion bombardment of such specimens removed the Co_3O_4 layer, and a Cu/Co ratio close to the bulk value was achieved.

A number of binary oxide catalysts have been studied by a set of physicochemical techniques including XPS. The catalysts $ZnO-SiO_2$ (with Zn/Si ratios from 85/15 to 15/85) are highly dispersed X-ray amorphous systems even at 800 °C [468]. Iron oxide with additions of potassium and chromium used in the conversion of ethylbenzene to styrene in the course of reaction became enriched in Cr_2O_3 and depleted of potassium owing to its diffusion into the bulk [469]. This was attended by the reduction of Fe_2O_3 to Fe_3O_4. In the binary system $V_2O_5-P_2O_5$ [470] calcined at 773 K in O_2, VO^{2+} ions segregate on the surface at a P/V ratio above unity. The opposite is observed at P/V < 1. Moreover, the surface cations form centers of various types, namely, Brönsted ones (P—OH) and redox ones (—V=O). On the basis of these data, a mechanism has been proposed for oxidizing butane on oxide V-P catalysts [470]. In the catalyst $SiO_2-Al_2O_3$ prepared by treating aluminium oxide with $SiCl_4$ vapour, the chemical state of the components changes [471]. By the data of XPS and quantum-chemical calculations, the precipitation of SiO_2 on Al_2O_3 produces a growth in the positive charge on Al. This explains the growth in the catalytic activity of these systems in the cracking of n-hexane whose maximum corresponds to 2% of Si/Al_2O_3 [471].

An analysis of the surface structure of mixed Hf-Zr oxide systems in connection with the manifestation of synergism of their effect in the conversion of secondary alcohols into alkenes enabled Davis [472] to conclude that binary oxide and metal alloy systems are similar. By data of XPS, AES, and ISS, the surface composition of these oxides is close to the bulk one, which makes it impossible to explain the sharp changes in the activity by the surface segregations. Since the catalytic properties of the bulk Hf-Zr oxides and of a system prepared by impregnating ZrO_2 with a hafnium-containing solution are close, Davis considers the synergism to be due not to the electron interaction of the components in the bulk, but to a surface ligand effect. But he gives no proofs of the electron interaction of the components on the surface.

Industrial catalysts are often mechanical mixtures of various components including oxides, therefore a quantitative analysis of such surfaces is of substantial interest. Paparazzo [473] analyzed mixtures of SiO_2, Al_2O_3, Fe_2O_3, and Cr_2O_3 employing XPS and AES. He used the position and shape of the O $1s$ and O KVV spectra and also the Me/O ratio to identify the individual oxides and determine their relative concentrations. The shifts in the O KVV spectra were larger than in the O $1s$ ones, but XPS allowed him to obtain more accurate quantitative data. He showed that the preparation of suitable mixtures makes it possible to model the properties on the surface and interface of real catalysts. An important factor determining the accuracy of the quantitative analysis was the absence of reactions between these oxides and of an appreciable phase redistribution. But these effects can be expected with various thermal treatments of composite oxide systems.

4.3   Zeolite Catalysts

When studying zeolite catalysts, two sets of problems closely related to the crystalline structure and nature of their catalytic action are of interest:

(a) the nature, structure, and reactivity of the acid centers and how to control them; and

(b) the nature and specific structure of the modifiers introduced into zeolites in the form of cations, oxides, reduced metals, organometallic complexes, etc.

The solution of these problems depends, in turn, on the determination of the electron structure of zeolites, the type of chemical bonds of the components, the composition of the surface and bulk of zeolite crystals, the chemical states of the transition elements, the sites of their localization, particularly their distribution between the surface and bulk, and on the determination of the dispersion and electron state of the reduced metals. When studying systems consisting of a zeolite and transition metal complex or of a zeolite and oxide, it is important to establish to what extent the inclusion compounds retain their initial properties and to what extent they have been modified under the influence of the zeolite matrix.

As applied to the investigation of zeolites, the surface sensitivity of EES and IS techniques can be appraised in two ways: (a) as a shortcoming because only the outer surface and near-surface regions are probed, and they may not reflect the properties of the "bulk", which in the given case is the reaction zone; and (b) as a merit when studying modified systems based on zeolites since the modification processes substantially redistribute the elements of the zeolite support and the active components, and this is first of all reflected in the EES and IS spectra. In zeolite systems with medium and narrow pores, which have at present gained a leading role in catalytical chemistry, owing to diffusion limitations it is just the near-surface layer that can be the reaction zone, while the centers localized in it can be the main catalytic centers. Already the first studies using XPS yielded data making it possible, on the one hand, to obtain direct information on the charge state of the framework elements and their bonding with the charge compensating cations, and on the other hand, to use the surface sensitivity of the technique to detect the migration of the reduced atoms of a metal to the outer surface [211, 291, 474]. Subsequently, XPS and SIMS were employed to study several surface characteristics of zeolites.

1. The state of the framework elements, the cation-zeolite framework bond, and the composition of the surface layer.
2. The state of the transition elements, their ability of being reduced and migrating.
3. The interaction of the highly dispersed metal particles with the zeolite support.
4. The formation of zeolite-transition metal complex or zeolite-oxide systems.
5. The formation and properties of bi- and polymetallic zeolite catalysts.

Separate stages of these investigations have been summarized in quite a few reviews [25-27, 197, 222, 291, 475, 476], among which Minachev and Shpiro [222, 475] devoted their main attention to high-silica (HS) zeolites of the pentasil type. Minachev *et al.* [197] and Shpiro *et al.* [476] analyzed the possibilities of employing EES and IS for studying zeolites. With this in view, we shall consider herein very briefly the results and conclusions of studies conducted in the last five or six years on the enumerated problems. We shall give our main attention to the new trends that manifested themselves in the application of EES techniques to zeolites, namely, to a more thorough analysis of the position and shape of the lines of the framework elements, the plotting of the concentration profiles, the quantitative determinations of the oxidation states, the studying of modified zeolite systems, and also to zeolite-like materials with a close crystalline structure. The nature of the interaction between a metal and its zeolite support is treated in Sec. 4.5, and bimetal zeolite catalysts, in Sec. 4.6.

4.3.1   State of Components and Surface Composition of Starting and Modified Zeolite Materials Containing No Transition Elements

The interest in studying the electron structure of zeolites (the effective charge of the separate elements or structural framework "blocks") is associated with the general theoretical concepts of acidity in zeolites. One of them relates the Brönsted and Lewis acidity to the electronegativity of the skeleton, i.e. is based on the suggestion that a zeolite is an ionic crystal or a solid electrolyte [477, 478]. Another theory of acidity is based on considering the properties of the acidic centers in their relation to their close environment formed from structural or nonstructural silicon or aluminium or the elements replacing them (the cluster approach) [479]. It was expected that an analysis of the binding energies of the levels of Si, Al, O, and Me^{n+} and of the density of states in the valence band will make it possible to find experimental confirmations of the ionic or covalent model of zeolites and answer the question on the relation between the structure and the charge state of the elements [27]. Assumptions were also advanced that the strong local electrostatic fields in zeolite cavities will substantially modify the charge state of elements in comparison with oxides. These predictions have meanwhile found no strong evidence, although up to recent times [480-482] attempts have been undertaken on the basis of repeated measurements and the analysis of XPS data to establish the electron structure of zeolites.

Table 14 presents the binding energy of the levels of Si, Al, O, and cations (Na^+, etc.) in zeolites differing in their structure (A, X, Y, mordenite, pentasil), in the Si/Al ratio, and in the degree of exchange of Na^+ for another cation or proton. Notwithstanding some quantitative differences between the latest data and those we obtained 15 years ago [26, 483], the same general trends can be seen: a smooth growth in E_b of Si $2p$ and O $1s$ with an increase in Si/Al, and the absence of appreciable changes in Al $2p$ with wide variation of Si/Al and the structure. Okamoto et al. [482] relate these trends, and also the changes in E_b of the O and Si levels in decationization (increase) and the introduction of more electronegative cations than Na^+ (decrease) to the changes in the acid-base properties of zeolites. The low value of E_b for O $1s$ in the zeolite NaA is considered to be due to the high basicity of the oxygen, and the high value of E_b for O $1s$ in the zeolite H-ZSM-5—to the high acidity. The absence of a shift of Al $2p$ is explained by the lower contribution of the Al—O bonds to the formation of the framework than the Si—O bonds. Even if we assume that the changes in E_b given by Okamoto et al. [482] do actually

Table 14. Binding Energy of the Levels of Framework Elements and Selected Cations in Synthetic Zeolites

Specimen	E_b, eV				References
	Si $2p$	Al $2p$	O $1s$	Na $1s$	
NaA	101.7	73.7	531.0	1071.0	482[*]
	100.9	73.20	530.2	1071.45	481[**]
	101.2	73.45	—	1071.65	480[***]
NaX	101.9	74.1	531.1	1071.9	482
	101.75	73.70	530.85	1072.05	481
	101.95	73.90	—	—	480
CaX	102.25	74.35	531.40	—	481
NaY	102.5	74,2	531.8	1072.1	482
	102.35	74.00	531.55	—	481
	102.65	74.25	—	—	480
NaM x = 10	102.5	74.1	532.3	1072.4	482
NaM x = 20	103.1	74.1	532.3	1072.3	482
ZSM-5	102.90	74.50	532.25	—	481
ZSM-5	103.6	74.80	532.6	—	476[*]
NaAlO$_2$	—	73.20	—	—	481
HNaY	103.1	74.6	532.5	1072.3	482
LiNaY	102.7	74.3	532.2	1072.3	482
KNaY	102.5	74.0	531.9	1072.1	482
CsNaY[4*]	102.5	74.0	531.9	1072.2	482
CaNaY[5*]	102.6	74.0	531.8	1071.9	482
CsNaY[6*]	102.3	74.0	531.5	1072.1	482
AAC[7*]	102.7	74.8	531.8	—	476
SiO$_2$	103.35	—	532.65	—	481
γ-Al$_2$O$_3$	—	73.6	530.35	—	481

[*] E_b of C $1s$ = 285.0 eV.
[**] E_b of C $1s$ = 284.4 eV.
[***] E_b of C $1s$ = 284.6 eV.
[4*] The degree of exchange is 8.4%; E_b of Cs $3d_{5/2}$ = 725.2 eV.
[5*] The degree of exchange is 34%; E_b of Cs $3d_{5/2}$ = 725.2 eV.
[6*] The degree of exchange is 58%, E_b of Cs $3d_{5/2}$ = 724.9 eV.
[7*] Amorphous aluminosilicate, SiO$_2$/Al$_2$O$_3$ = 0.4.

occur [although if we take Si $2p$ instead of O $1s$ as the reference level, there will in general be no shifts among the cation-decationized forms of the zeolites X and Y), we are immediately confronted with a new problem. The latter consists in finding to what extent these shifts are related to a change in the acid-base properties because in SiO_2, which does not have a high acidity, the maximal values of O $1s$ and Si $2p$ are observed. Barr [481] arrived at the conclusion that there are "structural" shifts of Si, O, and Al in a transition from zeolite A to X and Y. He is prone to ascribe definite characteristic values of E_b, the valence band spectra, and the energy losses to each structure. Barr, however, fails to analyse these data theoretically or compare them with what has been obtained for amorphous aluminosilicates. At the same time, as can be seen from Table 14, very close spectra are observed for the latter. It is also not clear how authentic the values of the Fermi level on the zeolites are, which Barr determined using the procedure of the complete charge compensation with the aid of a standard electron emitter [54]. The complications involved in applying this procedure for determining the "floating" Fermi level for insulators has been discussed in Chap. 2.

In our opinion, although a new analysis and systematization of the data on spectra parameters of the framework elements for the main types of zeolites will undoubtedly be useful (especially with thorough calibration and the use of monochromatic radiation [482]), they will hardly enable one to arrive at fundamentally new conclusions on the electron structure of zeolites and its relation to their acid-base properties. This will require a thorough analysis of XPS data and the results of MASNMP studies of ^1H, ^{29}Si, and ^{27}Al, IR spectroscopy, and theoretical calculations in the relevant publications [479, 484, 723]. Here special attention must be given to the procedure of zeolite heat treatment, because the acid-base properties of zeolites depend just as much on this treatment as on the initial composition and structure.

In some cases, by measuring the binding energy of the framework levels, we can apparently solve the opposite problem, i.e. identify the structural type of a zeolite or determine the degree of its amorphization. Figure 52 shows the changes in E_b for the Si $2p$ and Al $2s$ lines observed during the prolonged reduction of an iron-containing type A zeolite in hydrogen [485]. The rather sharp increase in the Si $2p$ binding energy after six hours of reduction points to destruction of the crystal lattice. If such a process involves the surface region, XPS is more sensitive than X-ray structural analysis techniques.

The highly informative nature of EES and SIMS for determining the surface composition of zeolite crystals is doubtless. But data obtained recently for separate zeolite single crystals by local techniques of microprobe, scanning

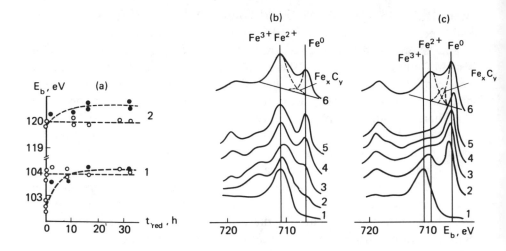

Fig. 52. Dependence of E_b of the Si $2p$ and Al $2s$ levels on the duration of reduction (a) and the changes in the Fe $2p_{3/2}$ spectra (b, c) during the reduction of the zeolites Fe/NaA and Fe/Na-ZVM in H_2 at 500 °C [485]:

a: 1 — Si $2p$; 2 — Al $2s$ (the black dots show Fe/NaA and the white dots, Fe/Na-ZVM); *b* and *c*: Fe $2p$ in Fe/NaA (*b*) and in Fe/Na-ZVM (*c*): 1 — initial specimen; 2 — 2 h; 3 — 8 h; 4 — 16 h; 5 — 32 h; 6 — 28 h + treatment in a mixture of CO and H_2, 3 h, 400 °C

electron microscopy, etc. reveal that the deviations of the surface composition from the bulk one for powders observed by XPS or FABMS may be due either to an actual change in the framework composition or to the fact that crystal intergrowths are analyzed. The latter have incompletely crystallized microphases of the type of aluminium oxide, amorphous SiO_2, etc. with different outcropping. Numerous data for initial, cationic, and decationized forms of zeolites with a silicate modulus ranging from 2 to 10 point to the uniform composition of such crystals [197, 476, 481, 482]. A different picture is observed for high-silica zeolites such as pentasil for which enrichment of the surface both in silicon and in aluminium has been noted. It follows from data of Cocatailo [486] obtained by microprobe analysis, MASNMR, and X-ray structural analysis that large single crystals of ZSM-5 are divided into zones (the edge, near-surface layer, nucleus) in which the Si/Al ratios in the framework differ substantially. Many authors [197, 476, 487-492] indicate a difference in the surface and bulk Si/Al ratios in studies performed by XPS, SIMS, and ISS. The repeated analysis of high-silica zeolites of the same batches resulted in an appreciable discrepancy of the results (up to 50%), which, as

Table 15. Surface Composition of Pentasil Type Zeolites Produced in the USSR [222]

	Specimen	Organic template	Form	SiO_2/Al_2O_3	
				Chem. analysis	XPS
1	ZVM[*]	—	Na	30.0	21.0
2	ZVM	—	Na	33.5	30-33
3	ZVM	—	NH_4	33.5	20-24
4	ZVM	—	Na, NH_4	46.0	37.0
5	ZVM	—	NH_4	45.5	50.0
6	ZSM-5	Monoethanolamine	Na	40.0	39.0
7	ZSM-5	Butanol	Na	42.0	46.0
8	ZSM-5	Butanol	Na	40.0	36.0
9	ZSM-5	Xylylenediamine	Na	40.0	41.0
10	ZSM-5	Ethanol	Na	40.0	26.0
11	ZVK-I[*]	TBA-Br	H, Na	63.0	61.0
12	ZVK-XI	Isopropanol	NH_4	71.0	98.0
13	ZSM-5[**]	Butanol	Na	97.0	71.0
14	ZSM-5	Normal propylamine	Na	70.0	50.0
15	S-130[***]	—	Na	260	120-140

[*] Designations of specimens of test batches.
[**] Produced in GDR.
[***] Produced in USA.

indicated above, is due to the nonuniform composition of individual catalyst particles. Some features of the change in the Si/Al ratio can nevertheless be observed depending on the nature of the template used in crystallization, the SiO_2/Al_2O_3 ratio, the synthesis conditions, and the subsequent treatment [197, 475, 486, 488] (Table 15). For instance, the growth in the concentration of Al in the decationization of zeolites synthesized without organic compounds (ZVM)* is probably due to dealumination and the subsequent migration of the extra-framework aluminium to the surface. XPS data reveal that these zeolites are less stable than those prepared with an organic template under conditions of carefully controlled synthesis.

---

* ZVM (ЦBM) is a zeolite designation used in Soviet literature for ZSM-5 synthesized without any organic template.

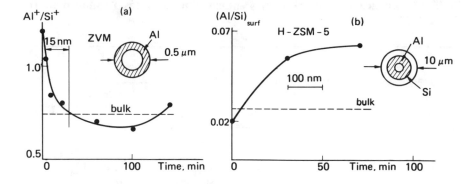

Fig. 53. Depth profiles of Al and Si in high-silica zeolites [475]:

1 — SIMS (1 keV), zeolite NH₄-ZVM; *2* — XPS with Ar⁺ ion etching (3 keV), zeolite H-ZSM-5

An attempt has been made to analyze the depth profile of Al by using SIMS and XPS in combination with ion etching (Fig. 53). For a specimen of H-ZVM whose outer surface is enriched in Al, the Al/Si ratio first diminishes quite rapidly, and then only slightly (SIMS), whereas for a specimen of Na-ZSM-5 initially enriched in Si, the ratio Al/Si changes in the opposite way, and when a layer depth of 100-200 nm is reached, the Al content becomes higher than in the bulk. Consequently, the outer edge (100-200 nm) of rather large crystals (10-20 μm) of a zeolite is enriched in Al. Although the changes in the Si/Al ratio may be partly due to prevalent sputtering of one of the components (silicon according to Okamoto *et al.* [482]), the different nature of the two zeolites, and also data of FABMS [500] indicate that the profiles reflect the real changes in the concentrations along the depth of the layer.

Let us now consider the most important factors determining the composition of local sections of pentasil crystals. The asymmetry of the Al $2p$ line often observed in pentasils [197, 222] may point to a significant contribution of a non-zeolite component to the total content of Al on the outer surface. Okamoto *et al.* [482] consider that for zeolites with Si/Al > 4, the fraction of the extra-framework Al is 50% of the total amount of Al in a zeolite; NaAlO₂ and Al(OH)₃ or Al₂O₃ are the main compounds. Since this conclusion is based only on an analysis of the Al $2p$ lineshape, the authenticity of the quantitative appraisal requires further confirmation. But there undoubtedly is a contribution of the nonlattice Al, and this may cause either enrichment of the surface in Al or its depletion when the Al migrates into channels. Judg-

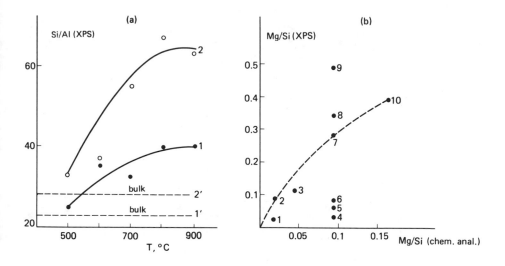

Fig. 54. Changes in the Si/Al ratio (XPS) depending on the temperature of calcination of H-ZVM zeolites (a) and in the Mg/Si ratio (XPS) depending on the MgO concentration, the preparation procedure, the modulus and temperature of calcination (b) [501]

a: 1 — H-ZVM ($x = 45.5$), 2 — H-ZVM (dealuminated, $x = 56$); 1, 2 — corresponding bulk values.
b: ($T_{cal} = 500$ °C), 1 — 0.66MgNa-ZVM ($x = 33.3$), 2 — 1.2%MgO/HNa-ZVM ($x = 33.3$), 3 — 3%MgO/H-ZVM (dealuminated $x = 56$), 4 — 6% MgO/H-ZVM ($x = 45.5$, $T_{cal} = 700$ °C), 5 — 6%MgO/H-ZVM ($x = 45.5$, $T_{cal} = 600$ °C), 6 — 6%MgO/H-ZVM ($x = 45.5$), 7 — 6%MgO/H-ZVM (dealuminated, $x = 56$), 8 — 6%MgO/H-ZVM (dealuminated, $x = 65$); 9 — 6% (3% + 3%) MgO/H-ZVM (dealuminated, $x = 56$), 10 — 10%MgO/H-ZVM (dealuminated, $x = 56$)

ing from Fig. 54, the high-temperature treatment of pentasil leads to the latter effect.

Of no less importance for the creation of a zeolite with a definite ratio of the outer and "inner" surface and a definite Si/Al distribution are the synthesis conditions (Si/Al in a gel, the duration of synthesis, the pH, etc.) [297, 486]. According to Derouane [297], who analyzed the surface and bulk composition of growing crystals by various techniques including XPS, the nature of Si and Al distribution is affected greatly by the composition of the initial gel (Fig. 55). A small number of nuclei form in a gel rich in Al, and the crystals grow because of the constant dissolution therein of the phase rich in Al. As a result, the composition of the crystals and gel constantly changes, the bulk of the crystals is enriched in Al, and their outer layers—in Si. At the end of crystallization the outer surface is additionally enriched, but already in Al (Fig. 55). In synthesis from a gel rich in Si, very minute crystals quite

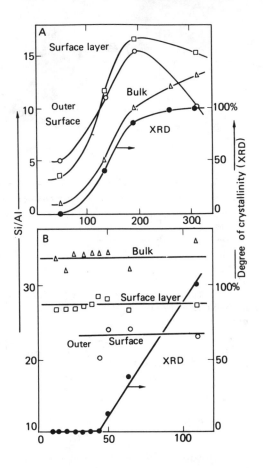

Fig. 55. Change in the composition of the bulk, surface layers, outer surface (Si/Al), and the crystallinity (XRD) during crystallization of pentasils [297]

homogeneous in composition form. The enrichment of their outer layers in Al is associated with the initial composition of the gel.

A general conclusion that can be drawn from the considered publications consists in that the nonuniform distribution of Al and Si in pentasil is rather a rule than an exception and may lead to the nonuniform distribution of the Brönsted (OH groups bonded to Al) and Lewis (trigonal Al, nonlattice Al, Si^+, etc.) acid centers. Since it is a quite complicated matter to predict the type of Si and Al distribution theoretically, the composition of the surface and bulk of crystals must be thoroughly monitored.

The conclusion on the nonuniform distribution of the silicon and aluminium holds to a still greater extent for dealuminated zeolites. Depending on the way dealumination is performed (treatment with HCl [222, 487, 494-496], HCl + hydrothermal treatment, ultrastabilization [497, 498], treatment with EDTA [495, 499] or SiCl₄ vapour [493, 499]), the content of Al on the surface of faujasites, mordenites, or pentasils may be higher or lower than in the bulk (Table 16). Depletion of a surface in Al is observed in the treatment of mordenite or pentasil with HCl [494, 500, 501]; at degrees of dealumination of 30-50%, the depletion is attended by an increase in the fraction of acid centers. An increase in the Al concentration on the surface is connected with its migration during thermal treatment under the action of a reactant from the channels

Table 16. Preferential Nature of Al Distribution in Dealuminated Zeolites

Specimen	Dealumination procedure	Al distribution	References
HY	Thermal and vapour	Enrichment of surface	495, 500
HY	EDTA	Depletion of surface	495, 498, 500
HY	SiCl₄	Enrichment of surface	495, 498
HY	Thermal and vapour + HCl	Homogeneous	495, 500
HMe	Thermal and vapour	Enrichment of surface	495, 500
HMe	Thermal and vapour + HCl	Homogeneous	496
HMe	Ditto	Depletion of surface	495, 500
HMe	Roasting + HCl	Great depletion of surface	494
H-ZSM-5	Thermal and vapour	Enrichment of surface	500
H-ZSM-5	Roasting + HCl	Insignificant depletion of surface	501, 502
H-ZSM-5	Roasting + HCl + roasting	Great depletion of surface	501, 502
H-ZSM-5	HCl	Depletion of surface	500-502

[495, 496]. The opposite process prevails for preliminarily dealuminated pentasil (Fig. 54) [501]—in a specimen calcined at 800 °C the Si/Al ratio on the surface doubles. It is interesting to note that there is also an increase in the O/Si ratio, in E_b of Al $2p$, and to a smaller extent in E_b of Si $2p$ and O $1s$, although the opposite ought to be expected by the model proposed by Barr [481].

Zeolites Modified by Compounds of Representative Elements. Other methods besides dealumination are used to contol the acid properties of high-silica zeolites, the shape and cross section of the channels, and in the long run the molecular-shape selectivity. These methods include the introduction of compounds of representative elements such as Mg, P, B, and Ga [500-502, 724]: (a) by impregnating the zeolites with the relevant compounds; (b) by ion exchange; or (c) by the direct incorporation of elements into the framework of zeolites or zeolite-like materials (ZLM) (silicates, phosphates). These methods should result in a different type of chemical bond between a compound and a zeolite and a different distribution thereof over a crystal. This, in turn, produces various catalytic properties. However, specific processes of redistribution of these elements, emergence from the framework, and exchange in the solid phase may substantially change the initial properties of a system. This is why the selection of optimal catalysts for processes such as the production of *para*-isomers of alkylaromatic compounds, and the aromatization of lower hydrocarbons is possible only after studying the formation of the surface of modified zeolites.

Systems consisting of MgO and pentasil prepared by impregnation or ion exchange with the use of initial or partly dealuminated pentasils were used as an example to see how the method of preparation, the concentration of the dopant, and the heat treatment conditions affect the composition of the surface layers, the state of the modifying components and oxygen [501, 502]. The most uniform distribution of Mg is achieved in ion exchange, although the state of an impregnated specimen with the same Mg content (1.2%) is similar. This is probably due to the reaction of MgO with the H centers of the zeolite. In specimens containing 3 to 10% of MgO, the major part of the dopant is on the outer surface, which is indicated by the high Mg/Si and Mg/Al ratios (XPS). The increase in Si/Al in such specimens is apparently associated with the preferential localization of MgO near Al and screening of the latter. When the silicate modulus is higher, more magnesium is concentrated on the outer surface, while an increase in the calcinating temperature produces the opposite effect (Fig. 54). At 700 °C, the Mg/Si ratio diminishes, which is explained by the migration of MgO into the channels. This process may be amplified owing to reaction of the oxide with the most acidic H centers, and this is why it is more intensive when the Si/Al ratio lowers. Indeed, the 0.5-0.7-eV drop in E_b of O $1s$ confirms the suppression of the surface layer acidity.

A similar mechanism of modifying ZSM-5 with phosphorus has been proposed by Rehman *et al.* [503] as a result of analyzing XPS data. The initial triphenylphosphine was evaporated from the gas phase, and then the specimens were calcined at various temperatures. Rehman *et al.* employed a model they developed [see Eq. (3.35)] for appraising the dispersion of the supported phase on the outer surface and its amount in the channels. By these estimates, the triphenylphosphine is initially on the outer surface in the form of large aggregates (10-30 nm), but after thermal decomposition, the phosphorus penetrates into the channels and by a mechanism of exchange with H^+ occupies cation positions. The mechanism of this unusual stabilization of the phosphorus ions has unfortunately remained unclear.

XPS has been used to study the chemical state and nature of distribution of Ga and the transition metal Zn, which are the most effective modifiers of zeolite catalysts employed in the aromatization of lower hydrocarbons [504, 505]. In specimens of 1.5% Zn/H-ZSM-5, the Zn/Si ratio decreased quite substantially after the running of catalytic experiments. Specimens containing 1.5 and 5% of Zn after calcination and reduction in H_2 were studied to reveal the causes of this phenomenon. In another specimen prepared by ion exchange and the subsequent impregnation of the zeolite with the unreacted solution, a considerable part of the zinc is on the outer surface, but after calcination the Zn/Si ratio dropped to one-third of its initial value because of diffusion of zinc into the channels, whereas additional reduction in H_2 did not lower the Zn/Si ratio. This is why the decrease in the surface concentration of Zn during the catalytic experiment is more likely associated with its migration into channels than with its volatilizing from the surface. It should be noted that heat treatments raise the binding energies of Zn $2p_{3/2}$ and O $1s$ and correspondingly reduce E_k of Zn LMM, which is due to the transition of the Zn to a more ionic state.

When studying Ga/H-pentasils, a "nondestructive" depth profile of Ga was obtained over a depth of 5-15 Å by determining the intensities of lines with a different E_k—Ga $2p$ and Ga $3d$ (Fig. 56). In the initial specimens prepared by impregnation, a maximum gradient of Ga concentration from the outer surface to the bulk is observed. When this and cation-exchange specimens (2% of Ga) were calcined, the Ga/Si ratio diminished. In specimens containing 5% of Ga, a high concentration of Ga was retained in the surface

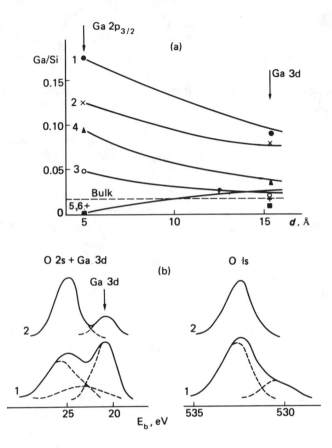

Fig. 56. Change in the Ga concentration over the surface layer in Ga/H-ZVM (a) and lineshapes of XPS spectra of O 1s + Ga 3d and O 1s (b) for 2% of Ga/H-ZVM and Ga-silicate [505]:

a: 1 — initial specimen, impregnation; 2 — initial specimen, ion exchange; 3 — calcined specimen, impregnation; 4 — calcined, ion exchange; 5 — 3 + H$_2$; 6 — 4 + H$_2$; b: 1 — initial 2% of Ga/H-ZVM, impregnation; 2 — initial Ga silicate

layer both in the initial state and after calcination. A comparison of these results with data of electron microscopy [725] has shown that specimens with 2% and especially 5% of Ga/H-ZSM-5 are characterized by the formation of large aggregates of Ga$_2$O$_3$ on the outer surface. This is indicated by the appearance of additional peaks in the O 1s and 2s spectra (Fig. 56). Hence, the great decrease in the Ga/Si ratio observed when these specimens were treated in H_2 is apparently due to reduction of part of the gallium to the metal state and its evaporation from the surface. The absence of shifts of Ga $2p$ or $3d$ is evidently explained by the remaining part of the Ga being in the Ga(III) state.

Yuan Shi-bin *et al.* [506] obtained other data on how hydrogen affects the state of gallium in Ga/H-ZSM-5. Treatment in H_2 was attended by a large shift in the Ga $2p$ line from 1119.6 to 1114.3 eV, and by TPR data the average degree of reduction of the Ga in this specimen was +1. The drop in the Ga/Si ratio was interpreted in this case as levelling out of the surface and bulk concentrations of the gallium oxide. It is interesting that treatment in H_2 increased the activity of Ga/H-ZSM-5 in propane aromatization [506]. Unlike specimens prepared by impregnation with Ga, in gallium silicate the Ga/Si ratio on the surface and in the bulk is close, while E_b of Ga $3d$ is higher by 0.8 eV than in Ga_2O_3 (Fig. 56). We can thus assume that the major part of the Ga is indeed contained in the framework of the silicate.

A comparison of the XPS data for zeolites of various structure and aluminophosphates AlPO-5 and AlPO-11 (E_b of Al $2p$, α') shows [507] that the effective charge of Al(III) in phosphates is close to what is observed in faujasites, while the Al/P ratio is close to the bulk one. Although E_b of P $2p$ in the two types of phosphates differs by 1.8 eV, the identical values of the Auger parameter point to the closeness of the charges on the phosphorus atoms.

The data obtained by Edgrell *et al.* [508] indicating a change in the Na $2s$ lineshape in faujasites subjected to very mild electron bombardment ($E_k \geqslant 6$ eV) were ascribed to the reduction of the Na^+ ions to the metal state. Although this process is possible theoretically, it has never been observed to date. In our opinion, the nonuniform charging that often occurs when the static charge of insulators is compensated by using a Flood Gun is a more plausible explanation.

4.3.2   Cation and Reduced Forms of Zeolites with Transition Elements

Within the limits of the electrostatic model, the high binding energy values of the core levels of cations in zeolites reflect the prevailing ionic nature of the cation-zeolite framework bond. This conclusion was made on the basis of correlations between E_b and the Pauling effective charge for representative (Na, K, Mg, Ca) and transition (Cr, Fe, Co, Ni, Cu) cations in faujasites and in individual compounds [25, 26]. Interactions of such a type are also typical of complex cations of Ru, Rh, and Pd [509, 510] whose binding energies of the 3d levels are higher than in the relevant complexes. Changes in the coordination state of transition elements owing to their localization in sites with various symmetry cause line broadening in the XPS spectra, but since the effective charges do not change appreciably, the chemical shifts are not large. XPS has revealed the ability of cations to be reduced or oxidized and migrate to the outer surface. In addition to the nature of the cations, this ability is determined by the type of structure, the degree of exchange, and the cation composition [25-27, 211, 291, 474]. The chemical shifts observed when the states of transition elements in zeolites change (see Table 2, p. 21) are often sufficient for quantitative analysis. Some metals (Ni, Cu, Ru, Rh, Pd) when subjected to heat treatment in vacuum, CO, or H_2, pass over into intermediate oxidation states [25-27, 511]. A noticeable mutual influence of cations on their reduction in zeolites has been established. For instance, the addition of Co or Cr to an Ni zeolite promotes its reduction, while the addition of copper retards the reduction of the Ni [211, 511]. An analysis of the dynamics of the increase in the intensity of the lines in the reduction of metals combined with the result of electron microscopy, XRD, and chemisorption have yielded a series of the mobility of metals in faujasites [25-27], Ag > Zn > Pd > Cu > Ni > Ru > Rh > Pt. Similar conclusions on the trends of migration of reduced metals in zeolites have been made when analyzing the data of AES and ISS [294, 512]. The ability of metals to migrate is influenced by the second component [25-27, 211], migration to the outer surface generally being retarded. The equality of R_s and R_b may indicate localization of the metals inside a zeolite structure, which for Ru [510, 513, 514], Rh [515, 516], and Pt [211, 509, 517] has been confirmed by independent measurements of the dispersion. At the same time only the constancy of R_s during reduction is not always a sufficient proof for such a conclusion. The formation of metal-zeolite structures in which the particles of the metals crystallize in the defective voids of the intercrystallite space is not attended by an appreciable change in the intensity of the metals [518]. In summarizing the results of studies of metal-zeolite catalysts by EES and IS performed in the 1970's, we can note that they related mainly to faujasites. Virtually no metal-high-silica zeolite systems were investigated. Moreover, a number of questions associated with the formation of the first type of systems also required more detailed elucidation.

With a view to the modern trends in the use of metal-zeolite systems in catalysis, we find it expedient to consider the following questions in greater detail:

(a) the structure of the metal complexes "assembled" in the zeolite cages;

(b) the intermediate states of the transition cations and the formation of highly dispersed clusters; and

(c) the formation and properties of systems consisting of a metal and high-silica zeolite.

4.3.3   Metal Complexes in Zeolites

Several types of metal complexes stabilized in zeolites have been studied by XPS, namely, metal-phthalocyanine [519-521], dimethylglyoximate [522], and *o*-phenylenediamine [519, 520] complexes. The most systematic data have been obtained for Y zeolites with anchored phthalocyanines of metals (MPc): Ni, Co, Fe, Cu, and Ru [519-521]. Studying of the metal-zeolite systems has shown that when clusters of metals form whose size is close to that of large voids the migration of the clusters is greatly impeded. We can expect by analogy that if a way will be found enabling one to synthesize a complex in a void and the size of the complex will exceed that of the entrance "windows", it will be firmly anchored in the matrix. Such requirements are satisfied by molecules of phthalocyanines, which have a size of 12-13 Å. Phtalocyanine molecules (Pc) were "assembled" for the first time in Y zeolites [523] by treating dehydrated cation (Ni, Co, Cu) forms of Y zeolite with the vapour of phthalonitrile (1,2-dicyanobenzene). Then XPS was employed to study a number of important questions relating to the identification and physicochemical properties of these systems, namely, (a) the structure of a complex; (b) the degree of complexing; (c) the nature of distribution over a zeolite crystal; (d) the interaction with the framework; and (e) thermal stability [519-521]. When the phthalonitrile vapour was admitted onto an ion-exchange specimen of 0.4 NiNaY, there was observed a shift in E_b of Ni $2p_{3/2}$ to 854.9 eV typical of the individual NiPc. Inspection of the values of E_b given in Table 17 reveals that the given treatment causes the relevant phthalocyanines to form in the zeolite matrix. Another confirmation of the formation of the phthalocyanine complexes is the decrease in the intensity of the satellite shakeup that is due to the reduction of the magnetic moment, especially on the Ni ions in the Pc. Splitting between the main and satellite peaks increases here, which reflect the trend observed when a more electronegative ligand (O) is replaced by a less electronegative one (N).

The reaction leading to the assembly of MePc in a zeolite does not proceed to the end because of steric hindrances for the ligands in individual structural

Table 17. Identification of MePc in a Y Zeolite and Appraisal of the Degree of Complexing [521]

Specimen	Metal content, mass %	E_b, eV	FWHM, eV	ΔE_{sat}, eV	$\dfrac{I_{peak}}{I_{sat}}$	$\dfrac{I_{Me}}{I_N}$	MePc, %	Number of atoms per elementary cell			
								total	extracted*	MePc	Me^{2+}_{resid}
NiY-13	—	855.6	2.4	5.7	1.63	—	—	—	—	—	—
NiY-30	—	855.7	2.3	5.9	1.70	—	—	—	—	—	—
NiY-40	—	855.8	3.0	6.0	1.57	—	—	—	—	—	—
NiY-60	—	856.2	2.3	6.0	1.46	—	—	—	—	—	—
NiPcY-13	1.0	855.2	2.2	6.2	1.40	0.96	30	4	1	1	2
NiPcY-30	1.7	854.5	1.8	7.3	1.88	0.43	67	8	2	4	2
NiPcY-40	2.0	854.9	1.9	6.8	2.26	0.44	66	11	4	5	2
NiPcY-60	3.1	854.9	1.9	7.5	3.24	0.37	78	16	4	9	3
NiPc	—	854.8	1.4	7.8	3.84	0.29	100	—	—	—	—
CoY-13	—	781.7	2.9	—	1.89	—	—	—	—	—	—
CoY-30	—	781.7	3.0	—	2.12	—	—	—	—	—	—
CoY-40	—	781.8	3.4	—	2.08	—	—	—	—	—	—
CoPcY-13	1.0	780.8	2.2	—	2.17	1.93	20	4	1	1	2
CoPcY-40	1.8	780.6	2.7	—	2.12	0.86	44	11	5	3	3
CoPcY-60	3.0	780.8	2.1	—	2.08	0.62	61	16	5	7	4
CoPc	—	780.8	2.6	—	—	0.38	100	—	—	—	—

* Extracted by reverse ion exchange with NaCl.

positions of the zeolite crystal. This conclusion can be made qualitatively for NiPcY by comparing the ratio of the peak and satellite intensities in the Ni $2p_{3/2}$ spectrum. A growth in the degree of exchange was attended by an increase in this ratio, which points to more complete complexing. The degree of binding the ions into complexes was appraised quantitatively by the ratio N/Me (see Table 17). For individual phthalocyanines, this ratio was taken equal to eight. The values of N/Me for zeolites are always under eight. The noncomplexed metal is present in the form of Me^{2+} cations, which is suggested by the absence of the lines typical of Me^{1+} or Me^{0}. The degrees of complexing found by comparing the values of N/Me and the total concentration of the cations (Table 17) show that with a growth in the degree of exchange the amount of PcMe formed per unit cell increases. For high degrees of exchange (60%) for Ni and Co, the fraction of MePc reaches 0.8. The number of free cations remaining after reverse exchange with NaCl and not entering into a reaction with phthalonitrile is two or three per unit cell for NiNaY. This corresponds to the number of Ni^{2+} cations localized at hardly accessible Si sites of the zeolite framework.

If a molecule of MePc is localized only in large cavities, the number of complexes in a unit cell should not exceed eight. Indeed, this is observed for practically all the specimens described in Table 18. Moreover, a comparison of the data of XPS and chemical analysis made it possible to directly appraise the distribution of Pc and the degree of aggregation. The latter owing to geometric factors can apparently occur only on the outer surface. The Me/Na ratios for the initial forms and after treatment with phthalonitrile are close within the limits of the error of determination. This indicates uniform distribution of the Pc over a crystal. The specimen NiPcY-13 was an exception here—the Ni/Na (Ni/Si) ratios for it exceeded the bulk ones five times. This was due to migration of the nickel to the outer surface in the thermovacuum treatment of the zeolite [474].

A special procedure was proposed for excluding the undesirable effects associated with the migration of the cations into unaccessible sites of the framework and for improving the selectivity of complexing. It consists in incorporating cations into a zeolite by the sublimation of carbonyl or metallocene compounds in the first step, and then "assembling" MePc by the reaction with phthalonitrile. This procedure also includes a step of washing off MePc from the outer surface of the zeolite crystals [524]. According to XPS and electron absorption spectroscopy, and also to chemical analysis [525], the phthalocyanines of Fe, Ni, Co, Ru, and Os obtained in this way are localized as individual molecules in zeolite cavities, while their fraction (Fe, Ni) of the total number

Table 18. State of Pd and Composition of the Surface Layer of High-Silica Zeolites [532, 534]

Treatment conditions	Zeolite	Pd^{2+} (comp.)		Pd^{2+} - O$_z$		Pd$^{1+(\delta+)}$ O - z		Pd^0		Pd/Si	Pd/Al	Si/Al
		E_b, eV	%	E_b, eV	%	E_b, eV	%	E_b, eV	%			
Initial[a]	Pd/H-ZVM	339.8	100	—	—	—	—	—	—	0.006	0.15	23.7
	Pd/H-DM	340.0	100	—	—	—	—	—	—	0.006	0.11	16.1
150°C, vacuum	Pd/H-DM	339.7	66	337.5	34	—	—	—	—	0.007	0.12	16.5
300°C, vacuum	Pd/H-ZVM	—	—	338.0	48	336.8[b]	52	—	—	0.005	0.14	24.0
	Pd/H-DM	—	—	337.4	51	336.4	49	—	—	0.008	0.12	14.9
450°C, vacuum	Pd/H-ZVM	—	—	338.2	40	336.9[b]	48	335.7	12	0.005	0.11	25.8
300°C, vacuum + H$_2$	Pd/H-ZVM	—	—	—	—	336.9[b]	66	335.5	34	0.005	0.10	20.0
(10 torr)	Pd/H-DM	—	—	—	—	336.6	100	—	—	0.006	0.11	17.3
450°C, vacuum + H$_2$	Pd/H-ZVM	—	—	—	—	336.9[b]	53	335.8	47	0.005	0.08	16.2
(10 torr)	Pd/H-DM	—	—	—	—	336.5	100	—	—	0.005	0.09	18.7
100-180°C, H$_2$, flow	Pd/H-ZVM	—	—	—	—	336.9[b]	30	335.9	70	0.005	0.11	25.8
	Pd/H-DM	—	—	338.0	31	336.5	69	—	—	0.005	0.08	14.8
300°C, H$_2$, flow	Pd/H-ZVM	—	—	—	—	—	—	335.8	100	0.005	0.11	23.8
	Pd/H-DM	—	—	—	—	336.7	56	335.4	44	0.004	0.04	16.0
450-500°C, H$_2$, flow	Pd/H-ZVM	—	—	—	—	—	—	335.7	100	0.004	0.05	12.2
	Pd/H-DM	—	—	—	—	336.6	52	335.6	48	0.003	0.05	14.8
500°C, H$_2$ + 300°C, O$_2$	Pd/H-ZVM	—	—	338.0	100	—	—	—	—	0.006	0.15	26.6

[a] SiO$_2$/Al$_2$O$_3$ for H-ZVM and H-DM is 33.5 and 28.0, respectively.

[b] Possibly, Pd—O—Pd clusters.

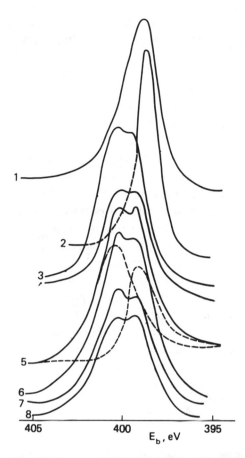

Fig. 57. Spectra of N 1s in MePcY and individual metal-phthalocyanines [521]:

1 — CoPc; *2* — NiPc; *3* — NiPcY-60; *4* — NiPcY-40;
5 — NiPcY-30; *6* — CoPcY-60; *7* — CoPcY-30;
8 — CuPcY-60

of cations reaches 90%. The dispersion of the MePc close to molecular dispersion reached in zeolites can facilitate the interaction between the Pc and the zeolite framework. Here both the central ion and the chelate ligands may participate in this interaction. The coincidence of the binding energy of Me $2p_{3/2}$ for an individual metal-phthalocyanine and MePcY signifies that the zeolite framework does not affect the state of the metal. On the contrary, the spectra of N 1s of the fixed complexes changed sharply (Fig. 57): instead of a narrow N 1s singlet with $E_b = 399.0$ eV and FWHM $= 1.9$ eV, either greatly broadened peaks (FWHM up to 3.5 eV) or a poorly resolved doublet are observed. Synthesis of the N $1s$ peaks yielded two components with binding energies of about 399 and 400.5 eV.

When analyzing the structure of the N $1s$ spectra, we can conclude that the low-energy component characterizes four central nitrogen atoms, whereas the high-energy component relates to four nitrogen mesic atoms. This is confirmed by the conclusion made above on the retaining of the close surroundings of the metal unchanged (E_b of N $1s$ = 399.0 eV). Protons of the structural OH groups may be a possible ligand for the additional coordination of the bridge N atoms. These protons form when H_2O decomposes on the noncomplexed cations of the metal. The data on E_b of N $1s$ (400.2 eV) obtained for the fragment N—H in the protonized form of porphyrin complexes agree with this assumption [526]. These conclusions were recently confirmed independently by Schultz-Ekloff et al. [527] who found changes in the shape of the electron absorption spectrum bands connected with the protonization of CoPc included in an NaX zeolite. Owing to the high dispersion of Pc in zeolites and the possible disturbance of the planar position of a molecule produced by steric factors, the thermal stability of the molecules drops somewhat in comparison with the crystalline state. But the complexes remain quite stable, expecially in a reducing or inert atmosphere. The interval within which the gradual destruction of metal-phthalocyanines occurs in zeolites, judging from an analysis of the parameters of the Me $2p_{3/2}$ and N $1s$ spectra and from the Me/N ratio is 100-300 °C in O_2, 400-550 °C in H_2, and over 500-550 °C in He.

4.3.4   Cations in Intermediate Oxidation States and Highly Dispersed Clusters

Metal complexes can be used as precursors for obtaining cations in intermediate oxidation states and zero-valent clusters in zeolites. Judging from the data of XPS, infrared spectroscopy, EPR, TPR, and electron microscopy, these cations and clusters are produced by stepwise heat treatments in vacuum, H_2, and O_2 of systems such as [Ru(NH$_3$)$_4$NO(OH)]-NaY [510, 513, 514, 528], [Rh(NH$_3$)$_5$Cl]-NaY or NaX [515, 516, 529-531], and [Pd(NH$_3$)$_4$]-HM, NaX [532, 533]. The formation of the Ru-Y systems with various treatments has been studied in detail by XPS. The data on Ru $3d_{5/2}$, Ru $3p_{3/2}$, and the N/Ru ratio reveal that a structure close to [Ru(NO)(NH$_3$)$_4$OH]Cl$_2$ is retained in the initial vacuum-treated specimen. This conclusion has been confirmed by infrared spectra in which there is observed a symmetric intensive absorption band of NO at $1875\,cm^{-1}$ (in the complex at $1850\,cm^{-1}$), and a less intensive absorption band of NH_3 with $\nu = 1345\,cm^{-1}$ (in the complex at $1300\,cm^{-1}$). Hence, after being incorporated into a zeolite, ruthenium retains a coordination bond with NO and NH_3. At the same time, the shift of the NO and NH_3 absorption bands into the high-frequency region with a corresponding increase in E_b of the N $1s$ line, and also the shift and broadening of the Ru $3p_{3/2}$ line observed for the complex cation in a zeolite in comparison with the individual complex apparently point to coordination rearrangements of the cation in the zeolite matrix owing to the replacement of Cl^- ions by an O^{2-} of the zeolite framework.

Stepwise thermal-vacuum treatment at 180-300 °C decomposes the complex and reduces the Ru^{3+} ions. The appearance of the Ru $3d_{5/2}$ line with $E_b = 282.3$ eV at 300 °C evidently indicates the transition $Ru^{3+} \rightarrow Ru^{2+}$. An increase in the duration of treatment at 300 °C causes an additional negative shift of the Ru lines by 1 eV. This probably reflects the reduction of Ru^{2+} to Ru^{1+} (Fig. 58). Further heat treatment in vacuum at 350 and 500 °C reduces the FWHM of the Ru lines with an insignificant change in their energy position and a drop in the N/Ru ratio. Consequently, activation of a zeolite in vacuum causes ruthenium to pass over to lower oxidation states, with no appreciable change in the concentration of the cations in the surface layer. Identification of the intermediate states of ruthenium was based on there being a linear relation between the degree of oxidation of the element in compounds with a close type of chemical bond and the chemical shift (see Fig. 21, p. 89).

By XPS and IR data, beginning from 350 °C, the Ru ions are virtually not bonded to NH_3 or NO and are in two oxidation states: Ru^{3+} and Ru^{1+}. Treatment in vacuum and H_2 at 450 °C yields a spectrum that characterizes one state of ruthenium (Fig. 58), which can formally also be ascribed to the Ru^{1+} ions. But a comparison of these results with data of Nijs et al. [528] suggests that the spectrum in the reduced zeolite belongs to Ru clusters in the zero-valent state. These clusters are dispersed within the zeolite cavities, while the positive shift of 0.9 eV relative to the metal is due to the reaction of Ru with the support. The absence of migration of Ru to the outer surface is confirmed by the Ru/Si ratio not increasing after reduction. These results agree very well with the data of Pedersen and Lunsford [514], and of Nijs et al. [528]. These authors by studying Ru-zeolites by TPR, electron microscopy, and XPS also concluded that highly dispersed Ru particles about 1.0 nm in size form in the zeolite cavities of reduced faujasites.

The process of formation of an Ru-zeolite by oxidative and oxidation-reduction treatment is of a differenet nature (Fig. 58). Treatment in O_2 at

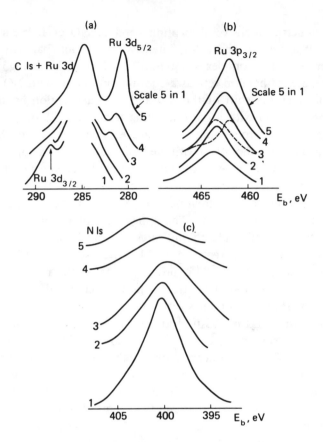

Fig. 58. Change in the XPS spectra of Ru $3d$ (a), Ru $3p_{3/2}$ (b), and N $1s$ during the preliminary treatment of RuNaY [510]:

1 — initial spectrum; *2* — 180 °C, vacuum; *3* — 350 °C, vacuum; *4* — 450 °C, vacuum + H_2; *5* — 500 °C, O_2 + H_2

300 °C leads to partial removal of the NH_3 and to the reduction of a small fraction of the Ru^{3+} ions. At 400 and 500 °C, complete destruction of the complex occurs, but unlike treatment in vacuum, the ruthenium is immediately reduced to a state close to the massive metal. The threefold growth in the Ru/Si ratio reflects the migration of Ru to the outer surface. This is explained by the reduction of the transition metal by ligands evolved in the course of treatment.

Subsequent oxidizing treatments (400 °C, O_2, 10 torr; 300 °C, air) lowered the Ru/Si ratio and increased the Ru $3d_{5/2}$ binding energy, which is apparently due to oxidation of the Ru to RuO_2 (see Table 2, p. 21).

As for RuNaY, the stepwise thermal treatment of RhNaY in vacuum at 180-500 °C decomposes the ammonia complex and reduces the rhodium. It is quite possible that the state Rh^{2+} appears at 180°, but it is unstable. The state Rh^{1+}, which a peak with $E_b \approx 308.5$ eV corresponds to [515], is apparently more stable. At 500 °C, its fraction exceeds 50%. Anderson and Scurrell [529] and Givens and Dillard [530] have confirmed the stabilization of Rh(I) in zeolites by using XPS. Unlike ruthenium, rhodium is not reduced when the complex decomposes in an oxidizing atmosphere, but subsequent treatment in H_2 readily transfers it to Rh^0. The degree of its reduction is close to 100% (300 °C, 1 atm, H_2). But neither at this temperature nor at 500 °C is migration of Rh^0 to the outer surface observed. The drop in Rh/Si and Rh/Al when RhNaY is directly reduced in H_2 (500 °C, 1 atm) evidently reflects a process of metal particle sintering, which prevails over migration to the outer surface [197, 222].

The distinctions in the formation of Ru- and Rh-zeolites are probably connected with the difference in the decomposition of complexes of the two metals. The high affinity of Ru to oxygen causes the formation of RuO_x particles that are mobile and migrate to the outer surface. But by employing stepwise reduction treatment, small clusters of these metals can be stabilized in the cavities of zeolites. Under mild conditions, the Ru clusters can be oxidized and reversibly reduced. More rigorous oxidation-reduction treatments can be used for Rh.

The closeness of E_b of the levels of cations in low oxidation states and highly dispersed clusters of Me^0 prevents their unambiguous identification. To solve this problem, Shpiro et al. [532] and Narayana et al. [533] employed EPR in addition to XPS. In Pd/H-DM specimens treated in vacuum and H_2 ($p = 10$ torr) at 450 °C or in a stream of H_2 at 300 °C, one state of Pd was observed characterized by E_b of Pd $3d_{5/2}$ equal to 336.6 eV (Table 18). It can be ascribed formally to Pd^{1+}, but this is not confirmed by EPR data. The broad singlet observed here may presumably be ascribed to clusters of $Pd_n^{\delta+}$ [533], although the mechanism of the superparamagnetism of such particles has meanwhile not been established. But we can exclude the formation of the cations Pd^{1+}, which appeared after reduction-oxidation treatment of Pd/H-DM (Fig. 59). The signal with $g_\parallel = 3.06$ and $g_\perp = 2.22$ (I) belongs to them. The adsorption of O_2 at 25 °C produces a new signal with $g_\parallel = 1.9887$ and $g_\perp = 2.0460$ (II). Judging from their great broadening in O_2, both signals belong to surface centers.

A signal in PdNaM close in its parameters is related to the cations Pd^{1+} localized in the center of 8- and 12-membered oxygen rings in the coordination of a pseudoplanar square [535]. The signal appearing in adsorption of oxygen

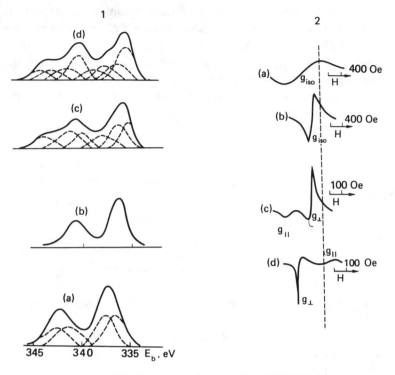

Fig. 59. XPS (1) and EPR (2) spectra of the zeolite Pd/DM [532]:

1 — Pd 3*d*; *a* — 300 °C, vacuum; *b* — *a* + 300 °C, H_2 + 450 °C, H_2 (20 torr); *c* — 300 °C, vacuum + 300 °C, H_2 + 550 °C, H_2 (20 torr) + 450 °C, O_2 (20 torr) + 450 °C, vacuum; *d* — *c* + 450 °C, O_2 (cooling in O_2, 25 °C); *2* — EPR: *a*, *b* — 300 °C, vacuum; *c* — 500 °C, H_2 (20 torr) + 400 °C, O_2 (20 torr) + 400 °C, vacuum; *d* — *c* + O_2, 25 °C

can be ascribed to the adducts $Pd(II)O_2^-$ or $Pd(II)O_2^{3-}$ [535]. But it may possibly also belong to $Pd^{3+} \ldots O_2$.

XPS data reveal (Fig. 59) that in the formation of Pd^{1+}, part of the palladium remains in the form of Pd^0, while another part is oxidized to Pd^{2+}. The fraction of the intermediate state of Pd by XPS data (40%) is higher than by EPR data (20%), which can be explained by the contribution from the signal of $Pd^{\delta+}$. When Pd^{3+} forms (12%), the relative content of Pd^{2+} and Pd^{1+} diminishes, but the fraction of Pd^0 grows. The latter may be due to disproportionation reactions:

$$3Pd^{1+} \rightarrow Pd^{3+} + 2Pd^0$$
$$3Pd^{2+} \rightarrow 2Pd^{3+} + Pd^0$$

The state of iron in various types of zeolites was studied by a set of techniques, namely, Messbauer spectroscopy, ISS, XPS, and SIMS in connection with the catalytic activity in the reaction of hydrogenation of CO [536]. In a reducing atmosphere, microcrystals of Fe^0 have been detected on the outer surface; when a second metal (Cu, Zn, Cr) was incorporated, no surface segregation of the components was found. XPS data on the state of Cr in H_2 and ultrastable forms of zeolite Y confirm the conclusion on the high ionicity of the cation-framework bond [537]. Surface ions of Cr(III) (XPS) are reduced to Cr(II) more easily than in the bulk (EPR). The same trend was noted for ions of Cr(VI) and Cr(V) in oxidized zeolites: in the bulk they are stable, and on the surface are reduced to Cr(III) under the effect of X-ray radiation.

4.3.5 State of Transition Elements and Surface Composition of Metal-High-Silica Zeolite Catalysts

We can expect substantial distinctions in the behaviour of these catalysts in comparison with faujasites due to the following features: (a) the channel structure of pentasil with a smaller opening size than in faujasites; (b) a low ion-exchange capacity; (c) the presence of strong and thermally stable Brönsted and Lewis acid centers; (d) nonuniform distribution of Al over a zeolite crystal; and (e) hydrophobic properties.

A number of authors [196, 475, 532, 538-540] have studied the catalysts Me (Pt, Pd, Rh-ZVM) prepared by the exchange of NH_4- or Na-forms (SiO_2/Al_2O_3 = 30-33.5) with aqueous solutions of metal ammines. Specimens prepared by impregnation were studied for comparison.

XPS data suggest that high binding energies of the $3d_{5/2}$ levels of Pd and Rh and the $4f_{7/2}$ level of Pt are retained after ion exchange. This may signify that the cations enter the pentasil channels. This is also confirmed by the lower surface concentration of the cations (Rh/Si) in specimens prepared by ion exchange in comparison with impregnated ones. But for all the specimens, the concentration on the surface is higher than in the bulk (Fig. 60). Indeed, for example, the size of the ammine of rhodium is 7.1 Å [531], which exceeds the size of the channels. In addition to diffusion hindrances, the concentration gradient may also be due to the higher surface concentration of Al in the given type of pentasil (see Fig. 53, p. 192).

Oxidizing treatment at 350-500 °C leading to the removal of water and ligands facilitates the migration of the cations into the channels, a considerable part thereof, as previously, remaining in the isolated state. Calcination is also

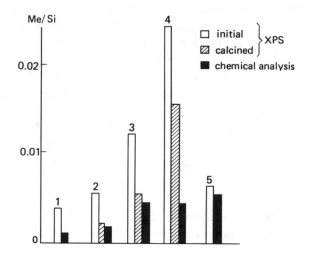

Fig. 60. Distribution of transition metal cations in pentasil type zeolites [475]:

1 — 0.25% Pt/H-ZVM (ion exchange); *2* — 0.5% Pt/H-ZVM (ion exchange + impregnation); *3* — 0.7% Rh/Na-ZVM (ion exchange); *4* — 0.7% Rh/Na-ZVM (impregnation); *5* — 1% Pd/Na-ZVM (ion exchange)

attended by a growth in the Si/Al ratio (for Pd/H-ZVM this was noted immediately after exchange). This may be caused by migration of the Al outside the framework or by "screening" of the Al because of localization of the cations near AlO_4^- tetrahedra. Weakening of the Al signal is possibly associated with the unusual inelastic scattering processes of photoelectrons within the molecular size channels, part of which have a zigzag shape. As a result, there is a decrease in the substrate signal intensity even with a small coverage by the supported component (cations) that is unusual for XPS [534].

Calcination is also attended by (a) reduction of part of the cations of Pt, Rh (air) or Pd (vacuum) to an intermediate and zero-valence state; and (b) exchange in the solid phase between the incorporated cations Na^+ and H^+. This process in $RhCl_3$/Na-ZVM prepared by impregnation resulted in a similar decrease in Rh/Al and an increase in Na/Al on the surface [538]. But as a whole, a more uniform distribution is achieved for calcined ion-exchange specimens than for impregnated ones, which is connected with the presence in the latter of large aggregates of the Me_xO_y type.

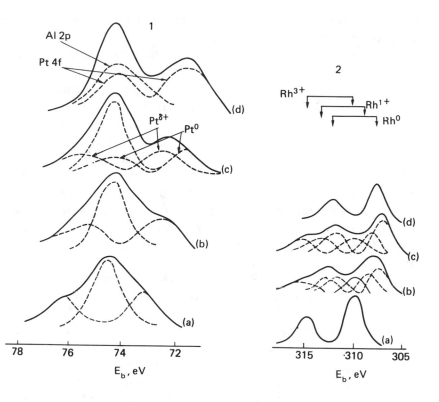

Fig. 61. Changes in the XPS spectra of Pt $4f$ + Al $2p$ (1) and Rh $3d$ (2) in treatments of the relevant forms of high-silica zeolites:

1 — 0.5% Pt/H-ZVM, spectra of Pt $4f$ + Al $2p$: *a* — initial specimens; *b* — 550 °C, air; *c* — 550 °C, air + 520 °C, H_2; *d* — 520 °C, H_2; *2* — spectra of Rh $3d$ for ion-exchanged 0.7% Rh/Na-ZVM in various treatments: *a* — 500 °C, air; *b* — *a* + 150 °C, H_2; *c* — *a* + 300 °C, H_2; *d* — *a* + 500 °C, H_2

Platinum. The increased surface concentration of Pt and the low content of Al made it possible to analyze the partly overlapping spectra of Pt $4f$ + Al $2p$ in specimens with a relatively low mass content of the metal [196, 539, 540] (Fig. 61). After calcination, 60-70% of the Pt is in the form of Pt^{2+} ions. Direct treatment in H_2 at 520 °C completely reduced the Pt to Pt^0, and also increased the Pt/Al ratio five times, which indicated the directed migration of the metal from the channels to the outer surface. Conversely, the reduction of a precalcined specimen at the same temperature did not change the Pt/Si or Pt/Al ratio, which indicated the preferential localization of Pt^0 within the zeolite structure. In this case, there is also observed a positive shift in the Pt

$4f_{7/2}$ spectra typical of small particles [226]. Close results have been obtained for specimens prepared by ion exchange or by a combination of the latter and impregnation. The Cl/Pt ratio after calcination of the second specimen did not exceed 1/5 [532, 534].

Palladium [532, 534]. Heat treatment in vacuum of specimens of Pd/H-ZVM and Pd/H-DM with a close modulus destroyed the complex (see Table 18). At 300 °C, about 50% of palladium is apparently in the form of isolated Pd^{2+} cations, while the other part can be related to an intermediate state of Pd. It is quite possible that in H-ZVM, where a peak of Pd $3d_{5/2}$ with E_b of about 337 eV is observed, the palladium forms bridge structures of the type of Pd^{2+}-O-Pd^{2+}. Such changes in the state of Pd are typical of Pd/Y [291], but in pentasil the cation forms are more stable. Indeed, an appreciable part of Pd in this state is retained even after treatment in H_2 (see Table 18). Only at 300 °C (H_2, flow) is palladium completely reduced in pentasil to Pd^0. Unlike H-ZVM, in H-DM palladium passes over completely into one state in mild reduction conditions. EPR studies (see Fig. 59) have shown that here we have to do with electron-deficient clusters of zero-valence palladium—$Pd^{\delta+}$.

A common feature in the properties of Pd in two types of high-silica zeolites is that at least under static conditions and in a flow at low temperature the major part of Pd remains localized inside the structure: the ratio Pd/Si changes only slightly, while the drop in Pd/Al with a simultaneous growth in Si/Al in H-ZVM and the absence of such changes in H-DM confirms the assumption that Al is screened [534]. Unlike this, for Pd applied onto H-NaZSM-5 from a solution of $Pd(acac)_2$, no shifts were observed in the XPS spectra that would indicate its reaction with the zeolite support [541]. This is simple to explain because the size of the Pd particles exceeded 100 Å.

Rhodium [538]. Treatment of calcined specimens of Rh/Na-ZVM at 150-500 °C in H_2 reduces the Rh^{3+} to Rh^{1+} and Rh^0 (see Fig. 61). As expected, the ability of rhodium to be reduced diminishes in the series Rh/Na-ZVM (impregnation) > Rh/Na-ZVM (ion exchange) > Rh/H-ZVM. In the last named specimen, about 30% of the Rh fails to be reduced in a stream of H_2 at 450 °C. As for platinum, the direct reduction in H_2 (450 °C) causes migration of Rh^0 to the outer surface, whereas oxidation-reduction treatment hardly changes the Rh/Si ratio in H-ZVM.

Ruthenium [542]. XPS and ISS data show that ruthenium in a specimen of 1.7% Ru/H-ZSM-5 is concentrated on the outer surface, but in the form of very highly dispersed particles. Melson and Zuckerman [542] associate the decrease in Ru/Si observed during calcination and subsequent reduction with the sintering of Ru^0. But having in view that the size of the particles by data of H_2 chemisorption was very small (4 Å), migration of Ru into the pentasil channels is more probable.

The considered XPS data enable us to conclude that the treatment procedure has a prevailing effect on the migration of noble metals in ZSM-5. It is confirmed by direct electron microscopy measurements [539, 540]. In specimens of 0.5% of Pt/H-ZVM, very highly dispersed particles with a prevailing size of 0.8 nm (it is difficult to observe smaller particles by this procedure) and a very narrow size distribution were observed. An analysis of the TEM patterns made it possible to conclude that Pt is localized in the channels. Conversely, in direct reduction, well faceted cubic crystals of Pt form with a predominating size of 10 nm and a wide size distribution. A major part of them is on the outer surface. When studying 0.5% Rh/H-ZMV in an electron microscope with a resolution of 4-6 Å, no Rh particles were observed, which evidently points to their small size (<1.0 nm).

Interesting data have been obtained for Pd/H-DM. Under conditions when XPS revealed one state of palladium, i.e. $Pd^{\delta+}$, TEM data indicate the formation of metallic particles with a predominating size of 2-3 nm (the maximum size is 6 nm). The absence of a pronounced habitus of the crystallites, their great disorder and high uniformity in size, and also their uniform population of the prismatic crystals of mordenite enable us to assume that they are "intergrowths" of clusters inside the zeolite structure. The intracrystalline cavities formed in dealuminizing may be the site of localization of the particles whose size exceeds the diameter of the channels (0.7 nm) of mordenite. A definite fraction of Pt (2.0 nm) may be localized in defective cavities of pentasil. Another explanation proposed for faujasites [543] is based on the recrystallization of the zeolite lattice, and it seems to be less probable for a very thermostable structure of pentasil.

Thus, the formation of a metal-high-silica-zeolite system is characterized by the following features:

(a) enrichment of the outer surface in cations when they are incorporated, which is especially great when employing impregnation;

(b) a great dependence of the distribution of cations and clusters of metals between the outer surface and bulk of crystals on the treatment procedure;

(c) difficult reduction, which is explained both by the higher thermal stability of the OH groups in pentasil and shifting of the equilibrium of the reaction $Me^{n+} + H_2 \rightleftarrows Me^0 + nH^+$ to the left by strong interaction of the cations with the framework;

(d) stabilization of the ions (Rh^+, Pd^+) and highly dispersed particles of the metals inside the channels and the formation of metal-zeolite structures;

(e) a certain deficiency of electron density (a positive shift) on the highly dispersed articles of the metals in the acid forms of pentasil and dealuminized mordenite.

Base Metals. The low exchange capacity of pentasils hinders the preparation of specimens containing transition elements only in cation positions. Wichterlova *et al.* [544, 727] instead of ion exchange successfully used for this purpose the procedure of incorporating cations of V, Cr, and Mo by a reaction in the solid phase, but the concentration of the incorporated metal is limited to 50% of the possible degree of exchange. An analysis of XPS and EPR data for Cr/H-ZSM-5 prepared by ion exchange shows that the Cr(V) cations are in an isolated state, and, moreover, part of them are contained in chromates. In specimens of 0.75% Cr/H-ZSM-5 close in composition and prepared by ion exchange and impregnation, the surface concentration of Cr is higher than in the bulk, while calcination causes the Cr(V) and Cr(VI) ions to penetrate into the zeolite channels. Reduction of such specimens in H_2 at 550 °C leads not only to the formation of Cr(III), but also to a very greatly reduced state of Cr whose Cr $2p_{3/2}$ spectra are close to Cr^0 [540].

Nguen Quang Huynh *et al.* [545] employed an original way of preparing a metal-pentasil system. Fe_3O_4 was incorporated into ZSM-5 during hydrothermal synthesis, the result being a catalyst with a special texture. Data of electron microscopy, XRD, and XPS revealed that the iron oxide is within the zeolite crystals, and reduction in H_2 at 350 °C does not change the catalyst texture.

In a number of cases, transition metals were incorporated into pentasil in concentrations greatly exceeding the ion-exchange capacity. This is typical of the Fisher-Tropsch synthesis catalysts in which the concentration of the active component is usually at least 10%. The zeolite nevertheless substantially affects properties of these systems such as the ability of a metal to be reduced, the surface concentration, dispersion, and catalytic activity [545-548]. The state of Co in specimens of 30% Co/Na-pentasil obtained by mixing the support and the basic carbonate of Co depends on the modulus of the zeolite, the treatment conditions, and the presence of a promoter [545, 546]. In Co-pentasil systems, the reduction of the cobalt is hindered when the modulus is lowered, preliminary calcination is employed, or when MgO is introduced. Calcination produces more difficultly reducible compounds (Co_3O_4 or spinel $MgCo_2O_4$), and the Co^{2+} ions migrate partly into the channels. The overall effect of all these factors lowers the degree of reduction of Co from 73 to 17%.

An analysis of E_b of Co $2p_{3/2}$, ΔE_{sat}, and I_{sat}, as well as of the O $1s$ spectra, allowed the surface structure of the catalyst to be represented as follows: zeolite-$Mg(OH)_2$-CoO-(CoO-MgO)-Co^0. The Co/Si ratios sharply diminish as a result of calcination and drop additionally during reduction, whereas Mg/Si changes only slightly, the catalyst surface being highly rich in magnesium oxide. The drop in the Co/Si ratio is apparently due to two competitive processes, namely, the sintering of Co and its migration into the tunnels. It is also quite possible that the Si in the initial specimens is partly screened by the phase of the cobalt oxide. Independent data on the dispersion of the cobalt oxide phase are needed for a more detailed consideration of the contribution of these effects and for appraising X_1, X_2, C_1, and C_2 [Eq. (3.35)].

XPS was employed to study the influence of various factors on the chemical state and degree of reduction of Fe in the sodium and hydrogen forms of pentasil with variation of the Fe concentration from 1 to 30%, of the preparation procedure, and of the treatment conditions [485, 547, 548]. In the initial zeolites, the surface concentration of Fe is close to the bulk one even at 10%, which is apparently explained by the competition of two effects—enrichment of the surface and aggregation of the oxide phase. As for cobalt, enrichment or the introduction of promoters such as MgO or MnO lowers the degree of reduction of Fe, whereas in the presence of K_2O iron is reduced more readily.

In the multicomponent catalysts $Fe-TiO_2-MnO-H$-pentasil [485], the iron is reduced much more readily than in the same composition on NaA (see Fig. 52, p. 190). This is explained by the different mechanism of catalyst structure alternation during reduction. For zeolite A, a considerable part of the Fe passes over into the cation form and as a result of destruction of the crystal lattice forms stable compounds with fragments of the zeolite. In pentasil, the Fe(III) oxide reacts more weakly with the zeolite and quite readily transforms into FeO, and then into Fe^0. An interesting, but not completely clear result is the decrease in E_b of Fe $2p_{3/2}$ (Fe^0) upon the prolonged treatment of Fe/Na-pentasil in H_2 (32 h). The negative shift by 0.5-0.6 eV relative to the massive metal may be associated with the electron interaction of the Fe^0 clusters with the zeolite, which is usually observed when the effect of strong metal-support interaction manifests itself (see Sec. 4.5).

4.4   Oxides on Supports

When studying supported catalysis, researchers display appreciable interest in information on the chemical state of the active components, their dispersion, and on how they are distributed on the support. These characteristics are in many ways determined by the interaction of supported component with its support, and for multicomponent systems also by the interaction of the components with one another. Metal-support interaction depends not only on the nature of the support, but also on how the catalysts in question are prepared and treated. This necessitates a thorough study of all the stages of formation of supported systems. Depending on the state of the components in the active form, supported catalysts can be divided into two groups: oxides and metals on supports. The first group also includes systems in which the deposited component is on the surface in the form of isolated cations.

In Chap. 3, we analyzed the possibilities of EES and IS for studying supported systems, considered the procedures for quantitative determinations of various chemical states, the dispersion of the deposited phase, and the nature of component distribution. In the present section, we shall attempt to systematize the data on the characteristics of the surface of supported oxides, which are of interest as catalysts of very important processes such as refining (hydrocracking and hydrodesulfurizing) of petroleum, the synthesis of methanol, dehydrogenation, aromatization, metathesis, and the partial and complete oxidation of hydrocarbons. These catalysts include first of all the systems Co, Ni, Mo, W, Co-Mo, Ni-Mo, Ni-W, Cr, and Re on traditional supports, namely, γ-Al$_2$O$_3$, SiO$_2$, amorphous aluminosilicate (AAS), and also the catalysts CuO-ZnO-Al$_2$O$_3$ and CuO-ZnO-Cr$_2$O$_3$ among others. It seems expedient to discuss the following matters:

(a) the dependence of the chemical state and dispersion of a supported phase on its concentration, the nature of the support, and the deposition method;

(b) the formation of surface compounds and their role in the formation of the active phase; and

(c) the formation of an active surface under the action of the reaction medium. These questions have been studied the most extensively for aluminium-cobalt-molybdenum catalysts (ACMC) of hydrodesulfurizing. The data of XPS and ISS for these systems together with the results of using other techniques have been discussed in detail in monographs and reviews [22, 23, 27, 132]. Without repeating the results and these works, we shall deal with some general conclusions and new information that develop or clarify the known models of the active surface of ACMC.

4.4.1   Supported Binary Oxides (Ni, Co, Mo, W)

Figure 62 presents dependences of the ratio of the surface intensities of the elements in the supported component and support on the total content of deposited oxides in XPS and ISS spectra for a number of oxides

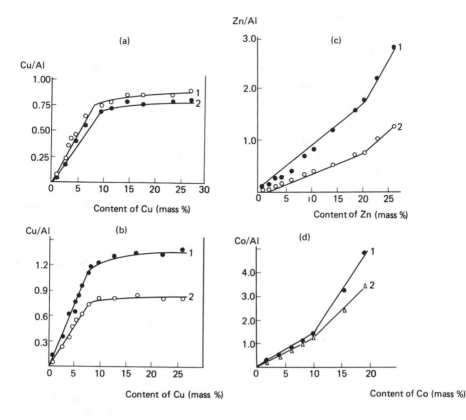

Fig. 62. Dependences of the surface ratios of the elements in a supported component and support on the total content of deposited oxides [104, 217, 278]:

a — CuO/Al$_2$O$_3$ (XPS: 1 — 400 °C; 2 — 600 °C; b — CuO/Al$_2$O$_3$ (ISS): 1 — 400 °C; 2 — 600 °C; c — ZnO/Al$_2$O$_3$ (ISS): 1 — 400 °C; 2 — 600 °C; d — CoO/Al$_2$O$_3$ (XPS): 1 — 400 °C; 2 — 600 °C, air

(Co, Ni, Cu, Zn, Mn, Mo, and W) on γ-Al$_2$O$_3$ within a broad interval of concentrations [132, 214, 216, 217, 278, 549-551]. All of them, notwithstanding the quantitative distinctions, have a common feature: the relations are approximated by broken lines, the initial slope of the straight lines in the second section may either grow or diminish. The shape of the curves, and also the distinctions registered by XPS and ISS for calcined and uncalcined specimens are due to the changes in the dispersion of the supported phases with a growth in the concentration, and also to the processes of migration of the components into vacant sites of the γ-Al$_2$O$_3$ lattice or of surface segregation of the oxide phase. It is difficult to determine unambiguously which of the processes prevails from the data of only one technique. For instance, a growth in the Co/Al or Cu/Al ratio (ISS) at high concentrations of the relevant oxides is interpreted as segregation on the surface of the oxide phases [214, 216], while a drop in these ratios attending calcination is interpreted as migration of the cations into the lattice of the aluminium oxides. Simple geometric reasoning shows that these changes can be explained in a different way, namely, a growth in the intensity can be explained as a result of dispersion of the supported phase, for example, when it spreads over the surface, while a drop in the intensity—as agglomeration of the oxide particles. Consequently, more reliable interpretation requires a comparison of data provided by different techniques such as XPS, ISS, and analytical electron microscopy, and also a more thorough analysis of other XPS spectra parameters: E_b, ΔE_{sat}, and $\Delta E_{s.o}$, which in a number of cases have characteristic values for massive and dispersed oxides, as well as surface compounds, and can reflect the interaction of a deposited phase with its support.

A common distinction of spectra of a supported phase from its bulk analogues is their broadening, which depends on the cation, its concentration, the support, the conditions of deposition and treatment [27]. Other conditions being equal, the broadening is more appreciable at a low oxide concentration and when the oxide is deposited on reactive supports such as γ-Al_2O_3. In other words, the broadening is associated with a high dispersion of the deposited phase. Moreover, it is due to nonequivalence of the energy states of the active elements and the nonuniform charging [22, 23, 27]. The latter depends on many factors, and it can be taken into account only by a comparative analysis of the FWHM of all the lines in the spectra, and also by lowering of the charging. Let us see how a set of XPS, ISS, and SIMS data can be employed for describing the surface structure of supported oxides using the well characterized Co, Ni, Mo, and W systems as examples.

The state of Co/Al_2O_3 has been studied in the greatest detail by Chin and Hercules [214]. They employed XPS, ISS, and SIMS to establish the trends of the change in the surface characteristics depending on the cobalt content and the calcination temperature (T_{cal}), and also the related changes in the ability of Co to be reduced. A comparison of E_b for Co $2p_{3/2}$ and ΔE_{sat}, as well as the XPS and ISS intensities, reveals that at low concentrations ($< 2\%$ Co) the entire cobalt when calcined migrated into the near-surface layer of the γ-Al_2O_3, where it occupies tetrahedral vacant positions (Co_t^{2+}) and forms $CoAl_2O_4$. When the concentration grows, part of the Co ions are localized in octahedral positions, and then also in the form of Co_3O_4. The intensity of the ISS signal was used as a criterion of the formation of Co_t^{2+}. Such a localization screens the signal, which is especially noticeable at high calcination temperatures (600 °C). The change in the slope of the curve in Fig. 62 is associated with a transition from a "monolayer" distribution to the phase Co_3O_4 (at a specific surface area of 90 m²/g it occurs at 10-12% of Co). A further growth in the intensity seems to be related to the formation of a highly dispersed oxide phase within the interval of 12-30% of Co in addition to the large crystals of Co_3O_4 (16-22 mm) observed by XRD. At the same time, such a low dispersion of the cobalt oxide is not typical of γ-Al_2O_3 and is explained by the low specific surface area of the support. Antoshin et al. [552] and Shpiro et al. [553] have compared the behavior of specimens of 6% of Co on γ-Al_2O_3 and SiO_2 with a specific surface area of 200-300 m²/g. The initial Co/Si ratio (XPS) was lower than the bulk one, while the maximal particle size, by appraisals using Kerkhof's model [185], was 3 nm. The cobalt oxide was X-ray amorphous on γ-Al_2O_3; thermal treatment in vacuum at 450 °C caused the formation of a mixed system containing $CoAl_2O_4$ and Co_3O_4. This was indicated by the values of E_b of Co $2p_{3/2}$, and also the cluster ions CoO $^+$, CoOH $^+$, and CoO_2^+ characterizing the oxide, and CoAl $^+$ and CoAlH $^+$ relating to the aluminate. XPS spectra with lower values of E_b of Co $2p_{3/2}$ typical of Co_3O_4 were observed when the support was SiO_2.

In Co-TiO_2-MgO-Kieselguhr specimens [554], interaction between the cobalt oxide and promoters has been observed that increases the dispersion of the supported species and hinders their reduction to Co^0. The change in E_b of O $1s$ presumably indicates the formation of a solid solution of CoO-MgO.

The state and distribution of Ni depend both on the support and on the preparation procedure. A transition from "monolayer" to "multilayer" distribution on γ-Al_2O_3 was observed at an NiO concentration of 17% [550]. At lower concentrations and higher temperatures of calcination, Ni^{2+} migrates to the tetrahedral positions of γ-Al_2O_3. The way of preparation has just as great an influence on the nature of Ni distribution and its reduction as the support. This is shown by XPS and ISS data for Ni/SiO_2 specimens prepared by impregnation and ion exchange, and also by anchoring $NiCl_2$ on silica preliminary modified with lithium [555-557]. Different relations between I_{Ni}/I_{Si} and the Ni content in the bulk have been obtained for these specimens (Fig. 63). For ion-exchangeable and anchored catalysts, they have the form of straight lines, which points to the constancy of Ni dispersion. Although the ambiguity associated with λ_{2p} did not allow Stakheev et al. [555] to calculate the Ni dispersion accurately, they presume that it is close to molecular disper-

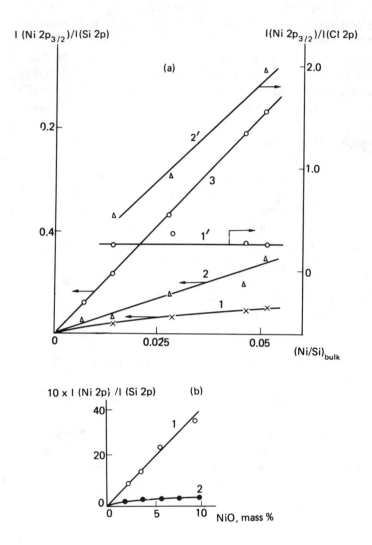

Fig. 63. Distribution of deposited nickel on the surface of silica:

a — Ni $2p_{3/2}$/Si $2p$ and Ni $2p_{3/2}$/Cl $2p$ versus Ni/Si (bulk) [556]; 1, $1'$ — impregnation; 2, $2'$ — fixing of NiCl$_2$/Si-O-Si; 3 — theoretical curve calculated from the monolayer module [185]; b — I(Ni $2p$)/I(Si $2p$) versus mass % of NiO [557]: 1 — ion exchange; 2 — impregnation

sion in an ion-exchangeable specimen. Conversely, in impregnated samples, the dispersion is much lower, and with an increase in the concentration it diminishes still more (see Fig. 63). Yuffa *et al.* [556] consider that the size of the NiO particles on such a specimen grows from 5 to 9 nm within a concentration interval from 1.3 to 4.6%. The experimental values of the intensities for an "anchored" catalyst are only 30% of the calculated ones [185], but judging from the binding energy of Ni $2p_{3/2}$ and having in view the preparation procedure, the formation of three-dimensional NiO crystals has a low probability. The differences between R_{exper} and R_{theor} are apparently associated with localization of part of the Ni^{2+} in the micropores and other hidden sites of the support structure. This is confirmed by the features of reduction of Ni [556, 557]. The Ni/Cl ratios differ substantially in the impregnated and anchored specimens: in the former, it is close to a stoichiometric one and does not depend on the Ni concentration, whereas in the latter, the Ni/Cl ratio diminishes with a growth in the nickel concentration. This points to a change in the ratio of the centers of single-point (I) and two-point (II) adsorption:

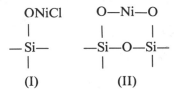

Within the interval from 1.3 to 4.6%, the fraction of the centers (II) grows from 0.1 to 0.7. One of the clue questions is that of the formation of nickel silicate at high calcination temperatures. According to data of Stakheev *et al.* [555] and Ocelli *et al.* [448], nickel silicate forms on SiO_2 and aluminosilicate with a high SiO_2 content; in an anchored catalyst at temperatures up to 450 °C, the formation of such a compound was not observed. The dispersion of Ni on AAS increases with a growth in the Al_2O_3 content [557, 558]. At high NiO concentrations (10%), the dispersion grows owing to the diminishing of the size of the main phase particles and to the appearance of a new finely dispersed phase of NiO at concentrations of the Al_2O_3 below 5%. The latter phase has been determined by analytical electron microscopy [557]. Within the interval of 1-5% of NiO, the growth in the dispersion is due to stronger interaction with the support rich in Al_2O_3 up to the formation of $NiAl_2O_4$. The size of NiO particles on such an aluminosilicate specimen, according to Kerkhof's model, does not exceed 2 nm [558, 559], which is consistent with the data of TEM [448].

The dispersion of MoO_3 on γ-Al_2O_3 depends on the oxide concentration and on T_{cal}: at 550 °C a linear relation is observed between I(Mo 3d)/I(Al 2p) and the surface density d_{Mo} [22] up to $d = 4$ atom/nm^2 [560]. Above this threshold value, "free" MoO_3 is present in addition to the epitaxial layer of this oxide. ISS data reveal that at a molybdenum concentration below 8% it can occupy only tetrahedral sites in γ-Al_2O_3, while at higher concentrations it can also occupy octahedral vacancies [216]. It follows from intensity calculations by Kerkhof's model that there is a monolayer distribution of Mo at MoO_3 concentrations of 12-14% [22]. Calcination of the catalyst or the use of solutions with a low pH for its preparation cause the dispersion of the MoO_3 to grow [549].

Many authors, including Barr [23], presume that a surface molybdate is the precursor of the active center in hydrodesulphurizing catalysts. Its formation has been proved by XPS, although its detailed structure requires further clarification [23]. The features of reduction of MoO_3 in a compact form and on γ-Al_2O_3 have been discussed in previous sections (see Figs. 21 and 22, pp. 89 and 90). The difficulty encountered in the quantitative determinations of the intermediate states of Mo is due to the ambiguities of ascribing Mo(V). Barr [23] indicates that the Mo(VI)$_t$ ions are reduced to Mo(V), and the Mo(VI)$_o$ ones, to Mo(IV). As expected, the ability of molybdenum to be reduced grows with the concentration of MoO_3. For instance, in specimens with 7% of MoO_3 at $T_{red} = 570$ °C, about 55% of the Mo remains in the form of Mo(VI), whereas in a specimen with 14% of MoO_3 only 28% [198] of the Mo remains in this form. Beginning from 700 °C, peaks with $E_b = 228.3 - 229.2$ eV belonging to Mo^0 are clearly registered in the Mo 3d spectra. The fraction of Mo^0 in specimens with a submonolayer coverage (7% of MoO_3) is 35-40% [198]. By many data [23], reduction results in breaking up of the monolayer and in a lower dispersion of molybdenum.

Possible surface structures for specimens of MoO_3/SiO_2 prepared by impregnation and anchoring of a pi-allyl complex have been proposed by analyzing XPS and ISS data [561]. In the first specimen, $R_{exper} > R_{theor}$, which is explained by segregation of the MoO_3 on the outer surface. In an anchored specimen, $R_{exper} < R_{theor}$, which, as for an "anchored" nickel specimen, is apparently associated with the localization of Mo in hidden sites, and not with the aggregation of the MoO_3 into large crystals. In addition to crystalline MoO_3, the formation of a surface molybdate and silicomolybdic acid is also postulated. The acid decomposes in calcination and is restored in the course of hydration [561]. As in the studies conducted by Horrell and Cocke [132], hydration screens the ISS signal from Mo [561]. The properties of MoO_3 on aluminosilicate depend on the SiO_2/Al_2O_3 ratio [562]. Unlike SiO_2, on which molybdenum oxide is present in the form of large crystals (15% of MoO_3), $Al_2(MoO_4)_3$ forms in specimens on aluminosilicate, especially at high MoO_3 concentrations and low or moderate contents of Al_2O_3. Practically monomolecular distribution of molybdenum ($R_{exper} \approx R_{theor}$) is observed in specimens with the concentration of MoO_3 under 15% and of Al_2O_3 over 30%. At lower Al_2O_3 concentrations, microcrystals (3-6 nm) of MoO_3 form on a surface.

Other supports are used in addition to SiO_2, Al_2O_3, and aluminosilicate for the preparation of highly dispersed molybdenum catalysts of metathesis, alkene disproportionation, etc. They include TiO_2, ZrO_2, MgO, and SnO_2 [563, 564]. When molybdenum was supported on ZnO, TiO_2, SiO_2, aluminosilicate, and Al_2O_3 [563], the binding energy of Mo $3d_{5/2}$ did not virtually change, although in this group of supports the FWHM of the Mo $3d$ lines more than doubled. This is evidently associated not only with the building up of a nonuniform charge on the surface, which may enhance when going over from semiconductors to insulators, but also with the energetic nonequivalence of the Mo species, which is maximal for Al_2O_3. When MoO_3/SnO_2 is treated in H_2, the SnO_2 is reduced and the Mo ions migrate into the bulk [565]. As a result, the major part of the Mo in the system MoO_xSn_{2-y} becomes encapsulated and thus becomes inactive in the disproportionation of propylene. The dispersion of the MoO_3 is substantially affected by preliminary modification of the support, for example, by Na or K ions [566]. At sodium concentrations up to 4 atom/nm^2, it is distributed over the surface of the support in a monolayer, while with a growth in the concentration or in T_{cal}, a part of the sodium migrates to the surface. Hence, a change in the sodium (or potassium) concentration appreciably modifies the dispersion of MoO_3.

The dependence of the ratio $I(W\ 4f)/I(Al\ 2p)$ on the tungsten content and its comparison with the theoretical ratio confirms the monolayer distribution of the supported component in calcined catalysts (Fig. 64a) [204]. A high dispersion is retained under mild reduction conditions, but it diminishes with elevation of the reduction temperature and the formation of a metal phase. Calcination of WO_3/Al_2O_3 at high temperatures (900-1100 K) causes dramatic changes in W/Al (ISS) (Fig. 64b)—a sharp maximum is observed at 1000 °C [567]. Although the binding energies of W $4f_{7/2}$ for WO_3, $Al_2(WO_4)_3$ and specimens of WO_3/Al_2O_3 are quite close, there are no doubts that at submonolayer coverages, the WO_3 is distributed in a practically monomolecular layer [204]. By analogy with MoO_3/Al_2O_3, one can expect the formation of a surface tungstate, or of ˙ polytungstate when the concentration of the

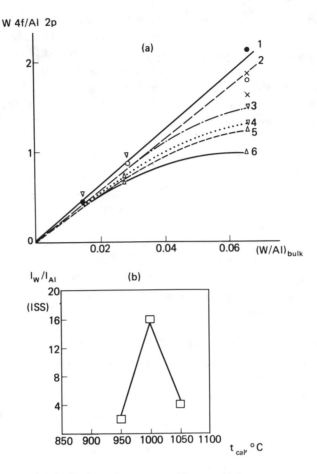

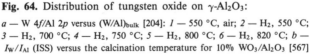

Fig. 64. Distribution of tungsten oxide on γ-Al$_2$O$_3$:

a — W 4f/Al 2p versus (W/Al)$_{bulk}$ [204]: 1 — 550 °C, air; 2 — H$_2$, 550 °C; 3 — H$_2$, 700 °C; 4 — H$_2$, 750 °C; 5 — H$_2$, 800 °C; 6 — H$_2$, 820 °C; b — I_W/I_{Al} (ISS) versus the calcination temperature for 10% WO$_3$/Al$_2$O$_3$ [567]

deposited phase is increased. The localization of W^{6+} ions in tetrahedral coordination in a specimen of 6% of WO$_3$/Al$_2$O$_3$ is confirmed by XPS and ISS data [452]. The structural changes occurring when the WO$_3$ concentration is increased can affect the subsequent behavior of the metal-oxide specimens in calcination and reduction. The XPS data confirm the extraordinary stability of W^{6+} on γ-Al$_2$O$_3$ [204]. At low concentrations, reduction is observed only at a temperature above 973 K, and a single phase forms, namely, W^0. The formation of W(IV) at high WO_3 concentrations is apparently due to inhomogeneity of the WO_3 layer even at submonolayer coverages [204]. The change in W/Al (ISS) was interpreted as a result of shrinking of the monolayer at temperatures close to the Tamman temperature [567], and of WO_3 segregation (a growth in the intensity) with the subsequent formation of $Al_2(WO_4)_3$ (a drop in the intensity) (Fig. 64b).

4.4.2 Ternary Ni-Mo, Co-Mo, and Ni-W Systems

The introduction of a second component (Ni or Co) substantially modifies the properties of Mo and W. The distribution of the components in Ni-Mo/Al_2O_3 catalysts depends on the sequence of impregnation [568] and the calcination temperature [569]. The depth profile obtained by ISS shows that at high temperatures (870 °C), nickel is present in the surface layer of the molybdate. The more detailed ISS studies [570, 571], the formation of a mixed Ni-Mo phase of the nickel isopolymolybdate type that decomposes at 500 °C was postulated. Beginning with work performed by Delannay *et al.* [572], the formation of a bimolecular Mo/Co layer in Co-Mo/Al_2O_3 catalysts has been confirmed repeatedly by ISS [132] and XPS [22, 23]. A comparison of ISS data for $CoMoO_4$ and Co-Mo/Al_2O_3 led Chin and Hercules [215] to the conclusion that cobalt molybdate forms only at cobalt concentration exceeding 7%, but subsequently this conclusion was extended to low Co concentrations [132]. At high temperatures, the bimolecular Mo/Co layer breaks up with lowering of the dispersion of Mo and migration of Co to the surface. Properties close to those described above are also exhibited by the system Ni-W/Al_2O_3 in which the nickel is also contained in the layer of the surface tungstate. Unlike cobalt, however, part of the nickel is present on the surface in the form of a well dispersed phase [573].

A procedure for obtaining concentration profiles in depth in Co-Mo/Al_2O_3 catalysts by ISS and XPS data has been proposed by Brinen *et al.* [296]. The analysis results show the components to be distributed nonuniformly over the layer. When Co and Mo are incorporated into γ-Al_2O_3, the Co/Al ratio on the surface is much higher than when these components are incorporated into boehmite. This result reveals that the preparation procedure can provide a definite depth profile of the components.

Since preliminarily reduced and sulfidized catalysts are active in a hydrodesulfurizing reaction, much attention has been given to studying how the genesis of catalysts and the conditions of their activation (including the reaction medium) affect the formation of the active centers [22, 23, 26, 132].

The correlations between the surface and catalytic properties of such systems will be discussed in Chap. 5. Here we shall only note some general conclusions:

(a) a combination of reduction and sulfidizing leads to the formation of MoS_2, and also of coordinationally unsaturated centers, which are assumed to have the following structure [23]:

(□ is an anion vacancy)

Molybdenum in lower oxidation states (III, 0) has not been detected in sulfidized specimens, although the close values of the binding energy for MoS_2 and Mo^0 do not completely exclude its presence. By the classical scheme, cobalt is partly reduced to the metal and is also sulfidized, possibly, with the formation of Co_9S_8 [23]. ISS data show the structure of sulfidized ACMC to be a monolayer of MoS_2 covered by the Co_9S_8. Moreover, unreduced Co in the form of an aluminate probably remains in the surface layer. The aluminate concentration can be varied by modifying the Al_2O_3 with alkaline (Na) or acidic (B) additives [22, 23]. One of the latest works performed by a set of techniques including XPS has studied the formation of active centers in sulfidized MoO_3/Al_2O_3 [574]. Direct sulfidizing forms particles of the type of MoS_2, whereas reduction with subsequent sulfidizing forms molybdenum oxysulfides or highly dispersed amorphous sulfides. The incomplete sulfidizing of functioning ACMC is indicated by Brown and Ternan [575], who employed the sulfidizing procedure *in situ*. According to these data, only half of the Mo is converted to MoS_2, while the remaining ions react with Al_2O_3 to form a molybdate or the complex $Al_2(MoO_2S_2)_3$. These data confirm the assumption that the active centers of hydrodesulfurizing include not only MoS_2 crystals, but also highly dispersed Mo-S-containing complexes [22, 23].

4.4.3   Chromium-Containing Catalysts

Notwithstanding the numerous studies of these systems in the 1960's and 1970's [576, 577], a number of matters associated with the chemical state of Cr in aluminochromium catalysts (ACC) have remained unclear. A detailed analysis of these catalysts containing 2.3, 7.8, and 20% of Cr_2O_3 by XPS [455, 578] revealed a number of new features in the behavior of these systems. In calcined catalysts with a low Cr content, like Mo and W, the distribution of the Cr is close to a monolayer one; with a high Cr content,

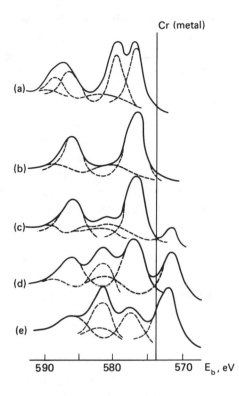

Fig. 65. Change in the XPS Cr $2p$ spectra with various treatments of aluminachromia catalysts:

a — 20% of Cr_2O_3, air, 550 °C; b — 20% of Cr_2O_3, air, 550 °C + H_2, 500 °C; c — b + 650 °C, H_2; d — 2.3% of Cr_2O_3, 550 °C, H_2; e — 1.8% of Cr_2O_3-0.62% of Pt, 500 °C, H_2 + 500 °C (n-hexane/H_2) + 500 °C, H_2

microcrystals of Cr_2O_3 (2.0 nm) form. The chromium in these specimens is in two states: Cr(VI) and Cr(III). The spectrum is described sufficiently accurately by these peaks, which excludes the presence of intermediate states— Cr(V) and Cr(IV) (Fig. 65). Brief treatment in H_2 is attended by the complete reduction of the chromium to Cr(III), an interesting feature being retained that was first observed by Okamoto et al. [579], namely, the spin-orbital splitting of Cr $2p_{3/2-1/2}$ diminishes with the dilution of Cr(III). No Cr(V) ions, which a Cr $2p_{3/2}$ binding energy of 579.4 eV was ascribed to [579], were observed. Reduction in H_2 at 550-650 °C caused the appearance of a new doublet with very low binding energies of 572.0-572.8 eV. Its parameters differ sharply from those of the Cr(II) spectrum obtained for Cr/SiO$_2$ [580]. The given signal was ascribed to Cr^0, although in comparison with the spectrum of the unsupported metal it has a lower E_b and larger $\Delta E_{s.o.}$. That such ascribing is correct is indicated by the observed dependence of the signal intensity on the Cr content and reduction temperature, and by its sharp increase in the presence of Pt (Fig. 65). Without discussing here the possible mechanism of the stabilization of Cr^0 on Al$_2$O$_3$ (see the following section), we shall note that its concentration is maximal for specimens with a low Cr content, i.e. whose surface accommodates highly dispersed clusters of Cr^{3+} instead of the α-Cr$_2$O$_3$ phase.

Unlike γ-Al$_2$O$_3$, considerable amounts of Cr(II) form on SiO$_2$ [580] or TiO$_2$ [581]. In the latter case, EPR and XPS data suggest that stepwise reduction of the Cr(VI) and Cr(V) ions to Cr(III) and Cr(II) occurs in an atmosphere of CO. The transition

$$Cr^{2+} (CO, \ 623 \ K) \rightleftarrows Cr^{3+} (H_2O, \ 873 \ K)$$

is reversible, i.e. the Cr is retained in a highly dispersed state (the β phase by EPR data) [581].

4.4.4   Supported Rhenium Catalysts

The state of rhenium in bimetal Pt-Re reforming catalysts is the subject of an extensive discussion [582, 583]. In this connection, and also because of the need to elucidate the forms of Re active in reactions of alkene metathesis, it is interesting to study the trends in the change in the chemical state of rhenium and its interaction with a support depending on various factors.

XPS was employed to study a series of rhenium-containing catalysts differing in the nature of the support and in the initial compound [205, 316, 450, 454, 584-586]. Figure 66 depicts the change in the experimental and theoretical values of the intensities for Re supported on γ-Al$_2$O$_3$ and θ-Al$_2$O$_3$. In the first case, at a rhenium concentration up to 5%, a distribution of Re^{7+} close to a monolayer one is observed on the surface, while a further growth in the concentration is attended by the formation of microcrystals of Re$_2$O$_7$ or ReO$_4^-$. These data are consistent with those obtained by the authors of the monolayer model [185], where up to Re/Al$_{bulk}$ = 0.02 the experimental and theoretical values of the intensities practically coincide. On θ-Al$_2$O$_3$, the value of $R_{theor} < R_{exper}$, which is probably due to the surface segregation of the rhenium oxide.

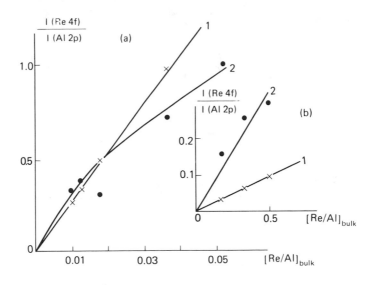

Fig. 66. Experimental and theoretical (simulated monolayer [185]) intensity ratios versus the component ratios for calcined aluminorhenium catalysts [205, 316]:

a — Re/γ-Al$_2$O$_3$: 1 — calculations; 2 — experiment; b — Re/γ-Al$_2$O$_3$: 1 — calculations; 2 — experiment

In specimens with 3.5-5% of Re/SiO$_2$, the values of R_{exper} and R_{theor} are approximately equal except for KReO$_4$, for which XRD revealed microcrystals 20-40 nm in size [450]. As for Mo or W, the question arises as to the nature of the surface complex of Re with Al$_2$O$_3$. The chemical shift in the Re $4f_{7/2}$ spectrum for NH$_4$ReO$_4$/Al$_2$O$_3$ in comparison with NH$_4$ReO$_4$/SiO$_2$ and the closeness of E_b to that observed for an individual mesoperrhenate suggest that already in the drying stage a mesoperrhenate structure of the AlReO$_5$ type forms on the surface. The most reliable evidence in favor of such a complex has been obtained by electron diffusion reflection spectroscopy [584]. At the same time, the quite ready reduction of Re^{7+} to Re^{6+} in calculation may also point to the presence of a free perrhenate that decomposes with the formation of ReO$_2$ and ReO$_3$. The reaction of Re^{7+} with F centers discovered in the structure of such catalysts is an alternative mechanism of Re^{6+} formation [584, 585]. At a higher calcination temperature of 500 °C, the entire rhenium is present in the form of Re^{7+}, whereas vacuum treatment transforms it into an intermediate state characterized by $E_b \approx 43$ eV. By the data of Table 13 (p. 178), this state is most likely Re^{4+}. The transition Re$^{7+} \rightleftarrows$ Re^{4+} is reversi-

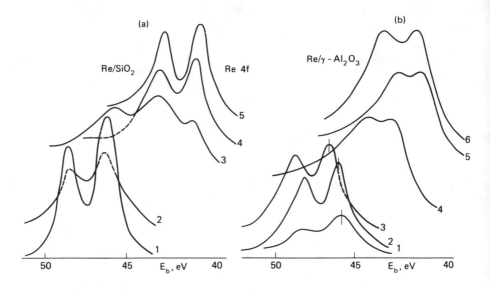

Fig. 67. Reduction of Re in supported catalysts [205, 450]:

a — Re/SiO₂: *1* — NH₄ReO₄; *2* — initial NH₄ReO₄/SiO₂; *3* — 200 °C, H₂, 10 torr; *4* — 500 °C, H₂, 10 torr; *5* — Re (metal; *b* — Re/γ-Al₂O₃: *1* — initial NH₄ReO₄/γ-Al₂O₃; *2* — Ba(ReO₄)₂; *3* — Ba₃(ReO₅)₂; *4* — 500 °C, H₂, 10 torr; *5* — 650 °C, H₂, 10 torr; *6* — 500 °C, H₂, 1 atm

ble: subsequent treatment in air returns the Re to the Re^{7+} state. The absence of noticeable changes in Re/Al in such treatments does not enable one to conclude (as for Mo or W) that Re ions migrate into the γ-Al₂O₃ lattice.

The treatment of rhenium catalysts in vacuum and H₂ leads to the formation of Re^0 (Fig. 67). Its appearance is noted already at 200 °C on both supports, but complete reduction on γ-Al₂O₃ is apparently hindered because of the strong reaction of intermediate Re forms with Al₂O₃. A similar conclusion has been made by Cimino *et al.* [455, 456], who have studied how the concentration of Re affects its reaction with TiO₂ and reduction. At low concentrations (up to 2.5% of Re), NH₄ReO₄ is dispersed quite well on the surface of the rutile and forms a perrhenate-like complex that is reduced in H₂ to Re^0. In the interval of 2.5-10% of Re, in addition to the two-dimensional phase, microcrystals of NH₄ReO₄ are present that are mainly reduced to ReO₂. The reaction of Re^{4+} with the support prevents the further reduction of Re. Finally, at Re concentrations of 10-14%, the phase NH₄ReO₄ is present on the surface. In this case, Re^0 forms according to the thermodynamics of reduction of the individual compound [455].

4.4.5   Copper-Containing Catalysts

The methodology of studying these commercially important systems is similar to that considered for hydrodesulfurizing catalysts. XPS and XAES are employed to study problems such as:

(a) the formation of the surface structure of simulating binary systems such as CuO-ZnO, CuO-Cr$_2$O$_3$, and CuO-Al$_2$O$_3$;

(b) the formation of the surface structure of real catalysts depending on their composition, the conditions of preparation in the individual steps preceding catalysis; and

(c) the evolution of catalysts caused by the reaction medium.

Such an approach makes it possible to elucidate the role of individual components in the formation of the active surface of catalysts and find the factors determining the surface properties and activity. One of the most controversial questions is that of the active forms of copper in the methanol synthesis and water-gas shift reactions. An analysis of CuO-ZnO systems from the standpoint of solid state chemistry allowed Klier [587] to advance the hypothesis that the Cu$^+$ ions localized on the Cu^0/ZnO interface and formed by the reduction of Cu^{2+} ions dissolved in ZnO are the active centers. On the face of it, this contradicts experimental data by which a high activity is exhibited by catalysts in which the main fraction of the copper is in the form of CuO aggregates that are reduced to microcrystals of Cu^0 ($\geqslant$100 Å) [587].

Let us analyze the XPS and XAES data obtained for model and real systems from these standpoints. The surface composition of Cu-Zn-Al catalysts depends greatly on how they are prepared [588]. If catalysts are prepared by the precipitation of two components—Cu and Zn—with the subsequent addition of the third one—Al(OH)$_3$, the surface concentration of the Al is very low. Conversely, in catalysts prepared by the coprecipitation of three components, an unusually high Al/Zn ratio was observed [589]. The drop in the satellite peak intensity in the Cu $2p_{3/2}$ spectrum in the presence of Al$_2$O$_3$ is interpreted as the formation of a solid solution of Cu^{2+} in ZnO. Moreover, it is assumed that compounds of the type of Cu-Al form. It should be noted that many of these conclusions have a purely tentative nature; Petrini and Garbassi [588], for instance, did not measure the Cu L_3VV spectrum at all.

Investigation of oxide and skeleton catalysts like Cu-Al, Cu-Zn, and Cu-Zn-Al shows that notwithstanding the difference in the preparation method, the chemical state of the components is quite close [589, 590]. In comparison with CuO or Cu-Al, the Cu $2p_{3/2}$ and Cu L_3VV spectra of the catalysts Cu-Zn and Cu-Zn-Al have other parameters, namely, a lower binding energy of the

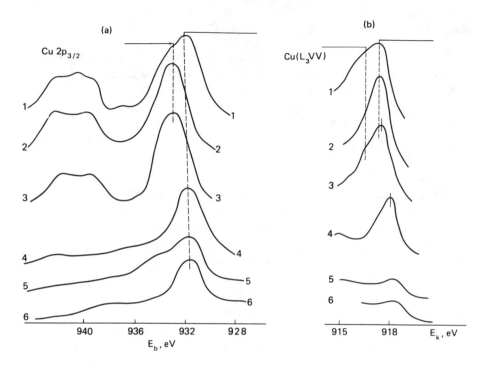

Fig. 68. Cu $2p_{3/2}$ (*a*) and Cu L_3VV (*b*) spectra of initial Cu-Al, Cu-Zn, and Cu-Al-Zn skeleton catalysts [590]:

1 — 57% CuO-5% ZnO-38% Al$_2$O$_3$; *2* — CuO; *3* — 31% CuO-24% ZnO-45% Al$_2$O$_3$; *4* — 41% CuO-33% ZnO-26% Al$_2$O$_3$; *5* — 34% CuO-46% ZnO-20% Al$_2$O$_3$; *6* — 29% CuO-71% ZnO

photoelectron peak, a lower satellite/peak ratio, and a higher kinetic energy of the Auger electron peak (Fig. 68). This is explained by the formation of more dispersed mixed clusters such as $Cu^{(2-\delta)+}\ldots Zn^{\delta+}\ldots O^{2-}$ in which the effective positive charge of the copper ions is lower than in the oxide. The surface composition of ternary systems prepared from Cu-Zn-Al alloys changes in a complicated way (Fig. 69): initially, with small changes in the copper concentration in the bulk, its concentration grows sharply on the surface, and then with appreciable changes in the bulk composition, the surface Cu/Zn ratio does not virtually change. In binary oxide systems CuO-ZnO, the surface is enriched in the dissolved component—Cu; when the copper concentration grows to at least 30%, the surface Cu/Zn ratio becomes lower than in the bulk. The observed changes in the intensities (Fig. 69) are associated both with a change in the composition due to the migration of Cu^{2+} into

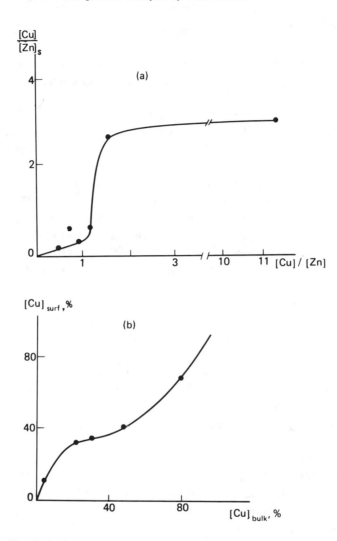

Fig. 69. Surface composition of skeleton (*a*) and oxide (*b*) Cu-Zn-Al catalysts [590]

ZnO or the surface segregation of CuO, and also to the change in the dispersion. Additional methods are evidently needed for a more detailed elucidation of the surface structure of these multiphase systems.

The chemical state of copper in catalysts differing in their composition and method of preparation has been studied in the greatest detail. Okamoto *et al.* [208, 593] have found a correlation between the concentration of Cu^+ formed in the reduction of binary CuO-ZnO systems in H_2 at 180 °C, the CuO content, and the catalytic activity in the water-gas shift reaction and the decomposition of methanol. The maximal Cu^+ concentration was observed at 30-40 mass % of CuO. At relatively low CuO concentrations (below 30%), the Cu^+ becomes stabilized even at high temperatures (250 °C) because of the formation of two-dimensional Cu^0-Cu^+ structures or similar mixed clusters dissolved in ZnO. The Cu^+ ions generated in the systems CuO-Cr_2O_3 [210] or CuO-ZnO-Cr_2O_3 [594] have the highest stability. For example, Cu^+ ions form readily in the vacuum treatment of CuO-Cr_2O_3 films [594], and they are not reduced in subsequent treatment in H_2 even at 270 °C. Such a high Cu^+ stability is associated with the formation of copper chromite. Chen *et al.* [591] arrived at a similar conclusion on the increase in the Cu^+ concentration in CuO-ZnO-Al_2O_3 systems when Me_2O_3 is added (Me = Cr, Y, Sc, etc.). The stabilization of the Cu^+ is due to the better compensation of the surplus valence with the aid of the Me_2O_3 whose cation radius is closer to that of Zn^{2+} than of Al^{3+}.

The different stability of Cu^+ in systems varying in their composition (CuO-ZnO, CuO-ZnO-Al_2O_3, CuO-Cr_2O_3, CuO-ZnO-Cr_2O_3-Al_2O_3, and others) and in the way they are prepared (coprecipitation, leaching of alloys) can be partly explained by the contradictory conclusions on the nature of the active centers (Cu^+ or Cu^0) that followed from the analysis of XPS and XAES spectra. Monnier *et al.* [594] and Okamoto *et al.* [208, 593] indicate that in both the synthesis of methanol and in the water-gas shift reaction, the Cu^+ ions are active; here the CO_2 does not affect the Cu^+ content in the reaction of methanol synthesis. In contrast to this, the Cu^+ ions formed in the stage of the reduction of CuO-ZnO-Al_2O_3 in H_2 or CO (up to 250 °C) after a reaction with the synthesis gas (73% H_2, 25% CO, 2% CO_2) at 250 °C completely transform into Cu^0 (the zinc remains in the form of Zn^{2+}) [209, 592]. The conclusion on the prevalent formation of Cu^0 in methanol synthesis is supported by data on the titration of the copper surface *in situ* using N_2O; a considerable amount of surface oxygen bound to Cu^{2+} was observed [595]. The concentration of the Cu^{2+} ions depended on the CO_2/CO ratio. Only the form, namely Cu^0, has been detected in industrial methanol synthesis catalysts in the stage of their reduction in H_2 by XPS [596] and EXAFS [597] data.

Depending on the catalyst composition (Cu-Al or Cu-Zn-Al), the formed Cu^0 (H_2, 250 °C, 1 atm) has a different reactivity with respect to CO, H_2O, and O_2. In the presence of zinc, the highly dispersed Cu^0 particles become

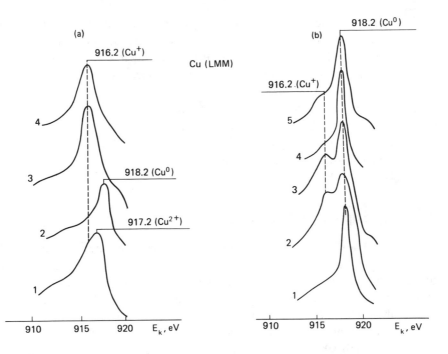

Fig. 70. Influence of different treatments on the state of the copper in Cu-Al-Zn (a) and Cu-Al (b) catalysts [589, 590]:

a — Cu-Al-Zn: 1 — H$_2$, 20 torr, 523 K; 2 — H$_2$, 1 atm, 523 K; 3 — 2 + H$_2$O (steam), 493 K; 4 — 2 + (H$_2$ + CO + H$_2$O), 493 K; b — Cu-Al: 1 — H$_2$, 1 atm, 523 K; 2 — 1 + H$_2$O (steam), 473 K; 3 — 1 + air, 298 K; 4 — 1 + (N$_2$ + CO + H$_2$O), 493 K; 5 — 4 + air, 298 K

stabilized; under the conditions of the water-gas shift reaction (CO + steam, 220 °C), these particles transform into Cu$^+$ (Fig. 70). Reaction of Cu^0 with H$_2$O also leads to this form of copper. Unlike this, skeleton copper (Cu-Al) has a lower dispersion and in the course of a reaction remains in the form of Cu^0. These data reveal that not only the presence of Cu^0, but also the reversibility of the transition Cu$^0 \rightleftarrows$ Cu$^+$ is important for catalysis. Moreover, an optimal electron density of the Cu^0 and Cu$^+$ is apparently achieved with a definite chemical environment and dispersion [208, 209]. The electron density depends on the genesis of the initial systems and the composition of the reaction mixture. Research in this direction will most likely give birth to a more universal scheme for describing the behavior of copper-containing catalysts in both reactions.

The surface structure of catalysts employed in the partial oxidation of hydrocarbons has been studied by a set of physicochemical methods including EXAFS and XPS [27, 598]. Since coordination of the ions is the most important characteristic determining the activity and selectivity of these systems, XPS is used as a supplementary technique.

Examples of such investigations are studies of the V_2O_5/SiO_2 catalyst of toluene oxidation to benzaldehyde [599] and of the MoO_3/SbO_4 catalyst of isobutene oxidation [600]. In both cases, XPS was employed to appraise the surface concentration and distribution of the applied phase.

4.5   Supported Metals

Many real catalysts contain highly dispersed particles of metals in low concentrations distributed on the surface of supports. The preparation of these systems includes a number of steps such as impregnation ion exchange, anchoring, decomposition of the initial compounds, and reduction, as a result of which the state of the metals, their dispersion and surface concentrations may change substantially. To find the laws of these changes and, consequently, the possibility of governing the properties of the supported metals, a number of fundamental problems have to be elucidated:

(a) the electronic and geometric structure of the deposited metal particles;

(b) the nature of the metal-support interaction;

(c) the influence of the nature of the support and the preparation and activation conditions on the electronic state and dispersion of metals; and

(d) the relation between the state and dispersion of metals and their catalytic properties.

In supported catalysts, a substantial part of the metals is on the surface, which is a prerequisite for studying them by surface sensitive techniques. In combination with structural techniques such as EXAFS [24, 601], RED [602], high resolution electron microscopy [603], and quantum-chemical calculations [604], the surface sensitive techniques can supply sufficiently complete information on the structure and composition of the surface of metal-supported systems. The present section is aimed at treating the modern notions on these problems formed on the basis of the data of EES, IS, and a number of other techniques. In Sec. 3.2, we analyzed the possibilities of XPS and AES for studying the electron structure of small metallic particles. There is a number of theoretical and experimental problems connected with the interpretation of spectra, and the changes in the electron structure can be assessed only qualitatively. At the same time, the variation of one of the parameters, e.g. the dispersion or support, enables one to arrive at more definite conclusions on the contribution of various factors to the EES spectra, including the metal-support interaction. As regards the chemical state and the distribution of metals on supports, they have been studied by many authors, a number of whose results was given in Sec. 1.3.

The main attention in this section will be given to the following matters:

(a) the chemical state of transition metals, their distribution and dispersion in reduced catalysts;

(b) the electron state of highly dispersed metal particles and the nature of metal-support interaction; and

(c) the effect of strong metal-support interaction for model and real systems.

The changes in the electron state of metals under the action of the reaction medium and correlations of the electron state with catalysis are discussed in Chap. 5.

4.5.1   Formation of a Metal Phase in Supported Catalysts

The state of the components and the distribution of the deposited metal precursor compounds depend on many factors: the nature and structure of the support, the concentration of the supported phase, the way of preparation and of the activation preceding reduction. As a result, centers that are not equivalent in their coordination and charge state are present on the surface and in the bulk. The entire set of centers can be divided conditionally into three types:

(1) atomically or molecularly dispersed centers localized on a surface, at vacant lattice sites, or in cavities and channels of porous crystals (zeolite) and exhibiting the greatest interaction with supports (I);

(2) microcrystals of the deposited phase that differ only slightly in their properties from individual compounds (II); and

(3) centers of type III intermediate as regards their dispersion and structure between the centers of types I and II.

The type I and III centers are typical of catalysts with a low metal content on γ-Al_2O_3, in zeolites, and for "grafted" catalysts in SiO_2. The type II centers prevail among impregnated specimens on SiO_2. The features of interaction of the metal predecessors with the supports and the distribution of the predecessors will affect the reduction and dispersion of the metal phase. Such results have been presented in Sec. 4.4 for reduced Re, Cu, Cr, Mo, and W on supports. This tells substantially on the completeness of reduction and dispersion of Group VIII metals. The differences in the forms of cobalt on γ-Al_2O_3 and SiO_2, also treated in Sec. 4.4, affect the formation of metal Co [214, 544]. At 350 °C (H_2, 5 torr), the degrees of Co reduction in specimens containing 6% of Co on γ-Al_2O_3 and SiO_2 do not virtually differ, which is explained by the initial reduction of large Co_3O_4 aggregates. But with elevation of the reduction temperatures T_r (in a stream of hydrogen), the degree of Co reduction of SiO_2 is much higher [552]:

Specimen	6% Co/Al_2O_3	6% Co/SiO_2	6% Co/Al_2O_3	6% Co/SiO_2
T_r, °C	350	350	450	450
α Co^0, %	24	21	28	54

The poorer reduction on Al_2O_3 is associated with the stronger interaction of Co^{2+} with the support leading to the formation of cobalt aluminate that lends itself with difficulty to reduction. Castner and Santilli [605] arrived at a similar conclusion. They employed XPS data to establish a series of reduction of Co(5%) at 480 °C, namely, SiO_2(100%) > TiO_2(85%) > K-Al_2O_3(50%) > Al_2O_3(20%). A close series has been obtained for nickel [606], namely, TiO_2 > SiO_2 > aluminasilicate > Al_2O_3 > MgO. On the same support, the laws of metal phase formation and its dispersion may depend greatly on the method of preparation [555-557]. This was observed for Ni/SiO_2 specimens prepared by impregnation and anchoring with $NiCl_2$ [556, 557]. The degrees of reduction of Ni were determined by two techniques: XPS and by measuring the magnetic susceptibility. Reduction in anchored specimens begins at lower temperatures, but complete reduction is not achieved at 450 °C (Fig. 71). Conversely, only metallic nickel is observed in an impregnated specimen under these conditions. The lower degree of reduction determined by XPS in comparison with magnetic granulometry at low temperatures is explained by the predominating reduction of Ni^{2+} localized in the near-surface areas of the structure—micropores or microcracks.

For higher temperatures, the data of the two techniques are highly consistent (Fig. 71). The data on the dispersion procured by the magnetic susceptibility technique reveal that the size of the Ni^0 particles in an anchored specimen is smaller (4.8-8.5 nm) than in an impregnated one (6-12 nm). In reduction at 280 °C, a sharp drop in the Ni/Si intensity ratio is observed, which cannot be explained only by the sintering of Ni^0 (Fig. 72). It is obviously associated with the unusually high mobility of the supported phase leading to the migration of Ni^{2+} into the bulk and its subsequent reduction. This effect may be

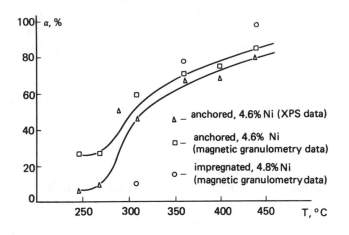

Fig. 71. Degree of nickel reduction (α) in anchored and impregnated catalysts versus the reduction temperature by data of XPS and magnetic granulometry [556, 557]

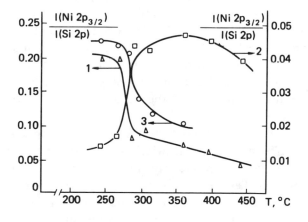

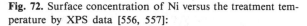

Fig. 72. Surface concentration of Ni versus the treatment temperature by XPS data [556, 557]:

1 — total Ni concentration (reduction); *2* — Ni^0 concentration (reduction); *3* — total Ni concentration (vacuum)

due to the partial breaking of the siloxane bonds of the silica. An analysis of the binding energy of Cl $2p$, Cl/Ni, and Cl/Si suggests that the chlorine after reduction is retained by the support: the Cl/Si ratio depends only slightly on T_r, but E_b decreases by 0.7 eV [556]. It is possible that the evolution of HCl is attended by additional breaking of the siloxane bonds and the formation of Si—Cl. Consequently, modification of the support with Li and Cl facilitates the decomposition of the silica, and at relatively moderate temperatures (440 °C) the migration of Ni into the microcracks apparently occurs together with its encapsulation owing to it being buried by SiO_2 fragments [555, 556].

The influence of the support has been observed in the reduction of other Group VIII metals—Ru and Rh [552, 553, 607, 608]. When Ru or Rh is treated on Al_2O_3 in vacuum and in H_2 at 200-300 °C, these elements are reduced to the intermediate state Me^{+1} similar to that described for the relevant zeolite forms (Sec. 3.3). Similar spectra due to the electron deficiency of the metal clusters were observed for the most dispersed particles of Ru^0 obtained from $Ru(CO)_{12}/Al_2O_3$ in H_2 at 500 °C [197]. A comparative study of two specimens of $Ru(OH)Cl_3/Al_2O_3$ prepared by impregnation and sorption at a definite pH reveals [608] that there are two types of centers on the surface, namely, a surface complex Cl_3-Ru-O (E_b of Ru $3d_{5/2} = 284.5$ eV) and a hydrated oxide $RuO_2(H_2O)_x$ ($E_b = 282$ eV). At 300 °C, mainly the oxide is reduced, whereas the surface complex is reduced only at 500 °C. On TiO_2, the Ru^{3+} ions are reduced quite readily. Already at 200 °C (H_2, 1 atm), the spectrum of Ru $3d_{5/2}$ practically coincides with that of metallic ruthenium. At 300-500 °C, the line becomes still narrower, while the Ru/Ti ratio drops to one-fourth of its former value. The latter is due to sintering of the Ru and its being buried by the reduced titanium oxide [609].

In specimens containing $RhCl_3$ (1-5%) on SiO_2, Al_2O_3, MgO, and TiO_2, the Rh $3d$ spectra differ only in the linewidth [607, 608]. At the same time, Rh^0 forms within various temperature intervals: on TiO_2 at 200 °C, narrow Rh $3d$ lines are observed characterizing Rh^0, on MgO at this temperature no reduction of Rh occurs at all, while at 500 °C about 20% of the Rh forming a compound of the type of $Rh_xMg_yO_z$ remains in the unreduced form. The spectrum of metallic Rh has a positive shift of 0.6 eV relative to the bulk metal. In the studied supports, the ability of Rh to become reduced grows in the following series: MgO < zeolite < Al_2O_3 < SiO_2 < TiO_2. There is no clear relation between the reducibility and the dispersion of rhodium. The latter becomes reduced readily on TiO_2 and by TEM data forms ultradispersed particles 0.5-2.0 nm in size [610]; a close dispersion is observed in a zeolite, in which the reduction of Rh goes on with difficulty [538]. The dispersion of Rh^0 is most likely associated with how Rh^{3+} and Rh^0 react with the supports.

From a practical standpoint, alumina-supported platinum systems containing 0.2-0.5% of Pt are of the greatest interest among catalysts containing Group VIII metals. But their studying by XPS is practically impossible because of the superposition of the most intensive line of the metal—Pt $4f$ and that of the support—Al $2p$. Another line, i.e. that of Pt $4d$, is less informative, and although recently a method has been proposed for determining the electron state of Pt by analyzing the Pt $4d$ spectrum [611], the lowest concentration of Pt in the specimens being analyzed by this method that can be determined is 1-3% [611-613]. In impregnated catalysts on SiO_2, MgO, and Al_2O_3, the major part of the platinum is in the Pt(II) state. Beginning from 100 °C with treatment in hydrogen, platinum is partly reduced to Pt(0) on MgO and SiO_2, while at 400 °C it chiefly passes over into the metal state. In specimens of 3% Pt/Al_2O_3 reduced at 500 °C, a positive shift of Pt $4f_{7/2}$ by 0.5 eV is observed, although the degree of reduction is close to 100% [612]. The dispersion of Pt determined by the "solubility" procedure [613] is 0.8 eV. Such a shift was not observed on SiO_2 and MgO at a Pt particle size of 4 nm.

The change in the intensities in reduction is due to two factors—sintering and redistribution of the particles. In the examples considered above, the data on the dispersion were obtained by independent techniques. As already noted in Sec. 3.3, attempts have been made to find a correlation between the XPS intensities and the dispersion of a metal [182, 284, 287]. Table 19 presents data on the dispersion based on an analysis of the XPS intensities for Pt/SiO_2 specimens obtained by ion exchange and impregnation [27]. The dispersion was also determined by the thermodesorption of hydrogen. A comparison of the experimental and calculated [185] intensities for calcined specimens yields a conclusion on the uniform distribution of Pt in ion-exchange specimens. The drop in the intensity during reduction is due to the change in the dispersion. The particle size found by thermodesorption correlates with the value of $\alpha_1 = c/\lambda$ [Eq. (3.23), Fig. 73] determined from the XPS intensities. When $\lambda(Pt\ 4f) = 1.0$ nm, the particle size found by the two techniques coincides. This relation between the intensity and dispersion can be used for other platinum specimens.

As for the Group VIII metals, the reduction of Re on γ-Al_2O_3 proceeds with greater difficulty than on SiO_2, which was explained by the presence of surface complexes of Re^{7+} and Re^{4+} on the first support [582, 584]. The positive shift of the Re $4f_{7/2}$ spectrum observed for the more dispersed Re^0

Table 19. Experimental and Theoretical [27, 185] Values of the Intensities and the Dispersion of Pt by the Thermodesorption of H_2 and XPS in Pt/SiO_2 Catalysts

Specimen	Treatment conditions	Pt/Si (exper.)[a]	Pt/Si (theor.)		$R = I_{exper}/I_{theor}$		$c(Pt)$[c]		$\alpha = c/\lambda(Pt)$		d, nm H_2
			λ_1[b] = 2.0 nm	λ_2 = 4.8 nm	λ_1 = 2.0 nm	λ_2 = 4.8 nm	$c_1(\lambda_1)$	$c_2(\lambda_2)$	$\alpha_1(\lambda_1)$	$\alpha_2(\lambda_2)$	
1.2% Pt/SiO_2 ion exchange	400°C, O_2	0.068	0.084	0.073	0.8[e]	0.93	0.55	0.5	1.1	1.0	1.02 (1.47)[d]
	400°C, H_2	0.056	0.084	0.073	0.67	0.77	2.5	1.8	5.0	3.6	1.82
	550°C, H_2	0.032	0.084	0.073	0.38	0.44	2.25	1.8	4.5	3.6	2.47
	700°C, H_2	0.034	0.084	0.073	0.4	0.46					
			0.099[e]		0.34[e]			3.0[e]	6.0[e]		
3.4% Pt/SiO_2 ion exchange	400°C, O_2	0.146	0.253	0.217	0.58	0.67	2.5	1.8	5.0	3.6	1.08 (1.43)[d]
	400°C, H_2	0.097	0.253	0.217	0.38	0.45	2.25	1.5	4.5	3.0	1.56
	550°C, H_2	0.101	0.253	0.217	0.40	0.47	2.25	1.5	4.5	3.0	1.81
	700°C, H_2	0.101	0.253	0.217	0.40	0.47					
			0.295[e]		0.34[e]			3.0[e]	6.0[e]		
3.4% Pt/SiO_2 impregnation	400°C, O_2	0.388	0.253	0.217	1.53	1.8	2.0	1.6	4.0	3.2	3.66
	400°C, H_2	0.106	0.253	0.217	0.42	0.49	5.0	4.5	10.0	9.0	3.47
	550°C, H_2	0.051	0.253	0.217	0.20	0.23	7.5	5.5	15.0	11.0	6.02
	700°C, H_2	0.038	0.253	0.217	0.15	0.17					
			0.295[e]		0.13[e]		8.5[e]		17.0[e]		

[a] In the bulk, Pt/Si = 0.0037, 0.011, and 0.011, respectively, S_{BET} = 200-290 m²/g (depending on the reduction temperature).

[b] $\lambda_1(SiO_2)$ = 2.0 nm, $\lambda(Pt)$ = 2.0 nm, $\lambda_2(SiO_2)$ = 4.8 nm, $\lambda(Pt)$ = 2.0 nm.

[c] $\lambda(Pt)$ = 2.0 nm, the particles are cubic.

[d] Repeated measurement.

[e] With account of decrease in S_{BET} to 200 m²/g.

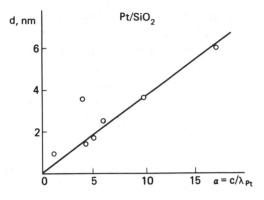

Fig. 73. Correlation between size of Pt particles determined by TDS of H_2 (d) and XPS (c) [27]

phase on Al_2O_3 is another, also quite common effect. The binding energy of Re $4f_{7/2}$ for the bulk metal is 40.6 eV, whereas the synthesis of Re $4f$ peaks for $HReO_4/Al_2O_3$ catalysts reduced in a stream of H_2 at 500 °C during 10 h yields two states: 40.8 eV (22%) and 41.5 eV (78%); for $KReO_4/Al_2O_3$ the relevant figures are 40.2 eV (64%) and 41.8 eV (36%) [614]. These specimens differ greatly in their dispersion: in the former, according to the chemisorption of hydrogen, it is close to 100%, while the average size of the Re^0 crystals by XRD data in the second specimen is 17 nm [614, 615].

As shown by Grünert *et al.* [204, 578] (see Sec. 4.4), Group VIB elements on Al_2O_3 under conditions of rigorous treatment in hydrogen are reduced to a zero-valent state. The tendencies of Cr, Mo, and W to become reduced are different. Cr^{3+} ions at low concentrations form the phase Cr_2O_3 whose reduction is hindered thermodynamically. There are no thermodynamic hindrances for the reduction of free MoO_3 and WO_3, and therefore at higher concentrations the degrees of reduction of these elements on Al_2O_3 grow. Conversely, at low concentrations of these compounds, owing to the strong reaction of Mo^{6+} and especially of W^{6+} with the support, they become reduced with great difficulty. Hence, XPS data show that metal oxide systems prepared in the traditional way, like specially synthesized systems based on carbonyls [564], can transform into supported metal systems. Unlike Group VIII metals in zeolites and on Al_2O_3, Group VIB metals on Al_2O_3 are featured by negative shifts in the spectra that reflect a different type of metal-support interaction (see Subsec. 4.5.5).

The formation and dispersion of Cu^+ and Cu^0 clusters are closely related to the interaction of the deposited components (CuO, ZnO) and their support (Al_2O_3) [208, 209]. The most accurate conclusions on the change in the electron density on copper follow from an analysis of the Auger parameter. Attempts to classify the interaction between copper and n- and p-type semiconductors by XPS and AES data have been made by Chen et al. [616]. In specimens of Cu/Cr_2O_3, the Auger parameter of Cu is larger than for pure copper, whereas for Cu/TiO_2 it is much lower. This is due both to the change in the electron density on Cu^0 and to the presence of a large amount of Cu^+ (50%) in Cu/TiO_2 that lowers E_k of Cu L_3VV by 2 eV. In CuO/MnO_2 specimens, depending on the reduction conditions, Cu^0 or Cu^+ forms and diffuses into the bulk of the support [617]. The various states of the copper give rise to different structures of Cu^0/MnO_x or copper manganite that exhibit quite a different activity in the hydrogenation of hydrocarbons.

An analysis of the cited EES and IS data suggests the trends in the reduction of dispersed metals and their distribution on a support. For specimens of 5% Me/SiO_2 prepared by impregnation, the interaction with the support is weak, and the reducibility in Group VIII grows in the series Co < Ni < Pt < Rh < Ru < Pd. Moreover, the ability of being reduced depends on how a component was deposited. Ni^{2+} ions on modified silica are readily reduced because of the anomalous coordination in the near-surface sites of the support, but when a silicate forms, reduction is sharply hindered. The latter conclusion holds for low-concentration specimens containing Co, Ru, Rh, or Re on γ-Al_2O_3 or MgO. A growth in the concentration is attended by the formation of type II centers, and reduction proceeds more readily. Heat treatment of catalysts prior to reduction may amplify the nonequivalence of the type I and II centers or level it out. For instance, the formation of surface compounds on Al_2O_3 and MgO and the migration of cations into the lattice hinder subsequent reduction, but facilitate the dispersion of the deposited precursor component and subsequently the formation of a more dispersed metal. For a concentration of 2-5%, a series of reduction on γ-Al_2O_3 can be compiled, namely, W < Mo $\leqslant$ Cr < Co < Re < Pt < Rh $\leqslant$ Ru.

The degrees of reduction obtained by XPS and bulk methods are well consistent, which confirms the reliability of quantitative determinations by XPS. It should be mentioned that here XPS can provide data on the intermediate oxidation states and on the influence of anions (especially Cl^-) on the degree of reduction and dispersion. The surface concentration of the residual Cl is higher on Al_2O_3 than on SiO_2 and changes in the series of metals Pd > Rh > Ru.

The data of XPS and other techniques employed in analyzing surfaces have been used to determine the dispersion of the unreduced and reduced phase within the range of sizes from 1 to 3 nm, where other techniques, e.g. XRD, have a low sensitivity. The linear relations $I_{Me}/I_{sup} = R_b$ observed for Re/Al$_2$O$_3$, W/Al$_2$O$_3$, Ni/SiO$_2$, Pt/SiO$_2$, etc. (see Figs. 63, 64, and 73) signify that the dispersion on particle distribution in this range of concentrations does not change, while the slope of the curves shows whether the surface is enriched in or depleted of a relevant component (at $d <$ 1-2 nm). The stronger interaction of metals with Al$_2$O$_3$ causes the prevalent formation of type I and III centers, which at high temperatures migrate to vacant sites of the lattice. This lowers the ratio I_{Me}/I_{sup}. For SiO$_2$ or TiO$_2$, this decrease may also be produced by the migration of support fragments resulting in the encapsulation of the deposited component.

4.5.2 Electron State of Highly Dispersed Particles. Metal-Support Interaction

Before considering electron emission spectroscopy data on this problem, let us briefly deal with the modern theoretical and experimental concepts on the structure of small clusters and the nature of metal-support interaction. The employment of EXAFS, high resolution electron microscopy, and other techniques revealed substantial distinctions in the geometric structure of small metal particles (elongation or contraction of the interatomic distances) and a change in the coordination number (generally lowering thereof). These changes are due to changes in the structure, the geometric shape of the particles, and adjusting of a cluster under the support lattice [599-602, 618]. On inert supports, the length of an Me—Me bond in small clusters is shorter than in the bulk of a metal, while a structure of the icosahedron type is more advantageous thermodynamically than close packing. With stronger metal-support interaction, the shape of the particles changes up to the formation of planar (virtually two-dimensional) structures in which all the atoms can interact with the substrate.

From a theoretical viewpoint, clusters containing several scores of atoms are intermediate as regards their physicochemical properties between isolated atoms and bulk metals [619]. Calculations suggest that characteristics such as the ionization potential and the electron affinity change smoothly when the number of atoms in a cluster grows, whereas others such as the density of states at the Fermi level and the width of the valence band change quite sharply [377, 604, 619, 620]. Calculations by the X$_\alpha$-SW SCF [377, 604].

Hartree-Fock-Slater-Dirac SCF [604], functional density [619], and MO (EHM, CNDO) methods yield different values of n at which the valence band of the bulk metal is achieved. The X_α-SW method predicts a bulk structure for clusters of Ag and Pd with $n \approx 10$; by EHM data for Ag and Pd clusters with $n \approx 80$, the valence band width is lower than in the relevant metals [620]. For all the studied metals, the width of the valence band (photoelectron spectroscopy) corresponding to the bulk was achieved at coverages exceeding 2×10^{15} atom/cm^2, which corresponds to an average cluster size of 1.5 nm ($n \approx 200$) [237, 238]. In other words, the experimental data are not consistent with the data of X_α-SW and are closer to the data of MO methods. Theoretical calculations also indicate a nonuniform distribution of the electron density on the surface and in the bulk of a cluster [377].

One of the most typical changes in the electron structure of a cluster caused by the support is the splitting of the d band of the cluster that increases the valence band width [604]. By EHM calculations [621] for an Ag$_5$ cluster on carbon, the valence band width is 3 eV in comparison with 1 eV for an isolated cluster. The interaction of Pd and Ru atoms with SiO$_2$ is considered as covalent between the d orbitals of a metal and the nonbonding p orbitals of oxygen. The formation of a metal-oxygen bond that splits the nonbonding and antibonding d orbitals is like the appearance of an impurity defect of a metal in the forbidden band of a semiconductor [604].

We can thus conceive cases when the interaction with supports will change the electron density on the atoms of a metal or it will not be attended by charge transfer. Of importance is the geometric arrangement of the metal and support atoms. As we have already noted, O^{2-} ions are often the closest neighbors of clusters with supports such as SiO$_2$ and Al$_2$O$_3$ (EXAFS data). But often irregular (Lewis or Brönsted) centers, reduced cations, etc. may be such neighbors as well.

The experimental studying of the electron structure of dispersed particles occupies an important place in the investigations of supported metal catalysts. Direct information on the electron state of metals and on metal support interaction is provided by electron spectroscopy and X-ray absorption spectroscopy, including EXAFS. In addition to real catalysts, in which the metal is dispersed on a porous support, thin films on the surface of inert substrates [226-246] are widely employed for this purpose, or atoms of the metal being studied are isolated in an alloy [226]. Sputtered on films are convenient objects for photoelectron spectroscopy, although, as will be shown on a later page, they do not transmit all the properties of supported catalysts. At very low coverages, the sputtered on metal is present on a surface in the form of isolated atoms or small clusters. With an increase in the coverage, the spectra experience changes pointing to the transition to a bulk metal.

Parallel studies of palladium as a sputtered film on SiO_2 by electron microscopy and photoelectron spectroscopy show that such spectra are observed for clusters with an average size of 2-3 nm (see Fig. 28, p. 101). At low coverages, there is no peak at the Fermi level, and the d band is much narrower. With a growth in the coverage, the increase in the valence band width is attended by a decrease in the binding energy of Pd $3d_{5/2}$ by 1.6 eV and in the FWHM from 2.5 to 1.6 eV, and by an increase in the intensity of this line. The intensity increases almost linearly with an increasing coverage; the departure from linearity above $\theta \approx 5 \times 10^{15}$ atom/cm^2 is associated with the circumstance that the particle size exceeds the inelastic mean free path of the photoelectrons.

The interpretation of the photoelectron spectra of small particles has been treated in Chap. 3. Although the physical aspect of the spectrum changes have not been elucidated to the end, it is quite clear that in definite cases they may be associated with the rehybridization of the valence orbitals and the transfer of the electron density. The first mechanism is more probable for weakly interacting systems like Au, Ag, Pt, or Pd sputtered onto carbon or SiO_2 [226]. The number of d electrons per atom in such systems is smaller than in the bulk owing to intra-atomic spd rehybridization. It can be seen directly from the spectra that a growth in the cluster size is attended by filling of the d band and an increase in the occupancy of the d orbitals near the Fermi level. If the pd orbitals of the support are not localized in the valence band of the metal, there will be no appreciable transfer of the electron density to the support. When the valence orbitals of the support and metal overlap, they may interact with an increase or decrease in the electron density on the metal [226]. These conclusions are based on data procured for evaporated films. Real supported metal catalysts have a number of features owing to which the degree of metal-support interaction and its contribution to the change in the electron structure of a cluster may be different.

The assumptions on the transfer of the electron density at the metal-support interface were advanced on the basis of studying the catalytic properties of these systems [326, 622], shifts in the vibration band frequencies of CO [623], the position of the absorption edge in X-ray absorption spectroscopy [326, 624], and so on. When metal particles are deposited on semiconductor oxides, the opposite direction of electron density transfer was presumed. Schwab [625] interpreted this as a result of levelling out of the metal and oxide Fermi levels because of the formation of Schottky-type barriers at the metal-semiconductor interface. The correctness of this theory for real supported metal systems, however, gives rise to serious doubts [326, 626] owing to the absence of an ideal interface, the dependence of the work function on the dispersion, and the negligible charge transfer, especially for metal/insulator systems. At the same time, with discovery of the effect of strong metal-support interaction [627], the question has again appeared on the transfer of the electron density between a semiconductor and metal (see 4.5.3).

Hence, theoretical and experimental data point to substantial differences in the electron state of small particles of metals that may be due both to their specific electron structure and to metal-support interaction. The latter most probably has a local nature, at least for metal-insulator systems (γ-Al_2O_3, SiO_2, zeolite). This is confirmed by data of EXAFS [600], and also of XPS [628, 629], which indicate interaction between the metal atoms and oxygen anions. An analysis of the interatomic O KVV Auger transitions and the valence band spectrum [628] reveals that the interaction of Pt with the electron system of the substrate [α-$Al_2O_3(001)$] manifests itself in the appearance of a new peak with $E_b = 10.4$ eV. Its intensity increased with a growth in the Pt concentration attended by a decrease in the intensity of the O $2p$ peak. This peak is ascribed to the formation of a Pt—O bond. The possibility of such bonding is confirmed additionally by the peak of the interatomic O KVV transition with $E_k = 518.5$ eV. The appearance of this transition is due to the interaction of a metal site (more likely Pt than Al) and O. The same interpretation has been proposed for sputtered systems Mo/SiO_2 and Mo/γ-Al_2O_3 in whose spectra peaks with $E_k = 518.4$ and 516.9 eV were observed.

Let us consider from these standpoints the data of EES obtained for real metal supported catalysts. The changes in the spectra of the core levels for highly dispersed metal particles on porous supports are similar to those observed for sputtered on films (compare Figs. 28 and 64; Table 20). Their main features are:

(a) broadening and positive shifts manifesting themselves at 100% degree of reduction of metals [27, 222, 614], their values depending on the support and dispersion (Fig. 74); and

(b) shifting of the peak of the density of states in the valence band, and lowering of photoemission near the Fermi level (Fig. 75).

In Chap. 3, we analyzed the contribution of the initial and final state effects to ΔE_b of small particles, which can be divided approximately by determining the Auger parameter. The value of ΔE_{ER} depends primarily on the size of the particles, therefore ΔE_{chem} can also be appraised by comparing the data for particles close in size on various supports. Inspection of Table 20 reveals that ΔE_{ER} for particles 1-2 nm in size does not generally exceed

Table 20. Shifts in the Inner Levels in Spectra of Deposited Metals

Metal	Support	Particle size, nm	ΔE_b, eV	ΔE_{chem}, eV[a]	% of state in total amount of the metal
Ru	H-NaY	1-2	1.0-1.2	(0.35-0.55)	> 90
	γ-Al$_2$O$_3$	2-3	0.6-0.8	(− 0.05 to 0.15)	
	SiO$_2$	3-5	0.4	(− 0.6)	
Rh	H-ZVM	1-2	0.8-1.0		50-60
	Na-ZVM	—	0.4		
	H-NaY	1-3	0.6		
	MgO	2-4	0.8		
	γ-Al$_2$O$_3$	2-4	0.4-0.6		
	SiO$_2$	2-8	0.2-0.4		
	TiO$_2$	1-3	-0.3	− 0.9	
Pd	DM	2-4	1.0	0.35	90
	H-ZVM	—	0.6		
	γ-Al$_2$O$_3$	2-4	0.4-0.8		
Pt	H-ZVM	0.8-1.4	0.8-1.0	0.2-0.4	60
	Ditto	10	0.4		
	H-NaY	2-6	0.4-0.6	− 0.2	
	γ-Al$_2$O$_3$	1-2	0.6	0	
	SiO$_2$	1-3	0.4-0.6		
	TiO$_2$	1-3	-0.3	− 0.8	
Re	γ-Al$_2$O$_3$	2-3	0.9-1.2		80
	SiO$_2$	10	0.4		
Cr	γ-Al$_2$O$_3$	b	− 2.0-2.5		20-30
Mo	γ-Al$_2$O$_3$	b	− 0.7-1.5		10-20
W	γ-Al$_2$O$_3$	b	− 0.8-1.3		10-20

[a] The figures in parentheses are approximate values of the "chemical part" of the shift evaluated with a view to ΔE_{ER}. The value of ΔE_{ER} for Rh/TiO$_2$ has been determined in [630], for Pt/TiO$_2$, in [231]; for Pd values have been taken from [228] (Pd/sapphire); Pt, from [226] (Pt/C); for Rh the values have been adopted for coatings of sputtered metals that are equivalent to the size of the deposited particles. This approximation has in view that ΔE_{ER} depends chiefly on the particle size and less on the support [226, 228, 231, 250].

[b] X-ray-amorphous.

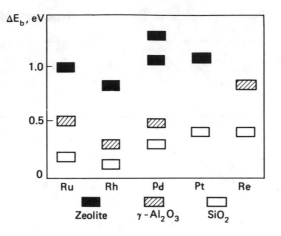

Fig. 74. Shifts in the spectra of the inner levels of small (1-3 nm) metal particles on supports [27, 222, 614]

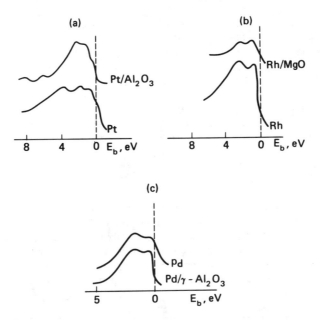

Fig. 75. Density of states and valence band width (differential XPS spectra) for small metal particles [27, 222, 614]:

a — Pt (1 nm)/Al$_2$O$_3$; b — Rh (2 nm)/MgO: c — Pd (2 nm)/Al$_2$O$_3$

0.5-0.7 eV and, consequently, positive shifts of at least 1 eV partly characterize the change in the electron state of metal clusters.

Two basic factors are responsible for positive shifts. The first factor is a size-induced one—lowering of the population of the d orbitals (a growth in the number of d vacancies) as a result of spd rehybridization with empty states above the Fermi edge, which is not attended by the transfer of the charge to the support [234]. Semikolenov *et al.* [631] attempted to find a correlation between the number of atoms (n) in a cluster and the chemical shift. For the cluster $P_2Pd_nX_2$, where $X_2 = (Si—CH_2CH_2PPH_2)$ and Si is the surface of silica gel, the binding energy of Pd $3d_{5/2}$ diminished from 337.3 to 335.6 eV when n increased from one to five, while at $n = 10$ the binding energy coincided with its value for the massive metal. The large width of the photoelectron lines suggested a certain lack of homogeneity of the clusters, although as a whole this result is qualitatively similar to the data obtained for sputtered on films. Also studied was the electron structure of clusters prepared by the decomposition of carbonyls: $Os_3(CO)_{12}$, $Os_6(CO)_{18}$, $Ru_3(CO)_{12}$ and $Ir(CO)_{12}$ on pyrolytic graphite in an electron beam [628]. When the number of atoms in a cluster decreased from six to three, the binding energy of the core levels relative to the massive metals grew by 2.5 eV. The first factor is more pronounced in the range of 1-2 nm, hence at very low coverages on model evaporated films noticeable shifts were observed for all the studied metals (Au, Pd, Pt, Ag) and depended only slightly on the substrate.

The second factor responsible for positive core level shifts is the interaction of a metal with electron-acceptor centers of the support that can result in charge transfer. In the particle size range of 1-3 nm, the shift depends substantially on the support (see Fig. 74). For Rh, for instance, its value grows in the series $SiO_2 < TiO_2 < Al_2O_3 <$ zeolite. For Ru, the shift grows in the series $MgO < SiO_2 < MgO\text{-}SiO_2 < Al_2O_3 < TiO_2\text{-}Al_2O_3 < SiO_2\text{-}Al_2O_3$ [632]. This suggests that the support plays an appreciable role in the formation of electron-deficient particles of the metal. Such a conclusion is consistent for a number of metals (Pt, Pd, Ir) with the above-mentioned data of XAS, EXAFS, RED, and infrared spectroscopy of absorbed CO. The electron deficiency of the small particles is confirmed by the valence band spectra (see Fig. 75) in which a decrease in the density of states near the Fermi level was observed for supports such as Al_2O_3 and MgO. The studied systems are wide-gap semiconductors, and it is hardly probably that the observed effects are associated with collective electron interaction leading to a change in the Fermi levels of the metal or oxide. The interaction is more probably of a local nature, however, and takes place between the most dispersed particles (atoms in the limit) of the metal and certain centers of the support. This is why the degree of electron deficiency (δ^+) and the fraction of the state $Me^{\delta+}$ diminishes in the series zeolite (H form) > Al_2O_3 > SiO_2. Lewis and Brönsted acid centers can be proposed as the electron-acceptor centers. The Lewis centers in zeolites and on γ-Al_2O_3 are trigonal ions of Al and Si, Al extra-lattice species and cations. The appreciable role played by the Brönsted acid centers in the interaction with a metal is indicated by the maximal positive shifts observed for Ru, Pd, and Pt introduced into the H forms of zeolites. But interaction with the transfer of an electron to a single Brönsted center of the Al-OH-Si type is not obvious from an energy viewpoint, primarily because of the high electron-donor nature of the alumino-oxygen tetrahedron having an ionization potential of 7 eV [479]. Quantum-chemical calculations of the cluster Si-OH-Ni [633] does not confirm a noticeable transfer of the charge between the Ni and OH groups. On the other hand, the arguments in favour of the interaction of the metal atoms and Brönsted centers follow from the data of infrared spectroscopy (the decrease in the intensity and change in the frequencies of vibrations of the most acidic OH groups [633]. An electron-acceptor center is possibly a more complicated structure consisting of combinations of Lewis and Brönsted centers. It should be noted that the trend to the diminishing of the ionization potential of AlO_4 occurring simultaneously with the growth in the proton-donor properties of a Brönsted center is observed when the Al-OH clusters are diluted, i.e. with a growth in Si/Al [479]. It is just in high-silica zeolites that the positive shifts in the spectra of metals are most pronounced. Primet and Ben Taarit [623] indicate that the zeolite framework may accept the charge excess from metal atoms.

In early publications [623, 624], where metal-support electron interaction was postulated, it was presumed that this interaction would result in "transmutation" of transition metals, i.e. in their acquiring the electron configuration of their left neighbor in the period (e.g. Pt $\rightarrow$ Ir, Pd $\rightarrow$ Rh, Ni $\rightarrow$ Co, etc.). On the other hand, the interaction of metal atoms with Lewis centers ought to oxidize them to unicharged ions [623]. But both of these assumptions have not been confirmed by available data. Studies of Pd/H-DM by XPS and ESR have shown that the mechanism of the formation of unicharged ions differs from the earlier proposed interaction $Pd^0 + Al^{3+} \rightarrow Pd^+ + Al^{2+}$ [532]. Moreover, clusters of $Pd^{\delta+}$ are present in the zeolite in addition to Pd^{+1}. A comparison of the results of XPS, XAS, and XANES with account taken of the ambiguity in appraising the charge by these techniques shows that the changes in the charge are less than 1 e^-/atom. For Pt/Al_2O_3, EXAFS yields a value of 0.1 e^-. In zeolites with a view to the change in ΔE_{ER}, the magnitude of the charge by XPS data is 0.3-0.5 e^-. A comparison of the valence band spectra of adjacent elements (e.g. small particles of Pt and bulk Ir) indicates the absence of complete analogy of the density of states. At the same time, the electron deficiency of small particles on supports with pronounced acceptor properties substantially affects the catalytic properties of the deposited metal systems in various reactions (see Chap. 5).

4.5.3   Strong Metal-Support Interaction

The nature of the support and the conditions of activation of supported metal catalysts greatly affect the structure of the metals and the catalytic properties. This manifests itself especially vividly for Group VIII metals on semiconductor oxides that can be reduced at moderate and high temperatures (TiO_2, ZnO, Nb_2O_5, ZrO_2, CeO_2, etc.). Tauster *et al.* [627] discovered that when the reduction temperature is raised from 200-300 to 500 °C, the chemisorption of H_2 and CO on metals drops sharply with no appreciable change in their dispersion. This effect, called strong metal-support interaction (SMSI), has the following phenomenological features [627, 634-637]:

(a) the suppression of the adsorption of H_2 and CO in high-temperature reduction (HTR);

(b) a sharp decrease in the catalytic activity in structure-sensitive reactions (hydrogenolysis, structural isomerization), and also in reactions of hydrogenation-dehydrogenation, H_2-D_2 exchange, etc.;

(c) complete or partial restoration of the initial adsorption and catalytic properties on treatment in O_2 with subsequent low-temperature reduction (LTR).

A large cycle of works has been devoted to studying the nature of strong metal-support interaction. Their results were summarized in the proceedings of symposiums [634, 635] and in reviews [636, 637]. All these works can conditionally be divided into two groups. The first one deals with sputtered on films, and unsupported mono- and polycrystalline specimens. These objects can be well characterized by EES and IS, but there is often no answer to the question as to the extent to which their properties are adequate to those of highly dispersed real systems. The second group of works deals with real catalytic systems; they characterize the catalytic and chemisorption properties sufficiently completely, but the data on the structure of the surface of these systems and on the electron state of the metals up to recently were incomplete and contradictory. In recent publications, which will be treated below, the employment of EES, IS, EXAFS, and high resolution electron microscopy also made it possible to describe the mechanism of strong metal-support interaction for highly dispersed supported systems.

An analysis of the experimental and theoretical data enabled us to single out five basic models proposed to explain strong metal-support interaction:

(1) Local (partly ionic) interaction between metal atoms and reduced Ti cations attended by the transfer of electron density from a cation to a metal [634, 638]. We have already mentioned that XPS has provided contradictory information on the charge on metal clusters in the state of strong metal-support interaction [639-643], namely a small negative shift in the spectra [639-641], and the absence of a change in E_b or a positive shift [642, 643] (Table 21). In these investigations, the changes in E_{ER} were not taken into consideration. By EXAFS data for Pt/TiO_2, a small charge transfer from TiO_2 to Pt has been identified [644]; this was not observed for Rh/TiO_2 [643].

(2) The formation of an intermetallic compound between the oxide cations reduced to Me^0 and the deposited metal [627, 645].

Table 21. Chemical Shifts in XPS Spectra for Me/TiO_x Specimens

Specimen	Treatment conditions	$\Delta E_b(Me^0)^a$, eV	ΔE_b(HTR-LTR)^b, eV	ΔE_{ER}^c, eV	ΔE_{chem}^d, eV	References
1% Pt/TiO_2	200°C, H_2	—	—	—	—	[639]
	550°C, H_2	—	− 0.4	—	—	[639]
4% Pt/TiO_2	210°C, H_2	—	—	—	—	[643]
	550°C, H_2	—	0	—	—	[643]
$Pt/SrTiO_3$ (100)	Ar^+, 2 keV + 450°C, vac.	+ 0.3	—	− 0.8	− 0.5	[231]
1% Pt/TiO_2	200°C, H_2	0	—	—	—	[610, 630]
	500°C, H_2	0	0	—	—	[610, 630]
2% Rh/TiO_2	200°C, H_2	—	—	—	—	[639]
	550°C, H_2	—	− 0.6	—	—	[639]
5% Rh/TiO_2	200°C, H_2	0	—	—	—	[610, 630]
	300°C, H_2	− 0.2	—	—	—	[610, 630]
	500°C, H_2	− 0.3	− 0.3	− 0.6	− 0.9	[610, 630]

[a] Shift relative to bulk metal.
[b] Difference in E_b for HTR and LTR.
[c] Found by Eq. (1.29).
[d] Shift due to initial state.

(3) Collective electron interaction at the Me/TiO_2 interface (between the metal conduction electrons and the impurity levels of Ti^{3+}). The formation of surface defects lowers the work function of TiO_2 from 5.5 to 4.6 eV (5.6 eV for Pt), and as a result of levelling out of the Fermi levels at the metal/semiconductor contact, the electrons should pass over from the oxide to the metal [646, 647]. This simulation is consistent qualitatively with the data on the electrical conductivity of Pt, Rh, and Ni on TiO_2 [647], although the latter is high enough already at low-temperature reduction. The shortcomings of the simulations based on collective interaction have been discussed above for real catalysts.

Two other models are geometric ones.

(4) The change in the morphology of the metal and support particles [648]. The particles of Pt or Rh on C or SiO_2 are globular, on Al_2O_3 are hemispherical, while on TiO_2 thin scaly crystals (planar structures of the "pillbox" or "raft" type) are observed. With high-temperature reduction, the TiO_2 transforms into Ti_4O_7 [648]. These data were obtained for sputtered films; when real catalysts were studied by TEM [649], no titanium suboxide was discovered.

(5) The model of encapsulation or decoration of metal particles by islets of the reduced oxide, which at high temperatures can diffuse to the surface [650-654]. Experimental proofs of the burying by TiO_x diffusion onto the metallic surface have been obtained chiefly for model specimens prepared by the sputtering of a metal onto TiO_x or vice versa. This mechanism does not contradict thermodynamics [652]: oxidation of the solid solution Pt-Ti already at low temperatures should result in the diffusion of TiO_x to the surface. But if at 500 K, the oxide diffuses over a distance of 10^{-5} nm during 24 hours, then at 900 K it diffuses over 1 nm. Encapsulation also explains the greater sensitivity of demanding reactions requiring large clusters to strong metal-support interaction.

It should be noted that a number of the cornerstones of these models were purely speculative when they were proposed (1978-1982). The possibility of migration of fragments of a reduced oxide to the surface of the metal was shown the most unambiguously. This circumstance in conjunction with the contradictory information on the electron metal-support interaction resulted in the interpretation of the adsorption and catalytic changes in the state of strong metal-support interaction from purely geometric considerations, i.e. as a consequence of the decrease in the number of surface centers without modification of their structure. This standpoint, however, is too simplified and fails to explain the entire set of changes in the catalytic properties, e.g. the change in the selectivity in the reaction of CO and H_2 [655-657], and the regenerating effect of oxygen at room temperature. This is why we find it very important to analyze the latest publications studying strong metal-support interaction by a set of techniques whose results make it possible to interpret this interaction as a complicated multistep process and also to develop, clarify, or limit the scope of application of the models considered above.

4.5.4   Effect of Strong Metal-Support Interaction for Model Catalytic Systems

The surface of Me/TiO_2 systems in the state of SMSI is simulated in various ways: by sputtering a metal film onto a single TiO_2 crystal or a preliminarily oxidized single Ti crystal with subsequent oxidation-reduction treatment or annealing-Ar^+ cycles, by the oxidation of Me-Ti intermetallics, by the preparation of "inverse" TiO_x/Me systems, etc. The ability of the metals to absorb H_2 or CO is used as the criterion of the manifestation of strong metal-support interaction. Most of these studies were aimed at obtaining direct proofs of the diffusion of Ti to the surface of a metal and determining the state of the Ti at the TiO_x/Me interface. The interaction of these components was studied in only a few cases. Two groups of researchers [646, 650, 653, 658] studied in detail the systems Rh/TiO_2, including those obtained by sputtering Rh onto $TiO_2(110)$, by XPS, AES, UPS, and LEED. For specimens reduced in H_2 or annealed in vacuum, proofs have been obtained of the migration of TiO_x (a small amount of Ti^{3+}) onto Rh [650]—the AES profile for Rh depending on the sputtering time passed through a peak, whereas for Ti and O it passed through a minimum. Such a specimen by UPS data did not absorb CO [658], but after sputtering produced a typical UPS spectrum of molecular CO. The capsulation model is supported by data of the static SIMS technique obtained for the first time for such systems by Belton *et al.* [651, 659]. They sputtered Rh onto a preliminarily oxidized surface of Ti(0001). Brief heating in H_2 (773 K) of the system $Rh/TiO_2/TiO_x/Ti$ lowers the chemisorption of H_2 by 75%; in sputtering, the Ti^+/Rh^+ ratio passes through a minimum coinciding with the increase in the amount of chemisorbed H_2. The possibility of the diffusion of oxidized Ti particles into a metal foil (Rh) has been shown by Ko and Gorte [660]. This diffusion is reversible, and when the specimen is cooled, the Ti emerges onto the surface again. When TiO_x/Rh is heated in oxygen, the oxide is irreversibly desorbed with Rh; the same is observed for the Ti/Pt system. This proves that partly oxidized titanium diffuses through the metal layer.

Annealing of TiO_2/Pt films in vacuum reduces the major part of the Ti^{4+}

to Ti^{3+} and results in an equilibrium coverage of TiO_x close to a monolayer one [661]. The accompanying increase in the binding energy of Pt $4f$ was explained by the formation of a Pt—Ti bond. Additional evidence in favour of the fact that the suppression of CO adsorption requires a coverage of the metal by TiO_x close to a monolayer one, i.e. to simulate complete encapsulation, have been procured by ISS [662]. Gorte et al. [662] controvert the conclusion of Levin et al. [663] according to whom only partial decoration of a metal surface is needed to suppress the chemisorption of CO. ISS has provided evidence of the reversible migration of Ti in addition to AES data [661]—when a flash (1100 K) is used, the titanium diffuses into the bulk of the platinum, whereas after cooling it again emerges to the surface.

The data on the state of Ti in reduced Me/TiO_2 systems are contradictory: Chien et al. [641] did not observe the reduction of Ti^{4+}; Sadeghi and Henrich [650] and Belton et al. [651] discovered the formation of Ti^{3+}, and only Takatani and Chung [664] point to the presence of Ti^{2+}. These differences are partly explained by the insufficient sensitivity of XPS to a surface. A considerable amount of titanium suboxide (TiO) is detected when more surface-sensitive techniques are employed—ISS [665] and ARXPS [666]. In the first case, the specimens were prepared by depositing an aerosol of H_2PtCl_6 onto TiO_2. The O/Ti ratio for the reduced specimens by ISS data was close to unity, whereas XPS mainly revealed Ti^{4+} and Ti^{3+} ions. ISS and XPS showed the reversibility of the encapsulation of TiO_x: in sputtering the oxide is removed, in H_2 (10^{-5} torr) it again migrates to the surface, while with subsequent treatment in O_2 (400 °C) the Ti/Pt ratio diminishes by more than 50%, which points to the "removal" of encapsulation. In catalysts consisting of a quite thick Pt film (its effective thickness is 22 Å) on $TiO_2(100)$ and reduced in H_2 (1 atm, 773 K). Tamura et al. [666] succeeded in separating a signal from Ti ions that had migrated to the metal surface; the latter is simultaneously a screening layer from the "bulk" TiO_2. This is observed especially clearly at low angles of photoemission:

θ	Ti^{2+} (%)	Ti^{3+} (%)	Ti^{4+} (%)
90	13	5	82
40	21	9	70
20	22	16	62

Levin et al. [667] provide additional conclusions on the reduced forms of Ti and their localization. They show that when Ti^{3+} ions are reduced in CO and H_2, these ions are arranged along the perimeter of TiO_x islets, while with a growth in the coverage the fraction of Ti^{3+} decreases. The absence of Ti^{4+}

reduction in TiO_x/Au specimens reveals that the preliminary adsorption of CO and H_2 on a transition metal (Rh) is needed for Ti^{3+} to form. Baker *et al.* [668] indicate substantial changes in the morphology of the oxide and metal layers. At 500 °C in the system Pt/TiO_2-graphite, "hexagonal" islets and dendrites form from the TiO_2 film, while the Pt localized on the TiO_2 changes its structure from compact globules to thin lamellas. Unlike this, the platinum localized on the graphite remains in the globular form at all treatment temperatures. It is interesting that in the region close to the Ti/C interface, the Pt particles acquire mobility at lower temperatures than on the "graphite" areas.

Hence, the surface segregation of TiO_x on a metal component is confirmed by a variety of EES and IS techniques for model specimens prepared in various ways. Such segregation has been predicted theoretically [652] and experimentally [359] in the oxidation of Pt-Ti intermetallics.

The mechanism of this process and the degree of interaction of the components are not so clear. Dwyer *et al.* [669] suggest that the migration of TiO_x to the surface of Pt at high-temperature reduction does not modify the electron state of the Pt, although the rate of CO hydrogenation grows by three orders of magnitude. In contrast to this, on the basis of data for specimens of Pt-TiO_2, Pt-Ti_2O_3, and Pt-TiO obtained by joint sputtering [670, 671], it was concluded that the surface intermetallic Pt_nTiO_x forms, where $n > 1$ and $x \approx 1$. It is obvious that the situation on the surface of model systems (thin films, single crystals) quite often differs from real ones (the coverages, dispersion of the metal, and the nature of the defects).

4.5.5   Effect of Strong Metal-Support Interaction for Real Highly Dispersed Catalysts

Already in the first publication on SMSI [627], the electron interaction between the metal atoms and the reduced oxide cations was postulated as the main cause of the effect. Subsequently, the results of calculations performed for Pt-Ti clusters [638, 672, 673] were cited as proofs of this hypothesis. By the simplest simulation, a Pt atom was incorporated at the site of an anion vacancy into a TiO_6 cluster. A partially ionic bond forms with transfer of the charge ($0.6e$) from Ti to Pt. Within MO limits, the charge passes from the Ti $3d$ impurity level near the bottom of the conduction band to a lower unoccupied MO consisting of a hybridized Pt $6s$-O $2p$ orbital. Further calculations [638, 673] revealed that the suppression of hydrogen chemisorption in such a system is explained by the fact that the only occupied antibonding Pt-O orbital is not suitable as regards its symmetry for interaction with the σ_u orbital of H_2, while there are no other active centers on lamellar metal particles in the state of SMSI. Although the conclusions arrived at by Horsley [638] in his calculations were disputed by Henrich [674] because of the simplification of the structure of the selected cluster, the consideration of more complicated structures did not change the basic conclusions on the nature of the interaction [672, 673]. That Ni/Ti clusters can be reduced more readily and are more stable than Ni/Si ones is indicated by calculations of the total energies and electron structure performed by nonempirical MO methods [675].

Experimental proofs of the electron and chemical nature of SMSI have been obtained when studying Me/TiO_2 catalysts by a combination of techniques including EES and IS, TEM, and isotope exchange [610, 630, 676, 677]. The transition of these systems to the state of strong metal-support interaction at 500 °C (H_2) was indicated by data on the change in the catalytic activity in reactions of ethane hydrogenolysis, benzene hydrogenation, and CO hydrogenation. These changes are reversible: the initial activity was restored after O_2-H_2 treatments at lower temperatures [610, 630].

XPS data (Table 21) show that at 200 °C rhodium and platinum (1-5%) are completely reduced to a zero-valent state. For specimens treated in H_2 at 500 °C, slight negative shifts (0.3 eV) in the Rh $3d$ and Pt $4f$ spectra are observed. Such shifts have been discovered for Ru/TiO_2 [678] and Pd/TiO_2 [679], but it was not clear whether they are associated with an increase in the electron density or with changes in E_{ER}. An appraisal of E_{ER} for specimens of Rh/TiO_2 (Tables 21 and 22) reveals that in the state of SMSI for the specimen BR 500 the value of E_{ER} drops by 0.6 eV, which leads to $\Delta E_{chem} = -0.9$ eV. A close value has been obtained for the specimen AR 500. Similar results for an Rh film on TiO_2 have been interpreted differently [680], namely, as a shift of the Fermi level to the bottom of the conduction band. At the same time, an analysis of the valence band spectra provided additional proofs of the increase in the electron density on clusters of a metal in the state of SMSI (see Fig. 25, p. 97).

A spectrum of a pure oxide is represented by the $2s$ (22.7 eV) and $2p$ (7.5 and 5.3 eV) states of oxygen. In the region above 5.3 eV, the intensity of photoemission is insignificant. When metals (Rh and Pt) are deposited, intensive photoemission of the d states of the metals is observed in this region. For 5% of Rh/TiO_2, bands with a binding energy of 2.3-2.8 and 1.1-1.7 eV can be clearly singled out. A comparison of the spectra obtained for various reduction temperatures shows very well the increase in the intensity in this region with elevation of the temperature. Since the Me/Ti ratio does not grow here, the observed changes can be associated with an increase in the electron density on the metal and with the contribution of the $3d$ states of Ti^{3+}.

Table 22. Electron State of Metals in Catalysts Supported on TiO$_2$ [610, 630]

Specimen	Additional treatment	Binding energy, eV			ΔE, eV		α'^{c}, eV
		Ti $2p_{3/2}$	Rh $3d_{5/2}{}^a$	Pt $4f_{7/2}{}^b$	Ti $2p_{3/2}$– Rh $3d_{5/2}$	Ti $2p_{3/2}$– Pt $4f_{7/2}$	
AR 500^d		459.0	307.0	—	152.0		645.7
AR 500	520°, H$_2$ (10^{-5} torr)	459.2	307.2		152.0		645.7
BR 200^e		459.2	307.4		151.8		644.6
BR 300		459.1	307.2		151.9		645.1
BR 500		459.1^e	307.1		151.8		645.8
CR 200		459.0	—	71.6	—	387.4	
CR 200	He$^+$, 1 keV, 2 h	459.5^e		72.0		387.5	
CR 500		459.0		71.3		387.7	
CR 500	He$^+$, 0.8-2 keV, 3 h	459.5^e		71.8		387.7	

^{a} For metallic Rh—307.4 eV.
^{b} For metallic Pt—71.6 eV.
^{c} $\alpha' = E_k(\text{Rh } 3d_{5/2}) - E_k(\text{Rh MLV})$.
^{d} In the specimen notation, A = 1% Rh/TiO$_2$, R = treatment in H$_2$, B = 5% Rh/TiO$_2$, C = 1% Pt/TiO$_2$, the figure is the temperature in °C.
^{e} 10% of Ti^{3+}.

An analysis of the Ti $2p$ spectra shows that in the specimens reduced at 500 °C, Ti^{3+} ions (5-10%) are present in addition to the Ti^{4+} ones. The spectra of the Ti *LMM* and *LMV* Auger transitions obtained by X-ray or electron excitation (Fig. 76) are still more sensitive to the defects in the TiO$_2$. The asymmetry of the line with $E_k \approx 420$ eV (an intra-atomic L_3VV transition) grows noticeably when the reduction temperature is raised from 300 to 500 °C. Moreover, for catalysts reduced at 500 °C, the shape of the line $L_3M_{23}V_{23}$ ($E_k = 380$ eV) changes. An appraisal of the contribution of the Ti^{3+} d states to the valence band spectrum (the spectrum was simulated by a broad peak in the region up to 3 eV below the Fermi level) suggests that at such concentrations the increase in the intensity near the Fermi level is mainly associated with the deposited metal. This is additionally confirmed by the spectra mea-

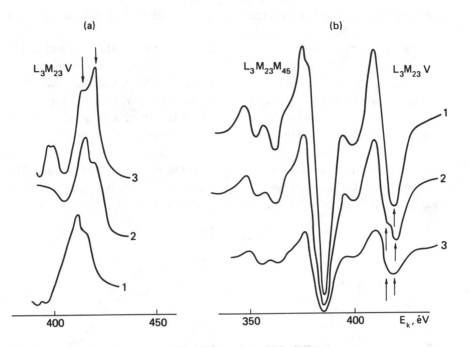

Fig. 76. Auger electron spectra of Rh/TiO₂ catalysts [610, 630]:

a — XAES spectra: 1 — 5% Rh/TiO₂, 200 °C, H₂; 2 — 5% Rh/TiO₂, 500 °C, H₂; 3 — 2 + Ar⁺, 1 keV; b — AES spectra: 1 — 1% Rh/TiO₂, 500 °C, H₂; 2 — 5% Rh/TiO₂, 500 °C, H₂; 3 — 5% Rh/TiO₂, 300 °C; H₂

sured after prolonged ion bombardment (see Fig. 25, p. 97). In this case, a deeper reduction of Ti^{4+} to Ti^{3+} and Ti^{2+} was observed, but an appreciable part of the rhodium was removed. The absence of a structure typical of the specimen BR 500 can be seen.

Attempts to analyze the valence levels of the system Rh/TiO$_x$ prepared by sputtering onto the surface of TiO₂ preliminarily reduced by ion bombardment to Ti₂O₃ have been made by Sadeghi and Heinrich [681]. A rather intensive Ti $3d$ peak was observed in the TiO$_x$ spectrum that was suppressed when Rh was sputtered on. Here the density of states at the Fermi level decreased somewhat, and a peak with $E_b = 2.7$ eV appeared. These changes were explained by the suppression of the Rh $3d$ photoemission because of encapsulation of the metal by TiO$_x$ and the formation of a hybridized Rh—Ti or Rh—Ti—O bond. Such a bond is similar to that in intermetallics (Ni₃Ti, Pt₃Ti) [223]. A common feature in the valence band spectra of real and evaporated Rh/TiO₂ catalysts is the presence of a structure with $E_b = 2.3$-2.8 eV; the differences are due to the different degree of reduction of the surface and encapsulation of the Rh particles.

Data of the most surface-sensitive technique—ISS (without considerable sputtering) show that the Rh/Ti ratio in real Rh/TiO$_2$ systems decreases 1.7 times from specimen BR 300 to BR 500. Data of XPS, XAES, and AES also point to the lower surface concentration of Rh and O with high-temperature reduction. These changes are less pronounced for Pt and range from 30 to 40%. The decrease in the Me/Ti ratio is due to lowering of the concentration of the metals in the upper surface layers.

More detailed information on the local structure of a surface and modification of Me/TiO$_2$ catalysts under the effect of Ar$^+$ ions has been evaluated from an analysis of SIMS data. Figure 77 shows the profiles of the changes

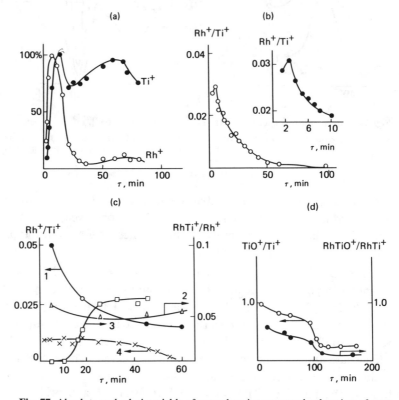

Fig. 77. Absolute and relative yields of secondary ions versus the duration of sputtering Rh/TiO$_2$ specimens (SIMS, Ar$^+$, 1 keV) [610, 630]:

a and *b* — 1% Rh/TiO$_2$, 500 °C, H$_2$; *c* — *1* and *2* — 5% Rh/TiO$_2$, 500 °C, H$_2$; *3* and *4* — 5% Rh/TiO$_2$, 300 °C, H$_2$; *d* — 5% Rh/TiO$_2$

in the secondary ion yields and their ratios depending on the duration of ion scattering. In addition to Ti^+ and Rh^+ ions, oxygen-containing fragments of TiO^+, Ti_2O^+, and $Ti_2O_3^+$ are observed, while for specimens reduced at 500 °C, quite intensive peaks of mixed clusters of the type of $RhTi^+$ and $RhTiO^+$ are registered already in the first measurements. In the specimens reduced at 200 and 300 °C, their appearance is noted only after 10-15 min of ion etching. The absolute and relative yields of the indicated ions depend on the duration of ion bombardment. An analysis of the secondary ion and cluster ratios suggests qualitative conclusions on the distribution of a component or the morphology of the deposited particles. The more rapid drop in Rh^+/Ti^+ in the specimen AR 500 in comparison with BR 500 may point to the higher dispersion or lamellar shape of the particles in the former catalyst. The growth in $RhTi^+/Rh^+$ and $RhTi^+/RhTiO^+$ (and also in Ti^+/TiO^+) witnesses the formation of a more defective structure TiO_x ($x < 2$) deficient in oxygen in which a larger number of the metal atoms can directly interact with the oxide cations.

After the prolonged sputtering of Ar^+ (1-3 h), a considerable reduction of Ti^{4+} to Ti^{3+} and to Ti^{2+} is observed (Fig. 77). An analysis of the Ti $2p$ spectra after the subsequent treatment of the specimens with oxygen suggests that only a part of the reduced Ti ions are on the surface. Hence, initial SIMS measurements enable one to obtain information on the local structure of the surface of Me/TiO_2 catalysts, whereas with more prolonged exposures account must be taken of the deep modification of the TiO_2 surface caused by ion bombardment.

Analysis of the set of spectral data indicates an appreciable change in the surface of Me/TiO_2 in a state of strong metal-support interaction. The direction of a shift in a core level indicates an increase in the electron density on dispersed particles or atoms of metals. This is confirmed by valence band spectra. Ti^{3+} ions (XPS), and also $RhTi^+$ and $RhTiO^+$ clusters reflect direct interaction between a metal and Ti^{3+}. The latter conclusion is consistent with direct data of EXAFS on the formation of an Rh—Ti bond (with a length of 0.267 nm) in ultradispersed catalysts having 0.3% of Rh/TiO_2 [682, 683]. Interaction may occur at the site of the metal-support interface and its enhancement requires diffusion of the metal into the oxide lattice or the migration of TiO_x onto the surface of metal particles. The small peak on the curve of Rh^+/Ti^+ against the time observed during the first two or three minutes of sputtering (Fig. 77), and also the diminishing of Me/Ti in the upper layers after high-temperature reduction confirm the occurrence of these processes, including the screening of part of the Rh by a thin layer of TiO_x. But these changes are not so significant as for sputtered on systems [659-667] and are apparently not the only cause of the changes in their catalytic activity.

Consequently, XPS, SIMS, and AES have provided proofs on the electron and chemical nature of strong metal-support interaction, namely, (a) an increase in the electron density on metal clusters; (b) the formation of a considerable number of defects: Ti^{3+} and V_O; and (c) direct metal-titanium interaction. Moreover, a decrease in the surface concentration of the deposited component is observed. The following scheme of modification of the surface layer of reduced Me/TiO_2 catalysts can be proposed on this basis:

(1) reduction $TiO_2 \overset{Sup}{\underset{Me}{\rightarrow}} TiO_x$ (Ti^{3+}, $\square$ O^-) ($x < 2$)

(2) mutual diffusion of TiO_x and $Me^0 \rightarrow Me^0$ $\square$ Ti^{3+}

(3) the incorporation of boundary atoms of a metal cluster at the site of oxygen vacancies $\boxed{Me^0}\, Ti^{3+}$, which by calculations [638] show to be profitable energetically;

(4) direct interaction $Me^0 + Ti^{3+} \rightarrow Me^{\delta-} Ti^{(3+\delta)+}$ ($\delta < 1$) with transfer of the electron density to metal atoms. The latter is also confirmed by calculations for clusters varying in size: $TiPtO_5^-$, $PtTi_4O_{16}^{-16}$, and $PtTi_4O_{15}^{-15}$ [638, 672]. When a metal atom (Pt) is incorporated at the site of an oxide defect, the Pt $6s$ orbital is below the Ti^{3+} $3d$ one, which causes the transfer of the charge from Ti to Pt. Such interaction becomes more probable with an increase in the dispersion. The morphology of small particles at a rhodium dispersion close to atomic reveals that the formation of Me-Ti clusters can occur by the reactions [683]:

$$[Rh(H_2O)_3(OH)_2(OTi)^{3+}] + 2H_2 \rightarrow 6H_2O + [Rh{-}Ti]^{3+}$$

$$[Rh(HOTi)]^{3+} + \frac{3}{2}H_2 \rightarrow [Rh{-}Ti]^{3+} + H_2O$$

Here complete dehydroxylation of the surface is presumed. Blasco *et al.* [684] and Gonzales-Elipe *et al.* [685] suggest that an important role in the electron interaction between a metal and an oxide cation is played by hydrogen. The latter in low-temperature reduction is captured by anion vacancies following a spillover mechanism with the formation of hydride complexes $(Ti{-}H)^{3+}$. High-temperature reduction is attended by the reaction

$$Rh + (Ti{-}H)^{3+} \rightarrow (Rh{-}Ti)^{3+} + H$$

with inverse spillover of hydrogen onto the metal.

The morphology of small Rh particles on TiO_2 [686] is such that "rows" of Rh atoms form directed along a definite plane of the TiO_2 lattice. This helps their subsequent incorporation into a vacancy. When a distance between the Rh rows of 0.325 nm is retained, there is a structural possibility for the formation of direct Rh—Ti bonds [683]. Such a mechanism is apparently correct for the highly dispersed systems described by Shpiro *et al.* [610] and Sakelsson *et al.* [682]. In the latter case, the size of the Rh clusters by TEM data was 0.5-1.5 nm. At a lower dispersion, the interaction is facilitated because of burying of the metal by TiO_x, which may even lead to physical blocking of the metal centers. But, judging from ISS and SIMS [610, 687], the capsulation effect in powdered Me/TiO_2 specimens is not so significant as for deposited films. The features of the kinetics of the low-temperature homomolecular isotope exchange of oxygen [677], namely, the suppression of the reaction with high-temperature reduction and its proceeding in two steps (slow and rapid) with low-temperature reduction confirm the proposed mechanism of strong metal-support interaction. They show that the promoting effect of the oxygen consists primarily in the breaking of Me—Ti bonds as a result of which the electron state of the metals changes. At high temperatures, the oxygen changes the TiO_x/Me ratio on the surface.

4.5.6 Effect of Strong Metal-Support Interaction for Other Catalysts

The effect of strong metal-support interaction (SMSI) manifests itself the most clearly for systems including a Group VIII metal and a semiconductor. An interaction mechanism with the formation of mixed clusters (Me-oxide cation) or even of chemical compounds of the type of intermetallics similar to that proposed for Me/TiO_2 has also been proposed for Pd/ZnO [645], Rh/ZnO [688], and Ni/Nb_2O_5 [689], as well as for simulated specimens such as a metal sputtered onto a nonstoichiometric oxide, e.g. Rh/SiO_x [690]. Investigations in which a semiconductor oxide, e.g. TiO_2, is a modifier of an insulator oxide are of substantial interest for establishing the role of electron and geometric factors in SMSI and also from a practical viewpoint. In one of these investigastions [690], XPS and infrared spectroscopy of adsorbed CO have shown that in a specimen of 7% $Ir/TiO_2/SiO_2$, the deposited phase retains its ability to be reduced to Ti^{3+}. When depositing Ti by building up layers from $TiCl_4$ on Al_2O_3 [317], the surface concentration of the Ti is higher than the bulk one, while the properties of the metal (Ir) become similar to Ir/TiO_2 at titanium coverages exceeding a monolayer. Judging from the easiness of reduction by Ar^+ ions, the mobility of the deposited phase and, consequently, the transition to the state of SMSI are facilitated when the Ti concentration grows. In some catalytic properties, catalysts prepared by anchoring Ti (or a rare-earth element) from organometallic complexes onto SiO_2 adjacent to ions of Group VIII metals are close to the catalysts Me/TiO_2 [691]. The chemical nature of SMSI is postulated on these grounds. When Pr and Pt ions were present on SiO_2, a positive shift was observed in the Pt spectrum ascribed to its interaction with the Pr ions [692]. Unlike this, the shift for Pd/La_2O_3 is negative (-0.7 eV) [693]. Here the magnitude of the shift grew unexpectedly with an increasing Pd content and increasing size of its particles. Fleish *et al.* [693] suggest that the large Pd particles are decorated by islets of La_2O_3 which are partly reduced to LaO. This leads to the charge flowing over onto the palladium. No direct proofs of such reduction have been obtained, however.

It should be noted that properties similar to SMSI are also ascribed to the traditional Me/Al_2O_3 or Me/SiO_2 catalysts. Here partial reduction of the support cations in the neighborhood of a metal atom is presumed [634-637]. The large negative shifts observed in the spectra of $Me(VI)/Al_2O_3$ [204, 578] may presumably be ascribed to the formation of type $Me^{\delta-}$-$Al^{\delta+}$ clusters or intermetallics owing to the reduction of Al^{3+} in the neighborhood of the highly dispersed metal particles. But these shifts may also be due partly to the contact charging observed in the sputtering of Cr, Cu, or Ag onto α-Al_2O_3 [694]. Although TPR underlied the assumption that Al is reduced in its oxide in the presence of Group VIII metals, direct proofs of the formation of Al(O) have meanwhile not been obtained. For instance, when a monolayer of Al is deposited on polycrystalline Rh foil, it transforms into Al^{3+}, and unlike Ti^{4+} is not reduced when the specimens are treated in hydrogen (750 K) [667]. Additional research is evidently needed to see how similar the mechanisms of metal-support interaction are for Me/TiO_2 and deeply reduced metal-oxide-insulator catalysts.

At the same time, a definite classification of local metal-support interaction can be proposed by analyzing EES and IS data. On supports with pronounced electron-acceptor centers, such interaction leads to the transfer of the charge to the support and to the formation of electron-deficient clusters. On supports containing electron-donor centers (of the Ti^{3+} type), the charge passes from the support onto the metal, and the electron density on the metal clusters grows. In rigorous reduction treatments, one can expect the transition of the first type of interaction to the second one because of the reduction of the oxide supports. The primary structure of the support and the defects forming in the genesis of a catalyst are quite significant in the realization of one of the interaction types. An attempt has been made to classify all the types of interaction on the basis of a model of vacancies by which a different type of interaction, including SMSI, may be realized depending on the type of cation sublattice of the oxide, the size of the vacancies, and the metal atom. Here there is no need of using semiconductor oxides; they may be replaced by additives having electron-donor properties.

Finally, mention must be made of systems deposited on nonoxide supports, for example, on polymers. At a high metal dispersion, here the chemical type of interaction may prevail that is characterized by the formation of fixed bonds with definite functional groups of the support [695, 696].

4.6   Bimetal Supported Catalysts

The interest in studying bimetal and polymetallic supported catalysts is due to two main factors:

(1) their practical importance as the most selective and stable catalytic systems; and

(2) the possibility of bringing to light the fundamental questions of catalysis: the role of the ligand and ensemble effects, and in a broader aspect, the promotion and poisoning of active centers.

A sufficiently reliable methodology has been developed to characterize the surface structure and composition of unsupported alloys by EES and IS (see Secs. 3.3 and 4.1). Matters are different for supported alloys. The difficulty of studying them is associated with the high dispersion of the components, their interaction with supports, and the inhomogeneity of the chemical states and distribution. These characteristics depend not only on the composition, but also on the method of preparation, the treatment conditions, and the catalytic reaction. This is why to date the investigations performed by EES and IS involving supported alloys are rather scarce. Other physicochemical techniques often have to be employed for the more authentic identification of alloys on supports.

In the present section, we shall briefly deal with the role of the ensemble and ligand effects in catalysis using alloys, an analysis of the literature on the identification of dispersed alloys by some modern techniques, and in connection with these problems we shall discuss the results of studying bimetal supported systems by EES and IS.

4.6.1   Ensemble and Ligand Effects in Catalysis with Supported and Unsupported Alloys

The distinctions in the catalytic properties of metals when alloyed are associated both with the geometry of an active center and with the effect of its closest neighbors on its electron structure. Strictly speaking, the two effects—geometric and electronic—are related. But in the literature, it is customary practice to treat them separately, giving preference either to the ensemble effect or to the ligand one. Beginning with the pioneering work of Sinfelt *et al.* [697], nickel-copper alloys are considered to be a classical example of the manifestation of the cluster effect. Indeed, the specific catalytic activity in structurally sensitive reactions such as the hydrogenolysis of ethane already with small additions of Cu drops by four orders of magnitude, whereas the activity in the dehydrogenation of cyclohexane is constant and diminishes only at a high copper content [326]. It should be noted that the surface composition of an Ni-Cu alloy depends on the preparation conditions, and a sharp change in the activity, e.g. in the hydrogenation of benzene, is due to the copper greatly enriching the surface already at low concentrations.

The contribution of the cluster and ligand effects is singled out by using CO adsorption: the intensity of the IR band or the TDS peak characterizes the first of the effects, while the shift of the band in the IR spectra (or the desorption temperature) may point to the ligand effect. We must note the noticeable contradiction of the results and their interpretation even for the most studied Ni-Cu and Pd-Ag alloys. The alloying of Ni with Cu or of Pd with Ag is attended by a decrease in the relative fraction of high-frequency CO bands characterizing bridge centers and by a growth in the band of "linear" CO; the shifts of the bands are negligible, while the desorption activation energy does not depend on the composition of the alloy [325]. Conversely, Dalmon *et al.* [698] note the shift of the bands and the change in the desorption activation energy on similar specimens. Consequently, even for well characterized alloy surfaces, no unambiguous information has meanwhile been obtained on the limits of manifestation of the cluster and ligand effects in the adsorption of CO.

The great changes in the activity and selectivity observed for Group VIII metals modified by additives such as Re, Sn, In, Pb, Cr, and Mo [325, 326] may suggest the ligand effect. In particular, it can be expected with a view to the strong perturbations of the electron structure in the formation of Pt-Me intermetallics. Indeed, by data of IR spectroscopy, the frequency of adsorption of CO on Pt lowers in the presence of Sn [325], but the behavior of Pt-Sn catalysts in various reactions depends substantially on the way of preparation, the Sn/Pt ratio, etc. [325]. This is explained by the diverse effect of the tin, namely, (a) an increase in the dispersion of the platinum; (b) the formation of an alloy; (c) blocking of the surface Pt centers, etc.

In Pt-Re/Al$_2$O$_3$, the maximum of the activity in the hydrogenolysis of cyclopentane was attributed to the interaction of Pt and Re, as a result of which the system acquires the properties of Os or Ir [326, 582]. The simplification of such an approach and the diverse effect of Re in reforming catalysts were noted by Ryashentseva and Minachev [583].

Metals that do not mix in the bulk (Os, Cu, Ru) and form quite heterogeneous clusters on supports nevertheless mutually affect the catalytic properties: in the suppression of hydrogenolysis, the hydrogenating and dehydrogenating activity of Cu-Os and Cu-Ru is retained [301, 328]. This is apparently due not only to the diminishing of the number of multiatomic centers of Ru, but also to the formation of mixed Me-Cu centers [328]. In other words, in this case too, we have to do with the ligand effect to a certain extent.

In reactions with the participation of oxygen or with its formation as an intermediate (the oxidation of hydrocarbons, Fisher-Tropsch synthesis, methanol synthesis, etc.), very substantial changes in the composition of the surface and the degree of oxidation of the metals are possible. This is probably why catalysts that are similar in composition but differ in the conditions of their preparation or in the reaction conditions exhibit different properties. Syntheses based on CO are convenient model reactions for establishing the contribution of the ensemble and ligand effects because they proceed on multiatomic clusters and their selectivity depends greatly on the energy binding various intermediates to the surface [699].

Consequently, the changes in the catalytic properties of bimetal catalysts are due to a multitude of reasons. The greatest clarity has meanwhile been achieved in characterizing the change in the geometry of an active center, especially in the presence of a component with little or no activity, and how this change affects catalysis. Much less evidence has been found for the ligand effect of the second component, although theoretical calculations and experimental data on the electron structure suggest that this effect can be expected for many alloys. Some authors correctly note the rather artificial nature of distinguishing the two effects: this is typical of highly dispersed clusters of metals on supports and for a large group of catalysts containing metal-cation groupings where the role of the second component consists not only in stabilization of the cluster of a metal, but also in creating additional adsorption centers. The elucidation of these matters is of a major importance for understanding the catalytic action of alloys.

**4.6.2   Physicochemical Characterization
of Dispersed Supported Alloys**

Since we are dealing with small particles, the criteria of formation of bulk alloys and their phase characteristics are difficult to use for these systems in the general case. Thermodynamic models of surface segregation can be applied in far from all cases because [700] (a) the formation of supported bimetal clusters is controlled by kinetic, and not thermodynamic, parameters; (b) bimetal clusters may form from metals insoluble in each other; (c) the values of the free energy cannot be employed; and (d) in deposited microcrystals, impurity atoms may be localized on edges, at apices, and at the metal-support interface. These arguments stress the need to thoroughly study supported bimetal particles, determine their surface composition and degree of homogeneity.

A variety of techniques are employed to identify supported bimetal clusters [700], namely, X-ray structural analysis, XAS, EXAFS, IR spectroscopy, Mössbauer spectroscopy, electron microscopy, and methods of surface analysis. The use of X-ray analysis is restricted by the large size of the particles ($>$ 2-5 nm) and the rather high concentrations, i.e. to the studying of model catalysts on SiO_2. Quite homogeneous bimetal Pt-Ir clusters have been identified by XRD in specimens of 10% (Pt-Ir)/SiO_2 reduced at 500 °C [327]. Mössbauer spectroscopy is quite effective for identifying bimetal clusters, but for a limited range of systems, e.g. Fe-Me (where Me is Ru, Rh, Pd, Ir, or Pt) on SiO_2 prepared by impregnation [701]. Only 20-50% of the iron by Mössbauer spectroscopy and XPS data was reduced to the metal with the formation of an alloy. An exception here was an Fe-Pd/SiO_2 specimen in which both metals were reduced virtually completely. In catalysts where the cluster $FeRh_4(CO)_{15}[NMe_4]_2$ was the precursor of a binary alloy, Mössbauer spectroscopy data reveal that a more dispersed particle distribution is achieved, but the particles contain Fe^{3+} ions interacting with the γ-Al_2O_3 [701].

The use of the procedure of an imaging in defocussed electron beams [702] permits the effective employment of TEM for identifying bimetal clusters and determining their size. TEM, EXAFS, and XANES were used to study a supported Pt-Ni catalyst [702]. All the techniques yield a consistent picture of the formation of bimetal clusters with a distorted face-centered cubic lattice and a size of 2.5-5.0 nm. Unlike compact alloys, no surface segregation of any of the components was observed in the dispersed bimetal clusters. The interatomic distances and coordination numbers determined by EXAFS [599, 600] enable one to judge not only on the existence of a bond between the two metals, but also to appraise the distribution of the components between the surface and bulk of the support, although, as noted by Zamaraev and Kochubei [601], there are some difficulties in interpreting the data. A comparison of the EXAFS results for 1% Pt-Ir on Al_2O_3 and SiO_2 has shown [703] that the structure of the bimetal particles differs for the two supports: on SiO_2 there is a "cherry" structure with a nucleus enriched in Ir and a shell of Pt, whereas on Al_2O_3 the particles are more homogeneous, the lowering of the Pt—Ir bond length pointing to interaction of the metals with the support.

EXAFS and TEM have also been employed to study bimetal catalysts consisting of components poorly soluble in the bulk, namely, Os-Ru [328], Ru-Cu, and Ru-Au [329], in connection with their interesting catalytic properties. In all cases, the distribution of the metals in a cluster was found to be quite inhomogeneous. In Pt-Re/Al_2O_3 reforming catalysts, EXAFS revealed no formation of a bimetal cluster [704]. But the rhenium apparently affects the state of the platinum: unlike a monometal specimen, already in the drying and calcination stage only O^{2-} ions are present in the coordination sphere of the platinum, while Cl^- ions are absent.

The use of IR spectroscopy to identify supported alloys is based on the changes in the ratio between the "bridge" and "linear" forms of CO adsorption, and also in the dipole-dipole interaction in the adsorbed layer. The mixed Pt-Cu, Pt-Pb, Pt-Sn, and Pt-Re catalysts have been studied in this way [700].

The considered examples show that attempts to characterize supported alloys meanwhile relate to a limited range of systems. The examples also reveal the involved nature and inhomogeneity of the surface structure of bimetal catalysts. The component distribution is affected by the support, this telling not only on the nature of surface enrichment, but also the composition of the core [700]. Since metals on supports are present in several valence states, XPS seems to be effective for studying bimetal systems [27]. We can note as an example the work performed by Ioffe *et al.* [705] who showed that in Pt-Me systems (where Me is Mo, W, or Re) anchored on SiO_2, metal-ion groupings form instead of purely bimetal clusters.

For a deeper understanding of the mechanism of catalysis on supported alloys, investigations are needed that are aimed at establishing the relation between the electron structure of alloys, surface segregation, and the local geometric structure of a surface. It is also necessary to establish the influence of the reaction medium on these characteristics and compare the properties of unsupported and supported alloys. In this connection, let us see how the following problems are solved with the aid of EES and IS:

(1) the chemical state of components and their mutual influence on reduction;

(2) the identification of bimetal clusters and the determination of their composition; and

(3) the structure of the surface of supported bimetal systems.

The systems being studied can be divided conditionally into two groups. The first one includes catalysts in which both components are chiefly in the metal state. The second one includes systems in which the promoters are not reduced or are readily oxidized under the reaction conditions.

4.6.3   Bimetal Supported Clusters

Ru-Co Supported Catalysts. Ruthenium and cobalt on γ-Al_2O_3 and SiO_2 are reduced within a different temperature interval, the degree of reduction of Co^0 on SiO_2 being higher [552]. Heat and vacuum treatment of mixed specimens results in decomposition of the initial compounds and partial reduction of Ru^{3+} to an intermediate state. With additional treatment in H_2 (20 torr), the cobalt is reduced, and the degree of reduction of Co^0 in mixed specimens is higher than in monometallic ones (Table 23).

The possible reasons for the more ready reduction of Co include (a) the activation of hydrogen on ruthenium and its spillover onto cobalt oxide; and (b) the interaction of Co and Ru to form an alloy. The increase in the ability of Ru to be reduced in the presence of Co must also be noted. This follows from the change in the FWHM of the Ru lines at 350 °C. This effect is apparently due to weakening of the interaction of Ru with the support (especially on γ-Al_2O_3), which is confirmed by the higher Ru/Co ratio on the surface in comparison with the calculated value.

More detailed data on the surface composition of catalysts and the interaction of their components have been obtained by SIMS [358, 553] (Fig. 78). The secondary positive ions Co^+, Ru^+, Al^+ (Si^+), and also $CoAl^+$ and $CoAlO^+$ ($CoSi^+$) clusters are registered in the mass spectra of specimens treated in vacuum and H_2; cluster ions of the type of $RuCo^+$ are additionally detected in reduced specimens, which suggests the formation of bimetal Ru-Co particles. The depth profiles of Me^+/Al^+ (Si^+) on the two supports change oppositely: the ratios diminish on γ-Al_2O_3 and grow on SiO_2. The Ru^+/Co^+ ratio on Al_2O_3 changes slightly with time, whereas this ratio grows in SiO_2, so that the steady-state value of Ru^+/Co^+ on SiO_2 is six times higher than on γ-Al_2O_3. The initial ratios Co^+/Al^+ and Co^+/Si^+ change in the same sequence as the relevant XPS intensities. The Co^+/Al^+ ratio on alumina in-

Table 23. Parameters of XPS Spectra and Chemical States of Metals in Bimetal Ru-Co Catalysts [358, 553]

Specimen	Treatment conditions	Binding energy, eV		FWHM, eV, Ru $3d_{5/2}$	Chemical states of metals					
		Co $2p_{3/2}$	Ru $3d_{5/2}$		Ru	%Ru^0	Co	%Co^0		
								bimetal	monometal
2%Ru-2.3%Co Al$_2$O$_3$	Initial	781.1	*	5.7	3+	0	2+	0	—
	350°C, vacuum	781.1	281.2	6.6	3+, +	—	2+	0	—
	350°C, H$_2$	781.2 777.2	280.4	3.4	0	100	2+	10	—
2%Ru-6%Co Al$_2$O$_3$	Initial	781.4	*	5.8	3+	0	2+	0	0
	450°C, vacuum	781.6	280.5	4.0	3+, 0	—	2+	0	0
	450°C, vacuum + H$_2$	781.6 778.6	280.6	—	0	100	2+, 0	45(55)**	28(26)**
	450°C, H$_2$, 1 atm	781.6 778.4	280.6	—	0	100	2+, 0	55-60	37
2%Ru-6%Co SiO$_2$	Initial	781.2	*	5.3	3+	0	2+	0	0
	450°C, vacuum	781.4	280.6	5.4	3+, 0	—	2+	0	0
	450°C, vacuum + H$_2$	781.4 778.4	280.5	4.6	0	100	2+, 0	70(79)**	55
	450°C, H$_2$, 1 atm	778.4	280.6	—	0	100	0	90	60

* The line is masked by a C 1s peak.
** By data of magnetic granulometry.

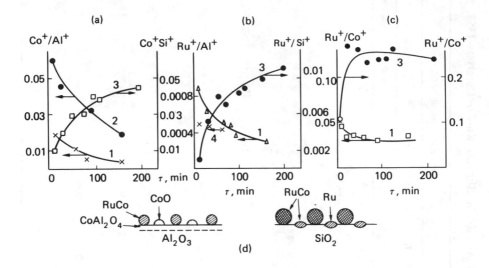

Fig. 78. Depth profiles of components (*a*, *b*, *c*) and presumed local structure of the surface (*d*) of supported Ru-Co catalysts [358, 553]:

a, *b*, *c*: *1* — 2% Ru-6% Co/Al₂O₃; *2* — 6% Co/Al₂O₃; *3* — 2% Ru-6% Co/SiO₂; *4* — 2% Ru/Al₂O₃. Specimens *2* and *4* were treated at 450 °C in vacuum, *1* and *3* in vacuum and H₂

creases in the series 6% Co < 6% Co-2% Ru < 2% Co-2% Ru. Moreover, in mixed catalysts, the $RuAl^+$ peak is less intensive. All this points to the stronger interaction of Co in bimetal specimens with γ-Al₂O₃ and to its incorporation into the lattice.

The unusual growth in Me^+/Si^+ in the initial period of sputtering is apparently associated with several factors, namely, (a) with partial screening of the metals by a thin layer of silica gel; (b) the larger size of the metal particles on SiO₂ in comparison with Al₂O₃, which was measured by the chemisorption of O₂; and (c) the preferential sputtering of SiO₂, although for pure supports the distinctions in the sputtering yields are not great enough for this factor to become determining [113]. The $RuCo^+$ ions, whose peak amplitudes are maximal at the beginning of sputtering, probably indicate the formation of microcrystals of an alloy. The $RuCo^+/Ru^+$ ratio on γ-Al₂O₃ is higher than on SiO₂, while the change in Ru^+/Co^+ with time is negligible, which may show the greater homogeneity of these particles on the aluminium oxide. On the other hand, the data of X-ray diffraction analysis suggest indirectly the formation of an Ru-Co alloy on SiO₂: the reflexes α- and β-Co were observed on a diffraction pattern of Co/SiO₂, while in the presence of Ru only β-Co was observed, in whose lattice the second component may dissolve. Judging from the Co^0/Ru^0 ratios on Al_2O_3 at 450 °C, the alloy ought to have the composition $Ru_{35}Co_{65}$ with complete alloying of the metals, and on SiO_2 it ought to have the composition $Ru_{70}Co_{30}$. In both cases, this is above the miscibility limit of Ru in Co [358]. This is why we can assume that only part of the Ru enters the alloy, or that the solubility of the components in microcrystals is higher than in the bulk. For Ru-Co/SiO$_2$, which contains larger particles, the first assumption is more probable [358].

The growth in Ru^+/Co^+ is possibly due to the varying composition of mixed clusters differing in their dispersion. They include large particles whose core is enriched in Ru and small particles with their shell enriched in Ru. Both cases have been observed for ruthenium-containing catalysts [328, 329]. The considered data suggest variants of the surface structures of Ru-Co/Al$_2$O$_3$ and RuCo/SiO$_2$ (Fig. 78). The unreduced cobalt is apparently present on the aluminium oxide in the form of an aluminate. Moreover, the presence of separate Co and Ru metal phases is possible, although the major part of the metals is present in the form of an alloy slightly enriched in Ru. On SiO$_2$ there is a phase of Ru^0 and also Co oxide in addition to alloyed particles. A feature of the structure of Ru-Co/SiO$_2$ is also that the metal particle is encapsulated or decorated by a thin layer of silica.

Complicated interaction of Ru and Os apparently occurs in catalysts prepared from carbonyl complexes on γ-Al$_2$O$_3$ [706]. By XPS and TEM, ruthenium in the reduced state forms Ru^0 particles with a size of 2-3 nm, while the major part of the osmium remains in the Os(II) state. The formation of bimetal Ru-Os clusters is hardly probable. Conversely, in systems containing readily reduced metals (Ag-Au/Al$_2$O$_3$) practically complete fusion occurs [413]. The structure of the Ag-Au particles differs from the bulk ones (EXAFS), although as in bulk alloys the surface is enriched in silver.

Bimetal Particles on Zeolites. When a second cation was incorporated into an NaY zeolite, the ability of Ni to be reduced appreciably changed [27, 197, 222]: Co-Ni > Cr-Ni > Ni > Cu-Ni, and the migration of Ni^0 to the outer surface was retarded. Additional studies of a Co-Ni zeolite by SIMS [553] suggest the formation of a Co-Ni alloy in the reduced specimens. Quite intensive $CoNi^+$ peaks were observed in the mass spectra; their amplitude with increasing sputtering time diminished somewhat, while in oxidation (500 °C) they virtually vanished (Fig. 79). By SIMS and XPS data, the Co/Ni ratio in a specimen reduced at 500 °C is greater than unity, which may point to the formation on the outer surface of an alloy rich in Ni that is more prone to migration. The degree of enrichment in Ni grows at 550 °C—here the

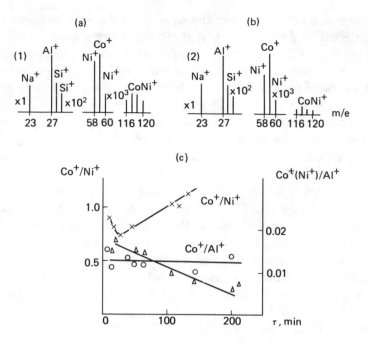

Fig. 79. Secondary ion mass spectra (*a*, *b*) and depth profiles (*c*) for specimen of 0.3CoO·3NiNaY [553]:

a — 450 °C, H_2; *b* — 450 °C, H_2 + 500 °C, O_2

Ni^0/Co^0 ratio (XPS) is three. If we assume that in sputtering the metals are initially removed from the outer surface, the growth in Co^+/Ni^+ signifies that within the cavities the bimetal clusters are enriched in Co (Fig. 79). It is also quite possible that part of the Ni^0 is in the form of a separate phase. At lower temperatures (400 °C) the migration of Ni is not so pronounced and, judging from the Ni^0/Co^0 ratio equal to unity, the composition of the bimetal particles localized chiefly in the cavities is close to the bulk value.

The treatment of Pd-Cu/DM in H_2 at 300 °C results in complete reduction of the metals, but shifts in Pd $3d_{5/2}$ (+0.6 eV) and Cu $2p_{3/2}$ (−0.9 eV) opposite in sign are observed in the spectra. These shifts are typical of unsupported Pd-Cu alloys [222]. Since no preferential migration of one of the components was observed here, we can assume that both metals are mainly contained in the alloy. As for the bulk alloy Cu-Pd close in composition (Sec. 4.1), the Cu/Pd ratio (XPS) in the zeolite is somewhat higher than the calculated value. By analogy with the data of electron microscopy for Pd/DM [475], we can assume that the particles of the alloy Pd-Cu are localized in the defective cavities of the intercrystallite space of the mordenite. The behavior of Pd-Cu and Ni-Co zeolites when oxidized (O_2, NO) is qualitatively similar to that observed for the relevant unsupported alloys: initially the preferential oxidation of the second component and its surface segregation occur, but under more rigorous conditions (300-500 °C, air) the complete oxidation of both components was observed. Reduction yielded an Me_1/Me_2 ratio (XPS) close to the initial one [197, 222].

4.6.4   Bimetal Platinum-Containing Catalysts

Additives of oxides of Group IVB (Sn, In) or VIB (Cr) metals to aluminium-platinum catalysts improve their activity, selectivity, and stability in various reforming processes [582], the dehydrogenation of higher alkanes [707], etc. Various opinions have been advanced on the mechanism of action of such additives, namely, synergism is related to the formation of intermetallic compounds, proceeding mainly from data for an SiO_2 support [708], or is ascribed to the stabilization of dispersed Pt particles on the surface of γ-Al_2O_3 modified by a promoter [611, 709]. The information on the state of an additive (chiefly Sn) obtained by XPS is also contradictory. Sexton et al. [710] indicate that tin is practically not reduced to Sn^0 on γ-Al_2O_3 even at high concentrations (6%) or with the use of the organometallic complex $Pt(SnCl[Et_4N]_6$; on SiO_2 under the same conditions, (Sn^0) exceeds 30%. Baronetti et al. [711] have found Sn^0 (40%) in low-metal-content specimens of Pt-Sn/Al_2O_3 (the total content of Sn and Pt was 0.6 mass %) prepared by joint impregnation. Many authors stressed the significance of the electron interaction of Pt and the modifier, but the direction of electron density transfer has not been established unambiguously.

To elucidate these problems, XPS was employed to study Pt-Sn and Pt-In catalysts on γ-Al_2O_3 and SiO_2 [612, 613], and also Pt-Cr on Al_2O_3 [578]. Notwithstanding the superposition of the Pt $4f$ and Al $2p$ lines, an attempt was made to determine not only the state of the promoter, but also that of Pt and the surface composition of the specimens.

An analysis of the position of the $3d_{5/2}$ level lines of In and Sn on Al_2O_3 reveals that in all the preparation stages, the main part of the promoters is in an oxidized state (Table 24). The state of In after calcination does not practically differ from the initial one in the nitrate, while after reduction it decreases by only 0.3 eV. The binding energy of Sn $3d_{5/2}$ corresponds to the states Sn(II)-Sn(IV), which are not distinguished by XPS spectra. At the same time, the

Table 24. Binding Energies of Pt and Promoter Lines, Relative Atomic Concentrations and Dispersion of Pt in Mono- and Bimetal Catalysts [612, 613]

Specimen	Treatment	Promoter	Pt $4f_{7/2}$ E_b, eV	Pt $4f_{7/2}$ ΔE, eV[a]	Pt $4d_{5/2}$ E_b, eV	Pt $4d_{5/2}$ ΔE, eV[a]	Sn(In) $3d_{5/2}$ E_b, eV	Sn(In) $3d_{5/2}$ ΔE, eV[a]	$\dfrac{Pt}{Al(Si)} \times 10^3$	$\dfrac{In(Sn)}{Al}$	$\dfrac{Cl}{Al} \times 10^3$	$\dfrac{In(Sn)}{Pt}$	$\dfrac{Cl}{Pt}$	γ_{Pt}^{b}
$\dfrac{3\%Pt\text{-}3\%In(Sn)}{Al_2O_3}$	Drying	In					444.5	+ 0.6	2.90	5.39	25.5	1.86	3.5	—
		Sn					487.2	+ 0.2	2.74	3.90	48.0	1.42	17.5	—
	Calcination	—							2.16	—	7.57	—	3.5	—
		In					445.2	− 0.4	2.49	5.89	8.96	2.36	3.6	—
		Sn					486.8	− 0.4	3.99	71.9	13.8	1.84	3.45	—
	Drying + reduction	—	72.0	—	315.4	—	—	—	—	—	—	—	—	—
		In	71.8	− 0.2	315.8	+ 0.8	445.0	—	1.40	5.90	4.28	4.2	3.05	0.23
		Sn	72.0	+ 0.4	315.8	+ 0.8	486.8	—	1.31	4.73	4.99	3.6	3.8	0.61
	Calcination + reduction	—	71.6	—	315.4	—	—	—	2.79	—	8.53	—	3.05	0.60
		In	72.0	+ 0.4	315.6	+ 0.2	445.3	− 0.3	2.83	4.44	8.61	1.6	3.0	0.84
		Sn	72.0	+ 0.4	315.4	0	486.8[c]	—	3.54	6.43	9.11	1.8	2.57	0.74

Sample	Element	Pt			In/Sn		I		Shift[a]			
$3\%Pt\text{-}3\%In(Sn)$ Calcination SiO_2	—	74.0	—	—	—	—	3.4	—	—	—	—	—
	In	74.5	—	—	445.8	—	1.0	—	2.38	—	—	—
	Sn	74.2	—	—	486.4	—	3.0	—	1.72	—	—	—
Calcination + reduction	—	71.6	—	—	—	—	2.8	—	—	—	—	—
	In	71.3 (50)[d]	—	—	443-9 (50)[d]	—	2.0	—	2.22	—	—	—
		72.3 (50)[d]			(50)[d]							
	Sn	71.5 (58)[d]	—	—	486.8 (45)[d]	—	3.0	—	1.45	—	—	—
		72.8 (42)[d]			484.0 (55)[d]							
Pt_2Sn_3 400 °C, vacuum	Sn	71.4	—	—	484.0	—	—	—	2.50	—	—	—
$PtIn_2$ Ditto	In	71.4	—	—	444.0	—	—	—	—	—	—	—

[a] A shift relative to the corresponding line position in a monometal specimen.
[b] The Pt dispersion is determined by the "solubility technique" [613].
[c] About 8-10% Sn^0.
[d] The percentage of a definite oxidation state.

synthesis of the peaks shows that about 8% of the tin are in the form of Sn^0, and the fraction of metallic indium does not exceed 5%. The position of the Pt $4d_{5/2}$ line changed only slightly with various treatments, whereas the Pt $4f_{7/2}$ line shifted noticeably. Its absence in the spectra of some specimens signifies that it overlaps completely with Al $2p$, which occurs at E_b of Pt $4f_{7/2} > 72$ eV. The relative surface concentrations of Pt, Cl, In, and Sn depend on the calcination stage. The reduction of preliminarily calcined specimens is attended by a growth in the surface concentration of Pt, while when the specimens are treated directly in H_2, this concentration decreases. The promoter concentration changes in the opposite way.

The results for Pt-In(Sn)/SiO_2 differ considerably (Table 24). When calcined specimens are reduced, two components can be seen quite clearly in the Sn $3d$ and In $3d$ spectra, and one of them relates to Me^0. The concentration of the reduced tin and indium exceeds 50%. The forms of Pt can be singled out from the Pt $4f$ spectra that are characterized by a binding energy of 71.2-71.5 eV and 72.3-72.8 eV and are also present in a ratio of about 1/1. When calcined specimens are reduced, the Pt/Sn ratio changes only slightly; a growth in Pt/Me was noted for Pt-In/SiO_2.

A comparison of the data for Al_2O_3 and SiO_2 indicates a different nature of interaction of the two components on these supports. The opposite shifts in the spectra of Pt and the promoters on Al_2O_3 may point to their direct interaction with the transfer of the electron density to the promoter. Judging from the spectra, this interaction appears already in the calcination stage. It should be noted that the magnitude of the additional shift in the Pt spectrum does not exceed 0.4 eV, although the total shift relative to Pt^0 is larger than 1 eV. It includes the change in E_{ER} and interaction with the support. It is hardly probable that the shift is due to modification of the Pt by chlorine, since the Cl/Pt ratio remains the same for both unpromoted and promoted samples while with a view to the content of the second component it drops. The conclusion on the increase in the electron deficiency of Pt in the presence of Sn and In is consistent with the data obtained on the same specimens by IR spectroscopy of adsorbed CO [707]. The presence of a certain amount of metallic tin and the increase in the Sn/Pt ratio in comparison with the calculated one may also suggest the formation of alloyed or intermetallic particles whose surface is enriched somewhat in the second component. Interaction of the components with the formation of bimetal clusters is typical of specimens on SiO_2 to a greater extent. As in unsupported intermetallics the surface is enriched in the second component. Two Pt forms are observed in a supported catalyst.

Having in view that at 500 °C platinum on SiO_2 is reduced completely, it can be assumed that the state with the larger binding energy relates to the platinum contained in the intermetallics.

Consequently, in reduced catalysts on γ-Al_2O_3 even at high contents of Sn or In, the latter are chiefly in the oxidized state and form mixed metal-ion clusters with the Pt. Bimetal clusters form on SiO_2. The substantial role of calcination in the formation of highly dispersed Pt must be stressed. The increase in the binding energy of the promoter levels in this stage is possibly due to the formation of compounds of the type of Pt(IV)-Sn(IV)-O-Cl, which has been presumed by Sachtler [582] and by Lieske and Völter [708]. The reduction of this compound yields clusters of the type of $Pt^{\delta+}$-Sn(II). At the same time, the influence of the alloy on the catalytic properties of Pt-Me/Al_2O_3 (Me = In or Sn) is apparently more appreciable if account is taken of the possibility of deeper reduction under the conditions of dehydrogenation or reforming reactions. Such a conclusion seems to be valid for Pt-Re catalysts. EXAFS and XPS data [704] reveal that in the initial stage of catalyst preparation a mixed Pt-Re complex forms; although during reduction in H_2 the average degree of oxidation is $+4$ [582, 704], the formation of Re^0 and, possibility, of the alloy Pt-Re is observed in a medium of hydrocarbons (C_4H_{10}/H_2, 500 °C) and in commercial reforming catalysts [712].

An additional example of the influence of carbon deposits on reduction are the data obtained for Pt-Cr reforming catalysts [455, 578]. Specimens containing 1.8% of Cr_2O_3 and 0.62% of Pt were reduced at 500 and 550 °C in an atmosphere of H_2, next a part of them were coked by treatment in a mixture of *n*-hexane and hydrogen at the same temperature. Complete reduction of the catalysts "without coke" and "with coke" yielded substantially different results (Fig. 80). In the first case, the degree of reduction of Cr to Cr^0 is 38% and in the second one is 67%; similar distinctions were observed for a specimen of Pt-Re-Cr/Al_2O_3: the degree of reduction of Cr^0 without coke was 27%, and with coke, 57%. We must also note the substantial promotion of the reduction of Cr by platinum, which can also be seen by comparing specimens of 1.8% Cr and 1.8% Cr-Pt reduced at 550 °C. Hence, a double catalytic effect of chromium reduction is observed that is associated with the presence of platinum and coke. The interaction between Pt and Cr is indicated by the positive shift in Cr $2p_{3/2}$ relative to Cr^0/Al_2O_3, which is especially noticeable for coked specimens. This may suggest the transfer of the charge from Cr to Pt or a change in the interaction with the support when an alloy forms. The formation of an alloy Pt-Cr/γ-Al_2O_3 is indicated by XRD data [713].

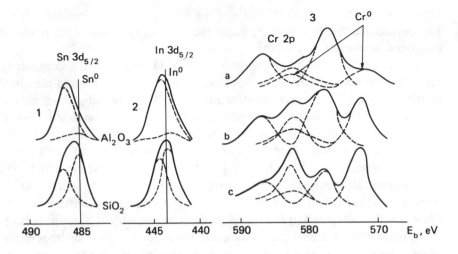

Fig. 80. Reduction of the second component in bimetal Pt catalysts [455, 612, 613]:

1 — Sn in 3% Pt-3% Sn/Al$_2$O$_3$(SiO$_2$); *2* — In in 3% Pt-3% In/Al$_2$O$_3$ (SiO$_2$); *3* — Cr in 0.62% Pt-1.2% Cr/Al$_2$O$_3$ (*a* — monometal specimen of 1.2% Cr/Al$_2$O$_3$); *b* — 0.62% Pt-1.2% Cr/Al$_2$O$_3$, 500 °C, H$_2$; *c* — *b* + 500 °C, *n*-hexane + 500 °C, H$_2$

The interaction of Pt with an oxide (CeO$_2$) is indicated by XPS and TPR data [714]. The mixed complex PtO-CeO$_2$ forms in the oxidized catalysts, whereas with rigorous reduction (920 °C, H$_2$), the platinum promotes the formation of a surface compound such as CeAlO$_3$.

The structure of reduced Pt-Me catalysts includes two main fragments, namely, highly dispersed Pt particles (clusters in the limit) stabilized on promoter ions, and alloy microcrystallites whose composition will depend on the support, the treatment conditions, and the catalytic reaction. Proceeding from data for the pure Pt-Sn and Pt-In phases, enrichment of the surface in the second component can be expected. Although the ratio between the two types of centers on Al$_2$O$_3$ and SiO$_2$ differs, a common feature is that in both cases there is a lower probability of formation of multiatomic Pt ensembles (a high dispersion or dilution by the second component) [274, 275, 582]. As a result, the reactions of C—C bond scission undesirable in reforming are suppressed. In an alkane hydrogenation reaction, not only the stability, but also the activity grows [707]. This can be associated with the ligand effect due to a certain drop in the electron density on Pt in the presence of an additive.

Other Me^0-$Me^{(n+)}$ Catalysts. The interaction of a noble metal with an oxide additive increases the dispersion and modification of the catalytic properties in reactions with the participation of H_2 or CO [715, 716]. In the catalyst Ru-MoO_3/Al_2O_3 [715], the chemical states Ru^{4+}, Ru^{3+}, Ru^{2+}, and Ru^0 have been detected, MoO_3 affecting the ability of Ru to be reduced. In the catalyst Pd-CuO/Al_2O_3 [716], interaction of the components changes the forms of adsorption of CO and oxygen and, in the long run, results in synergism in the low-temperature oxidation of CO.

An effective way of passivating the poisoning effect of metals (Ni, V, Fe) deposited on the surface of cracking catalysts from heavy oil products is their treatment with compounds containing antimony or tin [717]. Consequently, the resulting compounds are similar in their properties to bimetal supported systems. Moreover, Parks et al. [718] explained the passivation of the undesirable dehydrogenating effect of Ni in the presence of Sb by interaction of the components, which by XPS and XRD data resulted in the formation of an alloy (500 °C, H_2) and in blocking of part of the Ni centers because of antimony surface segregation. The concentration of the metals in these systems was much higher (5%) than in real ones (0.1-0.6%). More detailed studies employing XPS and SIMS [218, 448, 558, 559, 719-722] revealed that the state of the nickel and its distribution on the surface of a cracking catalyst depends both on the concentration of the deposited phase and the calcination and reduction conditions, and on the SiO_2/Al_2O_3 ratio in the matrix. In specimens rich in SiO_2 (including special additives of Si-containing minerals), the formation of nickel silicate was observed, which facilitates the "deactivation" of the nickel [448, 722]. Conversely, in specimens rich in Al_2O_3, it is presumed that nickel aluminate forms [558, 559], although a part of the Ni is reduced to Ni^0. The deposition of antimony onto such a specimen leads to a practically monolayer distribution of the passivator, treatment in H_2 not changing the state of the antimony, i.e. Sb(V) [558, 559]. Having in view that in the course of cracking the catalyst is reduced in the reaction mixture for a short time with subsequent oxidizing regeneration, a passivation mechanism associated not with the formation of an alloy, but with attenuating the interaction of the Ni with the support [558, 559] seems to be more probable. The antimony is localized at the Ni-catalyst interface which leads to a growth in the size of the particles calculated by Kerkhof's model [185]. The catalytic properties are also possibly modified because of Ni-Sb_2O_5 interaction, like strong metal-support interaction.

Hence, the active components and promoters interact in different ways in two-component supported catalysts. Sometimes, bimetal clusters form on a surface, whose presence is confirmed by data of XPS, SIMS, EXAFS, or XRD. Their behavior (Ru-Co, Ni-Co, Pd-Cu) is qualitatively similar to what is observed for unsupported alloys, which enables one to predict the change in the properties of bimetal clusters on supports proceeding from data for unsupported alloys. There are also substantial distinctions, however. They are connected with the presence on a surface of unreduced components, separate phases of a metal, and bimetal particles inhomogeneous in their composition. Another type of systems consists of metal clusters stabilized on ions or charged promoter clusters. The probability of the mutual transition of the two types of centers depends, in addition to other factors, on the reaction medium. Under the conditions of the reaction $CO + H_2$, a substantial part of the cobalt in Ru-Co catalysts transfers into the ionic state owing to the formation of oxide-like and carbide-like structures [358]. Under the conditions of synthesis of allyl acetate ($C_3H_6 + O_2 + CH_3COOH$), the transition of Pd^0 to the state $Pd^{\delta +}$ is observed in Pd-Cu/DM catalysts, the copper promoting this change [222]. As already indicated, in a medium of hydrocarbons (reforming, aromatization), the promoters may be reduced more effectively (Re, Sn, In, Cr), and, consequently, the fraction of bimetal clusters grows. The relation between the real structure of a surface and its catalytic properties will be discussed in the following chapter.

5 Relation between the Characteristics of a Surface and Its Catalytic Properties

When dealing with separate classes of catalysts in the preceding chapters, we repeatedly noted which of the properties studied by EES and IS are of the greatest importance for revealing the nature of the activity of catalysts and the mechanism of their action. We also pointed out the significant role of the reaction medium in the formation of an active surface. At the same time, the detailed uncovering of the mechanism of catalysis and the optimization on this basis of catalytic activity and selectivity require comparative studies of the main characteristics of a surface and the catalytic properties on the same objects. Such studies should also include measurements of the reactivity of a surface relative to individual reagents and reaction mixtures. The latter, however, is possible only for model objects within a limited interval of low pressures (10^{-7}-10^{-1} torr) [1, 2, 10, 18, 20, 24, 141]. These investigations are of a paramount importance for studying the mechanism of elementary events of catalysis. But in pursuing the main goal of our monograph, we shall attempt to give an idea of publications describing the results of studying the surface of real catalysts by electron and ion spectroscopies in combination with their catalytic properties. One of the main ways of establishing the relation between these parameters is the finding of correlations between the valence state of elements, the effective charge, the composition of a surface, and the catalytic activity and selectivity. Some examples of quite simple correlations of the type of "degree of reduction-activity" or "composition of surface-activity" are given in monographs concerning the application of EES and IS in catalysis [6, 7, 9, 15-27]. These monographs also explain the reasons for the differences in the activity (including deactivation) of catalysts that are close in composition but differ in the technology of their preparation and service. Such approaches probably have the right to exist and may sometimes tell specifically how a preparation method or process conditions must be changed to optimize catalytic properties.

At the same time, it is more fruitful to systematically study the structure and composition of a surface in the preparation stages preceding catalysis, and also in separate stages of a catalytic reaction (for example, in the initial stage when nonsteady-state activity is observed, in the steady state, and in deactivation or reactivation) in connection with the catalytic properties. The pretreatment and measuring procedures described in Chap. 2 enable one to conduct such research on serial instruments in principle.

The present chapter on the basis of several examples considers the correlations between the surface and catalytic properties established for various classes of catalysts and analyzes how valuable they are for establishing the nature of the activity of the catalytic systems of selected industrial or new prospective processes. Examples of the latter are the reactions of conversion of synthesis gas (CO/H_2) into methanol or other organic products (the Fischer-Tropsch synthesis), reactions of hydrocarbons and, primarily, the activation of stable molecules of the lower hydrocarbons, and hydrodesulfurizing and reforming of petroleum fractions. While understanding that such a separation is conditional, we nevertheless decided to retain the same structure of this small chapter as of Chap. 4.

5.1   Catalysts Based on Alloys and Intermetallics

The cluster and ligand effects in catalysis using alloys were dealt with in detail in Chap. 4. An important conclusion in the latest studies of the reactivity of thin-film bimetal catalysts by LEED, HREELS, XPS, AES, etc. [370-386] consists in that the formation of an alloy on a surface does not obey the laws of mutual solubility or thermodynamics of the formation of bulk solid solutions or intermetallics. Moreover, the conditional nature of the concept of an "active" and "inactive" component of an alloy has become clear. At different surface coverage, the second component may either directly react with the first one to form the same short-range order as in the structure of an intermetallic, or simply dilute or block the centers of the first component. The nature of interaction of the two components on a surface affects their reactivity, e.g. with respect to CO [341-345], differently and, consequently, may affect the degree of synergism of their action. With a bulk structure of a solid solution or intermetallic, of importance is the composition of the first and following surface layers, which may change additionally in the course of the reaction.

Single crystal alloys $Pt_xNi_{1-x}(111)$ have a higher activity in the hydrogenation of 1.3-butadiene than pure Pt(111) at conversions under 50% [282]. They also have a much higher selectivity in the formation of butenes. But with an increase in the conversion, the initial selectivity rapidly drops on $Pt_{0.78}Ni_{0.22}$ and does not change on $Pt_{0.5}Ni_{0.5}$. One of the assumptions explaining these differences is based on the existence of correlation between the selectivity and the contraction of the first layer of Pt-Pt: for $Pt_{0.78}Ni_{0.22}$, the thickness of this layer is $d = 2.72$ Å, for $Pt_{0.5}Ni_{0.5}$ it is $d = 2.65$ Å, and for $Pt_{0.1}Ni_{0.9}$ it is $d = 2.52$ Å. However, this is not consistent with the behavior of these alloys in reactions of hydrogenation of benzene or cyclopropane, where the first and third specimens have a close activity, while the activity of the second specimen is low. It is apparently necessary to take into account the differences in the concentration of Ni in the second layer [282] and the different degree of its interaction with Pt. In our early studies [728, 729], we discovered a strong increase in the catalytic activity of alloys of Pd with Rh or Ru in a reaction of dehydrogenation of cyclohexane, which increased with a growth in the duration of treatment of the alloys with the reaction mixture. The catalytic effect was attended by a sharp increase in the concentration on the surface of rhodium or ruthenium — the more refractory components than palladium. Such a segregation can be understood if we take into account the results of calculations [283]. The latter show that the presence of strongly adsorbed carbon on the surface of Pd-Rh or Pd-Ru alloys will cause inversion of segregation because ΔH_{ads} of carbon on ruthenium is 0.5 eV higher than on Rh and 1.0 eV higher than on Pd. It is also quite possible that the nature of segregation of the components on the surface also changes substantially for Pt-Ni alloys in the presence of hydrocarbons. This leads to various catalytic activity.

Other typical correlations of the catalytic activity and surface composition have been obtained for systems based on hydrides of intermetallics, which were subjected to oxidative segregation [398-405] (Fig. 81). The activity of catalysts based on the intermetallic hydrides Zr-Ni-H, Zr-Co-H, Zr-Cr-H, Zr-Mo-H, and Zr-W-H in reactions of hydrogenation and isomerization of 1-hexene, aromatization of cyclohexane and normal hexane is much higher than in the relevant intermetallics, while sometimes it exceeds the activity of the relevant pairs of metals supported on zirconium oxide. XPS data reveal that this is due to the segregation of the active component on the surface in oxidizing treatment. For the system Zr-Ni-H, the activity grows on additional treatment in a mixture of hydrocarbons, which is attended by the additional diffusion of Ni to the surface [398, 399]. A similar effect is typical of the catalysts Th-Ni in the reaction of CO and H_2 [401]. Examination of Fig. 81b reveals that a qualitative correlation is observed between the rate of methanation and the surface concentration of Ni determined by ISS. The absence of a quantitative relation between these parameters is due to the surface layer containing phases

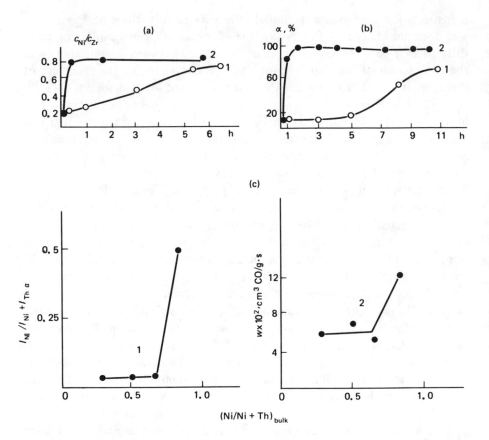

Fig. 81. Correlation between the catalytic activity of intermetallics and their hydrides and the surface composition:

a — change in the composition of the surface of $ZrNiH_{2.8}$ (350 °C) in time: 1 — under conditions of a catalytic reaction; 2 — in a stream of air;

b — dependence of the catalytic activity of $ZrNiH_{2.8}$ in the conversion of toluene on the time: 1 — in a stream of hydrogen; 2 — after preliminary oxidation in air at 350 °C [399];

c — dependence of the ratio of the ISS signal intensities [Ni]/([Ni] + [Th]) (1) and the rate of hydrogenation of CO (2) on the bulk composition of Ni_xTh_y alloys [401]

differing in their activity, namely, phases lean (Th_7Ni_3) and rich ($ThNi_5$) in nickel [401].

We have noted that the surface composition of Fe/Mn catalysts of the synthesis of alkenes from CO and H_2 when reduced in hydrogen and treated with the reaction mixture depends on the conditions of preparation (alloys, coprecipitated oxides) and the ratio Fe/Mn. Alloy systems are characterized by the surface becoming lean in Fe up to the formation of an MnO film with dissolved metallic Fe [411]. Similar systems of Fe incorporated into a matrix of Ti form when a TiFe alloy is treated with ammonia [730]. The surface of the alloy RuTi is completely covered by Ti nitride. As a result, these two catalysts differ sharply in the catalytic activity of ammonia synthesis [730]. Such qualitative correlations are undoubtedly helpful, although they do not allow one to determine the structure of the active centers to the end.

In a more detailed investigation performed by AES, XPS, UPS, and ELS, Biwer and Bernasek [410] attempted to study the electron structure of nitrided FeTi — a catalyst employed in ammonia synthesis, and to explain its increased activity. As in other investigations, they observed the disintegration of the FeTi lattice and the formation of TiN. But by the data of all the employed techniques, they did not succeed in detecting a substantial transfer of the charge between the Fe and TiN. The position of the inner levels of Fe, Ti, and N remained the same as in the separate phases. The absence of electron interaction was also noted for the catalyst Fe/TiO_2 reduced at 980 K and containing Fe particles about 10 nm in size [731]. Biwer and Bernasek [410] considered that the growth in the catalytic activity of FeTi is due to the structural changes associated with the morphology of TiN, in particular with the degree of exposure of Ti on the surface itself and the resulting activation of the N_2.

In a series of investigations performed by Lambert and his co-workers [383-386] that we described in Sec. 4.1, comparative studies of thin films and oxidized polycrystalline alloys were performed. They made it possible to arrive at interesting conclusions on the nature of centers responsible for the extremely high activity of systems consisting of copper and rare earth metals in the synthesis of methanol. The systems most probably include atoms of copper localized at the Cu/REM oxide interface, the latter having an appreciable deficiency of oxygen. It is also possible that rare-earth metals (La, Ce) are present in the oxides in an unusual bivalent state. The catalysts Cu-REM of the lutecium subgroup (Gd, Tb, Dy, Ho, etc.) obtained by deposition of the relevant intermetallics exhibit synergism of their catalytic effect in the reaction of o-p conversion of hydrogen, which in the opinion of Zhavoronkova et al. [732] is due to the direct electron interaction of two metals. But judging from the data of XPS, the rare-earth metals under the studied conditions (vacuum of 10^{-6}-10^{-7} torr) are partly oxidized so that when discussing the nature of the activity attention must be given to the formation of Cu centers at the interface with the REM oxide.

Consequently, many solid solutions and intermetallics under conditions of even low-temperature reactions are not monophase systems, and in addition to the direct interaction of two metals, metal-oxide interaction occurs, which can substantially modify the catalytic activity of these systems. This conclusion holds to a still greater extent for catalysts of the Raney type whose activation or deactivation is associated with segregation of additives on the surface, encapsulation of active particles by Al oxides, etc. For example, the surface segregation of Mo, Fe, and Ti in the leaching or treatment in H_2 of modified Raney nickel catalysts stabilizes the active structure of the catalyst of hydrogenation of anthraquinone into hydrogen peroxide [335]. Conversely, nickel in the Ni-Al-Ti system is enveloped with Ti oxide in the high-temperature treatment with hydrogen, which lowers its activity in the hydrogenation of CO to C_1-C_4 hydrocarbons [733]. In this case, an effect similar to SMSI is observed. Alloys of Cu promoted by Cr and Mo gradually became deactivated in the process of reduction of acetonitrile [734]. But no surface segregation of the additives was observed for them in the course of the reaction. The surface oxidation of the nickel and the additional leaching of the aluminium are a more probable cause of deactivation.

5.2   Oxide Systems

The nature of the activity of oxide catalytic systems (or oxide precursors) has been studied in the greatest detail by EES and IS in the reactions of hydrodesulfurization of hydrocarbons, conversion of synthesis gas, and the low-temperature water-gas shift reaction of CO by steam. Two universal relations were established in the reaction of hydrodesulfurization that have been discussed in a number of reviews and monographs [21-25, 27] (Fig. 82): (a) an extremal dependence of the catalytic activity on the ratio [Co]/[Co] + [Mo]. A peak is observed at a value of $[Co]/[Mo]_{surf}$ of 0.2-0.3 regardless of the conditions of preparation of the aluminocobalt-molybdenum catalysts. The second relation is a volcano-shaped curve of the change in the activity depending on the S/Mo ratio. Studies of the precursors of the catalysts Mo/Al_2O_3 and Co/Al_2O_3 after their reduction and sulfidizing have shown that the peak in the activity depending on the composition of the surface may be associated with its structure, in particular with the amount of cobalt, which should transfer into a sulfide to ensure the required promotion of the Mo. The local structure of the active surface of the hydrodesulfurizing catalysts includes the dispersed particles of molybdenum sulfide partly covering the cobalt sulfide of the type of Co_9S_8, the latter, in turn, being on a "pad" of $CoAl_2O_4$ [23]. The presence of Co^0 or Mo^0 in active catalysts has not been

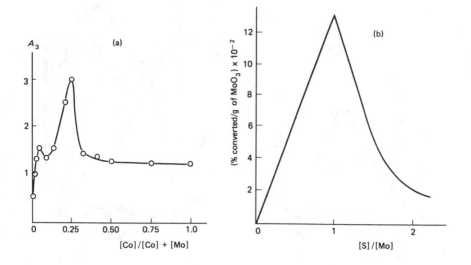

Fig. 82. Correlation between the activity of Mo-containing catalysts in a hydrodesulfurization reaction, the composition of the surface (a) and the degree of sulfurization (b) [21-25, 27]:

a — dependence of the specific catalytic activity (conversion in $\% \cdot m^{-2}$ in the hydrogenolysis of thiophene on the ratio $[Co]/([Co] + [Mo])$; b — dependence of the specific catalytic activity of Mo ($\%$ conv./g of MoO_3) in the reaction of hydrodesulfurization of thiophene at 400 °C on the degree of preliminary sulfurization of Mo in catalysts MoO_3/Al_2O_3

proved; this is rather difficult to ascertain by XPS because E_b of the inner levels of the relevant metals and sulfides are close. The dispersion of the molybdenum sulfide also plays a key role in ensuring a high catalytic activity. By the data of XPS, laser-Raman spectroscopy, and NO adsorption [735], an inverse relation is observed between the turnover frequency (TOF) in the reaction of thiophene hydrodesulfurizing and the dispersion of the sulfidized Mo provided that preliminary sulfidization was conducted in a mixture of H_2S and H_2. If sulfidization is conducted in H_2S alone, the TOF is higher and does not virtually depend on the dispersion. The different relations are due to the fact that the ratio between the two types of coordination-unsaturated Mo^{4+} centers changes in the catalyst. The first type has two coordination sites, and the second has three. The latter center has a higher activity. As a whole, the TOF is determined both by the number of coordination-unsaturated Mo-S centers and by their ratio. This explains the volcano-shaped dependence of the activity on the degree of sulfidizing observed in earlier works [23, 560]. These new results detail the structure of the active centers of alumicobalt-molybdenum catalysts in hydrodesulfurization reactions. By data of Kon-

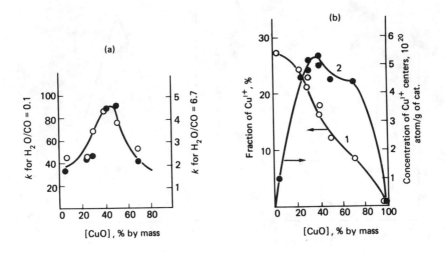

Fig. 83. Correlation between the content of Cu^{1+} ions in CuO-ZnO catalysts and their activity in the conversion of CO with steam [208]:

a — dependence of rate constant (molecule/atom$\cdot m_2$ Cu^{1+} $\cdot s$) per ion of Cu^{1+} on the content of CuO;

b — fraction (1) and concentration (2) of Cu^{1+} ions in catalysts CuO-ZnO activated in a stream of 5% H_2/Ar with heating to 300 °C on the CuO content

dratyeva *et al.* [736], the promoting effect of Co(MoS$_2$) or Ni(WS$_2$) is of an electron nature: XPS of the valence bands of the catalysts and calculations by the EHM method point to rising of the Fermi level when promoters are introduced, the position of the Fermi level and the catalytic activity of the deposited sulfides correlating with each other.

We have already noted the interest in studying the state of copper in catalysts of methanol synthesis and the water-gas shift reaction. Figure 83 depicts the correlation between the activity related to one Cu^+ ion and the ratio CuO/ZnO [208]. The maximal activity is achieved at 30-40% of CuO; here the number of Cu^+ ions formed after reduction of the catalyst is also maximal. The accuracy of the quantitative determination of Cu^+ is evidently not high because the intensity of the spectra is low and the main Cu^+ peak in Cu *LVV* is superposed onto the shoulder of the more intensive Cu^0 peak. Nevertheless, the conclusion that Cu^+ is involved into an active center of the water-gas shift reaction can be considered as proved. This state was clearly identified in skeleton Cu catalysts in the stages of their reaction with CO, H_2O, and also the reaction mixture of CO + H_2O (steam) under the conditions of catalysis (see Fig. 70). It can be seen (compare Figs. 70 and 84) that

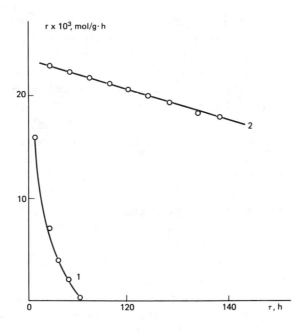

Fig. 84. Dependence of the activity of skeleton copper (Cu-Al) (*1*) and Cu-Zn-Al (*2*) on the duration of the water-gas shift reaction $CO + H_2O$ (vap.) [591]

the different stability of the activity of skeleton copper and the catalyst Cu-Zn-Al is associated exactly with the ability of Cu to form monovalent copper in the course of a reaction. A high activity is achieved if the Cu^0 has a high dispersion and is readily oxidized to Cu^+. The appearance of Cu^+ in the C $1s$ spectrum is attended by a peak with $E_b = 288.4$ eV probably corresponding to a formate complex [589]. These data together with kinetic studies of the water-gas shift reaction on skeleton and oxide Cu catalysts [737] speak up in favor of an associative mechanism with the participation of the formate complex of Cu^+ as an intermediate.

As previously, optimization (lowering) of the amount of copper in a catalyst needed to maintain a high activity is a very important practical task. It follows from an analysis of the dependence of the specific catalytic activity in the water-gas shift reaction on the content of Cu/Zn_{sur} (see Fig. 69) and on the change in the binding energy of Cu $2p$ that mixed $Cu^{(2-\delta)+}Zn^{\delta+}$ clusters in which the copper ions have a deficiency of a positive charge are the precursors of the active copper center. We can assume that by increasing the fraction of such groups (by varying the way of preparing them), we can lower the fraction of the "ballast" copper and, consequently, its total content. The matter of the active state of copper in the synthesis of methanol is not so clear. The more reducing atmosphere of the reaction reduces the major part of the copper to the zero-valent state [592], and only in chromites does part of the copper remain in the state Cu^{+1} [594]. All this allows us to presume that zero-valent copper is active in methanol synthesis. At the same time, the sensitivity of XPS is insufficient to completely exclude the presence of Cu^{+1} at the interface of Cu^0 and a solid solution of Cu^{2+} in ZnO assumed to be the active center of this reaction [587, 592]. The latest data of XPS and XAES obtained for cement-based copper containing catalysts (of the composition $CuO\text{-}CaAl_2O_4\text{-}CaAl_4O_7$) used in the hydrogenation of butyric aldehyde reveal [738] that Cu^{+1} can not only form in the course of a reaction, but can also facilitate the dissociation of hydrogen on metallic copper. The activity of these systems is higher than that of the binary oxides CuO-ZnO in which the copper was reduced completely to the metallic state.

The cations of oxides in intermediate valence states are considered as the active centers of quite a number of reactions. In an alkene metathesis reaction, the active carbene complex is assumed to form exclusively on ions of an intermediate valence [Re(VI, IV), Mo(V, IV, II), W(IV, II)], or even on zero-valent atoms of the metals ([460] and references therein). A detailed investigation of how the catalytic activity and state of tungsten and molybdenum depend on the activation conditions (medium, temperature, duration of treatment) of the catalysts WO_3/Al_2O_3 and MoO_3/Al_2O_3 [198, 204, 451] made it possible to propose a way of forming carbenes with the participation of coordination-unsaturated cations in their higher states of oxidation. Figure 85 shows the dependence of the turnover frequency in a propylene metathesis reaction on the conditions of activation of two WO_3/Al_2O_3 catalysts containing different amounts of WO_3. For the specimen with a low WO_3 content, high-temperature treatment in argon leads to a higher activity than treatment in H_2, the activity sharply diminishing with elevation of the reduction temperature. For 23% WO_3/Al_2O_3, the theoretical coverage approaches a monolayer one and at 923 K in hydrogen a higher activity is reached than for treatment in argon, although the activity drops again with further elevation of the reduction temperature. By the XPS data obtained on the same specimens [451], the antibate dependence of the activity on the reduction temperature T_{red} indicates that the W^0 formed at high temperatures does not participate in a metathesis reaction. The growth in the activity when the 23% WO_3/Al_2O_3 catalyst is reduced at 923 K is connected with the transition of a part of the W^{6+} ($\sim 10\%$) into

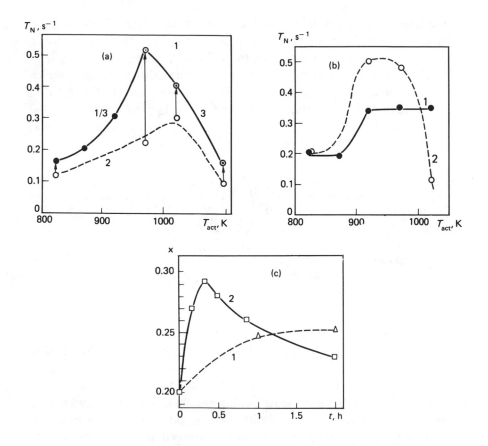

Fig. 85. Influence of the preliminary treatment conditions on the catalytic activity (turnover frequency) for 11.3% (mol) of WO_3/Al_2O_3 [451] (a) and 23.3% (mol) of WO_3/Al_2O_3 (b) and (c)

a — length of activation period, 2 h, activation medium • — Ar (curve 1), ○ — H_2 (curve 2), ⊙ — H_2 with additional 2 h Ar treatment (curve 3 partially identical with curve 1) (the arrows indicate the effect of the Ar treatment); b — variation of the activation temperature; length of activation period, 2 h; activation medium • — Ar (curve 1), ○ — H_2 (curve 2);
c — variation of the length of the activation period; activation medium H_2; activation temperature: ▲ — 923 K (curve 1), □ — 973 K (curve 2), $t = 0$ corresponds to a 2 h activation in Ar at 923 K or 973 K

W^{4+}, which when T_{red} is raised reduces to the metal. Hence, the catalysts WO_3/Al_2O_3 have two types of centers; the first is W^{6+} whose activity grows substantially (by one or two orders of magnitude) in the high-temperature treatment of the specimens in an argon atmosphere [451]. It is quite possible that this is due to the removal of the OH^- groups neighboring with the cations and to the increase in the coordination-unsaturation of the W^{6+} [451]. The second type of centers is W^{4+} ions whose appearance appreciably raises the activity of the specimen with the high WO_3 content. The turnover frequency achieved on these centers is estimated to be $3\text{-}4\,s^{-1}$, which indicates their exceedingly high reactivity. However, on the main part of the centers — the cations W^{6+}, the reaction proceeds at a sufficiently high rate (TOF $= 0.5\,s^{-1}$).

By analogy with the schemes considered in the literature [739, 740], two schemes of the formation of a carbene complex on these catalysts have been proposed:

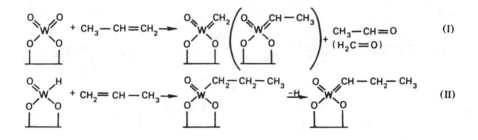

The final form of the complexes is similar.

The reduction of Cr^{3+} to Cr^0 in alumina-chromia catalysts discovered by Grünert *et al.* necessitated a more attentive analysis of the data on the aromatization of normal hexane on typical catalysts [741]. It was found that the Cr^0 most probably does not participate in this reaction, while the development of the catalyst in time is associated with the removal of water. The formation of Cr(0) plays a more important role in the conversion of hydrocarbons on the catalysts Pt-Cr/Al$_2$O$_3$ and Pt-Cr/H-ZSM-5 (see Sec. 5.4).

The possibility of the participation of an unusual form of copper — Cu^{3+} — in the catalysis of the decomposition of N_2O has been discussed by Christopher and Swamy [742]. The copper in this state, judging from the data of chemical analysis, is stabilized in a spinel of the composition $La_{1.85}Sr_{0.15}CuO_{3.95}$. This catalyst is more active than specimens containing Cu^{2+} (La_2CuO_4) and Cu^{1+} ($CuAlO_2$). But Cu^{3+} was not detected in XPS spectra, which was explained by its reduction to Cu^{2+} in a vacuum of 10^{-9} torr [742]. In another publication devoted to studying the nature of the superconductivity of such systems, Steiner *et al.* succeeded in detecting Cu^{3+} by XPS according to the shift of the Cu $2p_{3/2}$ line [743].

Mixed oxides are used as precursors or active components of Fischer-Tropsch synthesis catalysts in addition to alloys. The identification of various intermediate compounds of metals with carbon-containing species is of sig-

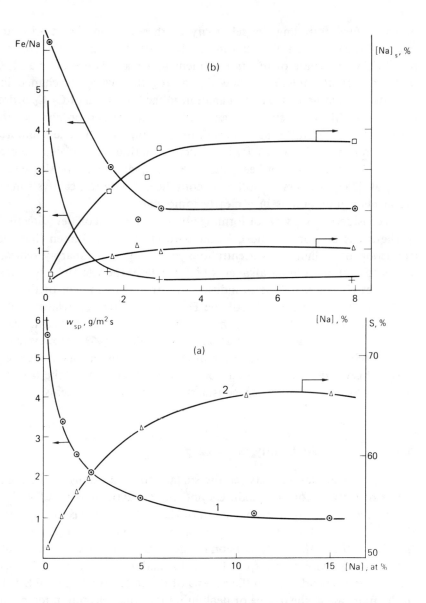

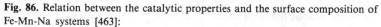

Fig. 86. Relation between the catalytic properties and the surface composition of Fe-Mn-Na systems [463]:

a — dependence of the activity (*1*) and selectivity (*2*) for C_2-C_4 alkenes on the Na content; *b* — dependence of the Fe/Na ratio on a surface and the surface concentration of Na on its content in the bulk for calcined (x, △) and reduced (⊙, ⊡) specimens

nificance for controlling the selectivity of these systems in addition to the trends of the reduction of metals in Fe-Mn oxides [463] discussed above. Employing the procedure of *in situ* treatment under a pressure of 15 atm, Copperthwaite *et al.* [461, 744] showed that together with the change in the selectivity of synthesis (from methanation to the formation of C_5-C_6 hydrocarbons) on Fe-Mn catalysts the shapes of the surface carbon change from a carbide state to one differing from carbide and graphite. Of no less importance for the control of the selectivity is the determination of how alkaline promotors act. Their role has not been elucidated completely even for single crystals of metals [332]. Matters are still more complicated with real catalysts in which the alkaline promotor is in a clearly ionic state.

In considering two ways of forming alkanes on Fe-Mn oxides (directly from synthesis gas or from alkenes), Kuznetsova *et al.* [463] showed that the introduction of sodium at concentrations exceeding 1.5% sharply retards the rate of formation of lower alkanes, which increases the selectivity with respect to alkenes (Fig. 86). XPS data indicate the strong enrichment of the surface in sodium oxides, the activity of the Fe-Mn-Na system correlating with the ratio Fe/Na on the surface, while the selectivity with respect to alkenes changes similarly to the surface concentration of the sodium [463]. However, no appreciable change in the electron state of the active components in the presence of sodium was detected. We shall revert to the discussion of the laws of the Fisher-Tropsch synthesis when considering metal-zeolite systems.

5.3   Zeolite Catalysts

Correlations between the surface concentrations of the main elements of a framework or modifiers and the catalytic activity, as well as the noticeable changes in the composition of the surface in the course of the reaction were observed in processes of alkylation of aromatic compounds by alcohols [501, 502, 745]. The conversion of methanol in the alkylation of toluene, and also selectivity with respect to the formation of *p*-isomers of xylene or ethyltoluene depend both on the genesis of the catalysts (the initial Si/Al ratio in the framework, the degree of dealumination, the calcination temperature), and on the reaction temperature [501, 502]. The influence of the calcination temperature on the $(Si/Al)_{sur}$ ratio and the catalytic activity is especially noticeable (Fig. 54, Table 25). Elevation of the temperature from 400 to 700 °C is attended by a lower conversion of the methanol and a simultaneous increase in the selectivity in the formation of *p*-isomers of xylene and ethyltoluene.

Table 25. Influence of the Catalyst Composition, Modification Method, and Calcination Temperature on the Catalytic Properties of H-ZVM in the Reaction of Toluene with Methanol [502] (T_r = 350 °C, $C_6H_5CH_3/CH_3OH$ = 2/1)

T_{calc}, °C	Conversion of CH_3OH, %	% of toluene disproportionation	Composition of arom. hyd. C_8H_{10}, %				$(Si/Al)_{sur}$	$(O/Si)_{sur}$
			o-xylene	m-xylene	p-xylene	ethylbenzene		
			H-ZVM (x = 45.5)					
400	100	2.6	19.0	51.0	28.5	1.5	—	—
500	100	2.6	22.5	47.0	29.0	1.5	25.5	1.4
600	100	6.0	18.0	49.0	31.0	2.0	35.9	2.7
700	100	2.2	18.0	47.0	34.0	1.0	39.9	2.1
800	98	0.5	18.0	42.0	39.0	1.0	39.5	2.0
900	95.5	—	25.0	30.0	45.0	—	39.7	2.1
500←900[a]	98.5	0.2	25.0	32.0	43.0	—	—	—
			H-ZVM[b] (x = 56)					
500	100	3.5	21.0	49.0	28.0	2.0	32.7	1.9
600	100	6.0	20.5	52.0	25.0	2.5	37.4	2.2
700	100	2.5	20.0	51.0	27.0	2.0	54.6	2.2
800	99	0.8	19.0	46.0	34.0	1.0	67.6	2.0
900	96	0.2	28.0	28.5	43.0	0.5	67.2	2.0
500←900[a]	98	0.3	28.0	29.5	42.0	0.5	—	—

[a] Regeneration at 500 °C.
[b] Dealuminated with HCl.

The growth in the *para*-selectivity correlates with the increase in Si/Al on the surface, registered by XPS. At 900 °C, the catalysts are irreversibly deactivated, and the ratio $(Si/Al)_{sur}$ changes only slightly. The noted change in the surface composition is apparently associated with the emergence of Al from the framework and its migration into the zeolite channels at high temperatures. In a preliminarily dealuminated specimen of H-ZVM', a part of the extraframework Al may remain on the surface after treatment with HCl, and then migrate into the channels. The increase in the *para*-selectivity can be explained by the reduction of the contribution of the acid centers on the outer surface to the proceeding of the reaction, and also by the change in the shape of the channels because of the migration of aluminium into them. At 800-900 °C, strong dehydroxylation of the zeolite framework occurs, which leads to deactivation of the catalyst.

The reaction medium additionally affects the composition of the surface, especially at 400 °C. First of all, water facilitates the dealumination of the framework, which was noted when studying the reactions of methylation [502] and ethylation [745] of toluene. Additives of MgO, whose distribution depended substantially on the method of preparation and the treatments [501, 502], are still more labile. As noted above, an optimal relation between MgO_{sur} and MgO_{bulk} is needed. The role of the surface located MgO consists in lowering the acidity of the outer surface of the zeolite crystals, while the role of the bulk located MgO consists in reduction of the overall acidity of the zeolite and in changing the cross section and shape of the channels up to their blocking. In the studied catalysts, such optimal activity and *para*-selectivity were achieved when introducing 6% of MgO by impregnation [502]. The reaction

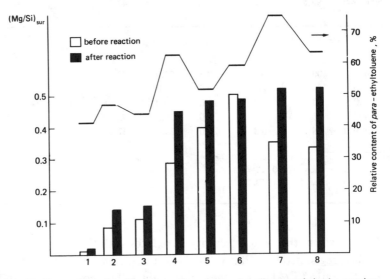

Fig. 87. Change in the Mg/Si ratio on the surface of zeolite crystals in the reaction of toluene and methanol and the para-selectivity of different MgO/H-ZVM systems [502]:
1 — 0.66 MgNa-ZVM; *2* — 6% MgO/H-ZVM ($x = 45.5$); *3* — 3% MgO/H-ZVM′ ($x = 56$); *4* — 6% MgO/H-ZVM′ ($x = 56$); *5* — 10% MgO/H-ZVM′ ($x = 56$); *6* — 6% MgO (3% MgO + 3% MgO)/H-ZVM′ ($x = 56$); *7* — 6% MgO/H-ZVM′ ($x = 65$); *8* — 6% MgO/H-ZVK ($x = 68$)

medium again substantially affects the distribution of the additive (Fig. 87). In the majority of cases, the surface concentration of the MgO increased after a reaction, and as for the H forms, this reduced the surface concentration of the aluminium. The diffusion mobility of the MgO in the reaction is connected with the way of its incorporation and the composition of the zeolite: in MgNa-ZVM and 6% MgO/H-ZVM obtained by twofold impregnation (3% MgO + 3% MgO) the changes in the ratios Mg/Si in the reaction are insignificant [502].

In reactions of lower hydrocarbon aromatization, the role of the modifiers (Ga, Zn, Cd, Pt, Rh) consists not only in changing the acidity of the zeolite support, but also in the direct participation in the composition of the active centers in individual states of the process. The data of IR diffuse reflectance spectroscopy [746] have shown that Zn is a strong Lewis center mainly localized in the zeolite channels, whereas the contribution of Ga to the Lewis acidity is less noticeable. To a considerable extent (if not completely), the gallium by the data of IR and EPR remains on the outer surface. On the other hand, it is exactly Ga silicate with the structure of ZSM-5 that is one of the most effective catalysts of propane aromatization [747]. These contradictions are partly eliminated if we analyze the data of XPS and catalysis obtained for Ga-zeolite catalysts differing in the way of their deposition and treatment (Figs. 56, 88) [505].

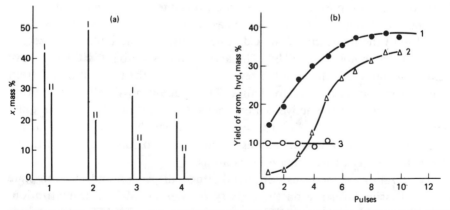

Fig. 88. Catalytic properties of Ga/H-ZVM in the aromatization of ethane and propane [505]:

a — dependence of the conversion of ethane (*I*) and the yield of aromatics (*II*) on the way of preparation of Ga/H-ZVM (reaction, 600 °C, 120 h^{-1}): *1* — ion exchange + impregnation; *2* — impregnation; *3* — mechanical mixture of Ga$_2$O$_3$ + H-ZVM; *4* — H-ZVM;

b — dependence of the yield of aromatics on the number of propane pulses (2% Ga/H-ZVM, T_r = 550 °C); *1* — ion exchange + impregnation; *2* — impregnation; *3* — H-ZVM

A glance at the diagram in Fig. 88 reveals that catalysts in which part of the gallium was incorporated by ion exchange have the greatest activity and selectivity in the aromatization of ethane; conversely, a mechanical mixture of Ga_2O_3/H-ZVM has the lowest activity that does not virtually differ from that of H-ZVM. As we have already noted (see Fig. 56), it is just on the former catalyst that the gallium is distributed more uniformly over the crystal, although part of it is on the outer surface. The development of the catalyst in the pulse mode (Fig. 88b) additionally confirms the possibility of migration of the Ga into the channels and of the increase in its catalytic effect. Additional experiments performed by XPS and HRTEM show that breaking up of the Ga_2O_3 aggregates in the course of the reaction and migration of part of the Ga into the zeolite matrix do occur in the course of a reaction [748]. What the valence state of the gallium is in the reaction is not clear. The absence of noticeable shifts of the Ga $2p_{3/2}$ and Ga $3d$ lines does not make it possible to reach a conclusion on the appreciable reduction of gallium by the reaction mixture [505, 748]. At the same time, a growth in the catalytic activity in the aromatization of propane after the treatment of Ga/H-ZSM-5 in H_2 was noted [506]. The situation with the drop in the ratio of the intensities Ga/Si with direct treatment in H_2 (Fig. 56) in the absence of shifts in the spectra reminds one of the oxidative dehydrogenation of ethane observed for the CdA catalyst [749]. In the latter case, reduction in H_2 also caused neither a shift in the Cd $3d_{5/2}$ spectrum nor a change in the shape of the Cd LVV line, although the Cd/Si ratio dropped two or three times. It is interesting that on subsequent treatment with air, the ratio again increased, and such a reversibility was observed until the reduction of the Cd concentration in the bulk of the zeolite crystals became noticeable [749]. The activity and selectivity in the oxidative dehydrogenation of ethane in the reversible stage of the removal of Cd from the outer surface also failed to change. This made it possible to assume that the composition of the active centers includes the cations Cd^{2+}. It is quite possible that part of the Ga also evaporates from the outer surface in the reaction, while the cations remaining in the mouths of the channels or in the channels themselves participate in catalysis.

Localization of the transition metals, their dispersion and electron state have a decisive influence on the activity of Me/HZSM-5 in aromatization. This is indicated by data of XPS, electron microscopy, EXAFS, and catalysis obtained for Pt/HZSM-5 and Rh/H-ZSM-5 [196, 465, 476, 540]. In both cases (Fig. 89), the yield of aromatics depends on the preliminary treatment of the metal-zeolites: zeolites preliminarily calcined and reduced in H_2 or simply calcined have a high activity, whereas zeolites directly treated with hydrogen are

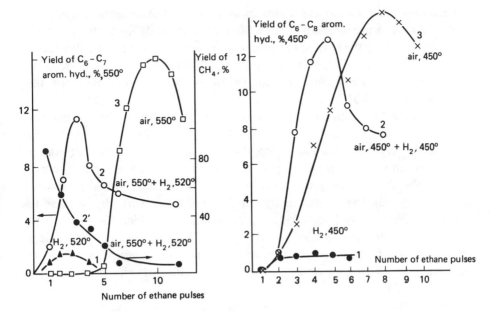

Fig. 89. Dependence of the yield of aromatics on the number of ethane pulses for three series of M/H-ZVM specimens [196]:

a — 0.5% of Pt/H-ZVM (T_r = 550 °C): *1* — H_2, 520 °C; *2* — air, 550 °C + H_2, 520 °C; *2'* — the same, yield of CH_4; *3* — air, 520 °C;

b — 0.5 Rh/H-ZVM (T_r = 450 °C): *1* — H_2, 450 °C; *2* — air, 450 °C + H_2, 450 °C; *3* — air, 450 °C

only slightly active. We have already noted that preliminary calcination is the key stage for the stabilization of highly dispersed platinum or rhodium in H-ZSM-5. We therefore have a correlation between the dispersion of the metals and the activity. As regards the high activity of the calcined specimens, it is associated with the presence in them of highly dispersed Pt or Rh formed as a result of the autoreduction by ammonia, and also the additional reduction of the metals by the reaction medium [196, 475]. To elucidate the relation between the dispersion of a metal (Pt) and its reactivity with respect to C_2-C_3 alkanes, the reaction of hydrogenolysis of ethane and propane was also studied [540]. The turnover frequency in this reaction changes only slightly with a growth in the platinum content in the zeolites that lowers the dispersion somewhat. But the transition from air-hydrogen to hydrogen treatment highly reduces the rate of hydrogenolysis and the TOF. EXAFS data obtained recently for 0.5% and 1.0% Pt/H-ZSM-5 [750] reveal that when employing air-hydrogen treatment, the Pt particles, although they differ in size and the number of atoms in a cluster (18 atoms in a cluster of 8 Å in 0.5% Pt/HSM-5 and 45 atoms in a cluster of 11 Å in 1.0% Pt/H-ZSM-5), in both cases they are coordinated by the oxygen of the zeolite framework, i.e. are probably localized inside the zeolite structure. In specimens directly treated in H_2, judging from the data of TEM and XPS, the major part of the Pt or Rh is on the outer surface [538-540].

The interaction with the zeolite framework causes an electron deficiency of the Pt or Rh particles reduced by hydrogen [538-540]. A comparison of the data of XPS and catalysis in individual stages of the reaction of ethane aromatization (after each ethane pulse) clearly shows that the positive charge (δ^+) on the Pt particles grows additionally in interaction with the reaction mixture (Fig. 90), the growth in the charge correlating with the increase in the yield of aromatics [196, 475, 540]. A specimen in which the Pt prior to catalysis has not been reduced transfers in the reaction into the zero-valent state (five pulses) and then acquires an excess positive charge. Only after its appearance does the reaction of aromatization proceed with a measurable yield (Fig. 90).

We can thus single out the following laws of the aromatization of alkanes on metal-zeolite systems:

(a) a high activity is achieved on catalysts containing metal clusters in channels in direct proximity to the Brönsted acid centers;

(b) metal clusters are deficient in electrons, and the degree of electron deficiency grows in the initial stages of the reaction parallel to the growth in the activity;

(c) there is a definite parallelism in the activity of the metallic particles in two competing reactions of hydrogenolysis and aromatization, although they are measured under different conditions. This is apparently associated with the fact that the initial stage of both reactions — dehydrogenation — proceeds more readily on electron deficient centers, while the further transformations of the dehydrogenated intermediate C_xH_y depend on the conditions of the process (temperature, partial hydrogen pressure), and the presence of acid centers.

The correlation between the activity and the electron deficiency of Pt is also intriguing. It was previously observed for Rh/Al_2O_3 in the aromatization of ethylene [751]. This can presumably be explained by the fact that in the initial stage of the reaction a carbon-containing layer is deposited on the surface. It modifies the state of the Pt and facilitates a growth in the activity. In later stages of the reaction, this layer transforms into an inactive one, blocks part of the centers, and gradually deactivates the catalyst (Figs. 89, 90). The

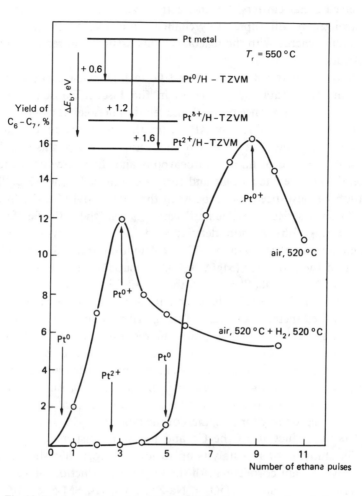

Fig. 90. Development of catalyst Pt/H-ZVM in the aromatization of ethane and the modification of the electron state of platinum in the reaction medium [196]

data of AES and measurements of $\Delta\Phi$ for C/Pt(III) show that as a result of the interaction of Pt atoms with periphery carbon atoms of the graphite islets, the work function of the Pt lowers, and its bond to H_2, C_6H_6, and CO weakens [752]. As a whole, however, the nature of the adsorption of these reactants changes slightly even at substantial surface coverages of the surface with carbon, i.e. a short-range effect is observed. In the presence of very highly dispersed metal clusters, the interaction with the carbon layer will possibly be stronger. It will appreciably change the charge on Pt, and also optimize its binding energy with the reactants. This assumption requires more detailed verification.

Studying of the state of Co and Fe in high-silica zeolites of the ZSM-5 type and their catalytic properties in the Fischer-Tropsch synthesis points [545-548] to the presence of a definite relation between the degree of reduction of the metal determined by XPS after the treatment of the catalysts with hydrogen and their activity. On the catalysts Co/pentasil and Co-MgO-pentasil, the yield of liquid hydrocarbons and the degree of reduction of Co depend on the zeolite module and correlate with each other (Fig. 91a); a direct relation has also been found between the total activity in the hydrogenation of CO on a series of Fe/pentasil catalysts and the ratio Fe^0/Si determined by XPS (Fig. 91b). A more detailed studying of these systems depending on the conditions of preparation (Co), calcination and reduction (Co, Fe), the nature of the promoter (MgO, K_2O), the support (Fe: zeolite, TiO_2, Al_2O_3, SiO_2, MgO) [545-548, 753] shows that in addition to the degree of reduction, the activity and especially the selectivity are determined by the dispersion of the deposited metal, the nature of its distribution on the supports (especially in zeolites), and the metal-support interaction. The preliminary calcination of Co-pentasil or Co-MgO-pentasil causes a sharp change in the selectivity: C_1-C_4 alkanes and alkenes form instead of liquid hydrocarbons [545]. By XPS data and measurements of the chemisorption of CO, this is associated with a decrease in the degree of reduction of Co and amount of chemisorbed CO. Calcination probably forms spinels of the type of Co_3O_4 and $MgCo_2O_4$, which hinders the reduction of the Co and lowers its dispersion.

The studying of Fe catalysts on various supports has shown [753] that if the activity again correlates with the degree of reduction of Fe and grows in the series $MgO < SiO_2 < TiO_2 < Na$-ZSM-5 $< H$-ZSM-5 $< Al_2O_3$, the selectivity of formation of C_2-C_4 alkenes changes practically in the reverse sequence: $Al_2O_3 < H$-ZSM-5 $< SiO_2 < Na$-ZSM-5 $< TiO_2 < MgO$. One of the controversial matters often discussed in the literature is that of the nature of the active centers in the synthesis of alkenes from CO and H_2 on Fe catalysts. The conclusion was arrived at from the data for Fe oxides that Fe_3O_4 clusters are the active group [754]. On the other hand, it is general knowledge that in a mixture of CO and H_2, surface and bulk carbides of Fe form whose predecessor is Fe^0 [744, 755, and references therein]. The data for Fe-zeolite systems reveal that Fe^0 can be a precursor of an active center for the obtaining of alkenes. Figure 92 shows how the changes in the activity and selectivity

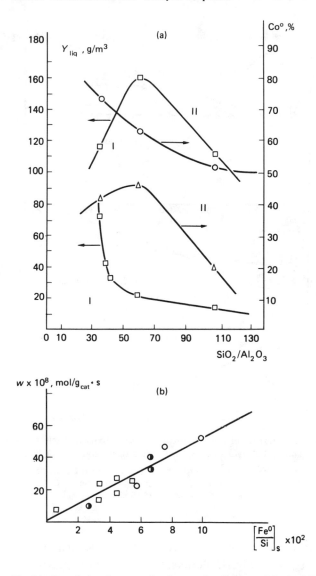

Fig. 91. Correlation between the degree of reduction of Co and Fe in pentasils and their catalytic activity in the Fischer-Tropsch synthesis [545-547]:

a — influence of zeolite module on the yield of liquid hydrocarbons and the degree of Co reduction (XPS) for Co-pentasil (*I*) and Co-MgO-pentasil (*II*) [547];
b — dependence of the activity in the hydrogenation of CO (240 °C) on the surface concentration of metallic iron; ⊙ — 10% Fe/ZVM, uncalcined; ◑ — 10% Fe/ZVM, calcined; □ — 10% Fe/ZVM + 5% K₂O

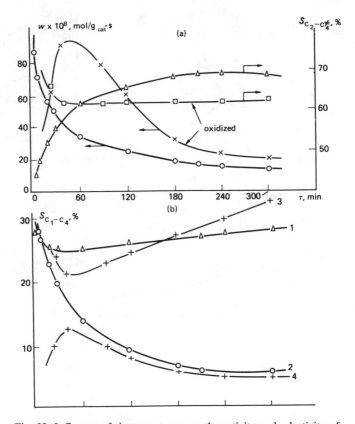

Fig. 92. Influence of time on stream on the activity and selectivity of the catalyst 10% Fe/Na-ZVM in the hydrogenation of CO [547]:

a — activity and selectivity with respect to C_2-C_4 alkenes: *1, 2* — reduced specimen; *3, 4* — unreduced specimen; b — selectivity with respect to C_1-C_4 alkanes: *1, 2* — reduced specimen; *3, 4* — unreduced specimen

depend on the duration of the reaction for unreduced and preliminarily reduced catalysts, while Fig. 93 shows how the Fe $2p_{3/2}$ spectra change in various stages of treatment. Prior to a reacton, the state of the Fe in the two catalysts and their catalytic properties sharply differ. Next the interaction of the unreduced catalyst with the reaction mixture leads to the formation of Fe^0, and with the reduced catalyst to a decrease in the degree of reduction of Fe associated with its transition into the oxide and "intermediate" forms. The latter also appears in the prolonged interaction of an unreduced specimen with the reaction mixture so that the spectra, activity, and selectivity of the

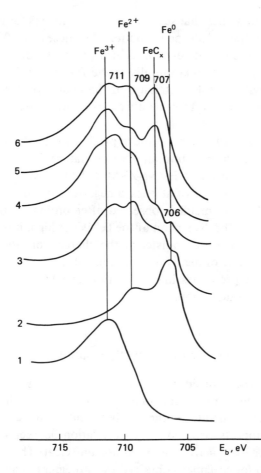

Fig. 93. Changes in the state of Fe in the catalyst 10% Fe-5% K_2O/Na-ZVM during preliminary treatments and the catalytic reaction [547]:

1 — calcined; *2* — H_2, 450 °C; *3* — *2* + CO/H_2, 1 h, 300 °C; *4* — *1* + Co/H_2, 1 h, 300 °C; *5* — *2* + CO/H_2, 6 h, 300 °C; *6* — *1* + 300 C, CO/H_2, 6 h

two catalysts in the steady state are close. With a view to the position of the Fe $2p_{3/2}$ line, it is most probable that the intermediate form of the Fe is a surface compound of the Fe_xC_y type. The presence of an intensive and broad C $1s$ peak with $E_b = 285$ eV hinders the separation of the other types of carbon, which are identified by electron spectroscopy techniques on the surface of pure metals [755].

It can nevertheless be assumed that surface carbide complexes formed from Fe^0 participate in the catalysis of alkene formation. The necessity of reducing Fe to Fe^0 in the initial stage of the reaction is also proved by data obtained for Fe-PO_4 systems [753]. The iron is initially in the framework in the form of Fe^{3+}, and this catalyst has a low activity; in the course of the reaction proceeding at 400 °C, Fe^0 appears (Fe evidently emerges from the framework), and the activity of the catalyst sharply grows.

The data given for Co and Fe point to the appreciable role of the support in forming the activity, which may have an electron nature. For example, in the Fe $2p_{3/2}$ spectra of the catalyst Fe/Na-ZVM + TiO_2 + MgO, which is sufficiently selective in the synthesis of alkenes, a negative shift in comparison with the spectrum of the pure metal is observed after prolonged reduction by hydrogen [485]. But since the dispersion of the Fe^0 is not high, it is difficult to uniquely ascribe this shift to the transfer of the charge from the support to the metal. The electron state of highly dispersed metals (Ru, Rh) on various supports and the catalytic properties in the reaction of CO and H_2 can be compared with greater substantiation.

5.4   Supported Mono- and Bimetal Systems

Practically when considering metal-zeolite systems, we have already mentioned the influence of metal-support interaction on the electron state of metals, the local structure of the catalyst, and catalytic properties. The results for other supports also show that correlation between the degree of reduction and the catalytic activity is often not universal. The degrees of reduction of Ni on SiO_2 for specimens obtained by impregnation and anchoring of $NiCl_2$ are quite close (see Fig. 71) [556], although the dispersion of the metal is somewhat higher for the second specimen. At the same time, the specific catalytic activity of two types of catalysts even in such a slightly hindered reaction as the hydrogenation of benzene differs by an order of magnitude [556]. These distinctions may be associated with the special hexagonal structure of the microcrystallites of Ni^0 on an anchored catalyst: XPS data explain the dependence of the catalytic activity of anchored specimens on the reduction temperature (Figs. 72 and 94) quite well [556]. Specimens in which the maximal part of the metallic support (Ni^0/Si) is available for reagents have the greatest activity (Fig. 72). The drop in the activity at T_{red} exceeding 400 °C is due not only to the sintering of the Ni^0 particles that is not so appreciable, but also to the encapsulation of the metallic phase by silica fragments.

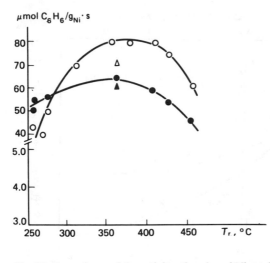

Fig. 94. Dependence of the activity of anchored Ni catalysts in the hydrogenation of benzene ($T_r = 100\ °C$) on the reduction temperature [556];

● — 4.6% Ni; △ — 3.9% Ni; ▲ — 2.6% Ni; ○ — 1.3% Ni

A comparison of the XPS data and measurements of the magnetic susceptibility enables one to explain the reasons for the deactivation of the catalyst Ni/SiO_2 in the benzene hydrogenation reaction in the prolonged treatment of the reaction mixture containing admixtures of H_2 and H_2O:

Stage	Before reaction	After reaction	After regeneration
α_{Ni^0}, % (XPS)	52	21	53
α_{Ni^0}, % (magnetic susceptibility)	56	56	57
d, nm	5.3	6.5	5.4

Here only surface oxidation of the nickel occurs without a change in its dispersion, which makes it possible to readily regenerate the initial activity by brief treatment of the specimen in H_2 or the pure reaction mixture.

The same catalyst reduced by $LiAlH_4$ exhibits an extreme dependence of the activity of hydrogenation in the liquid phase (1-hexene, cyclopentadiene, cyclohexene, cyclopentene, phenylacetylene) on the amount of reducing agent (the ratio Al/Ni) [756]. This correlates neither with the basic characteristics of nickel (the degree of reduction, dispersion) nor with the surface ratios Al/Si or Cl/Si. The dispersion changes slightly, while the other quantities increase

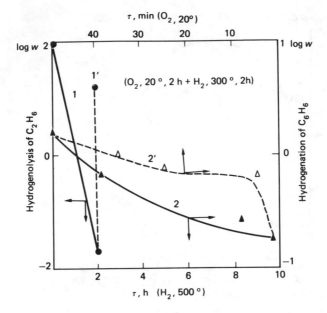

Fig. 95. Influence of high-temperature reduction and treatment with oxygen on the activity of 1% Rh/TiO₂ in the reactions of ethane hydrogenolysis (*1*, *1*) and benzene hydrogenation (*2*, *2*) [630]

with a growth in Al/Ni. All this points to a more complicated relation between the catalytic activity and the structure of the catalyst. Assuming that the hydride groups Ni-H and Li-H are active in catalysis, the drop in the activity at large Al/Ni ratios is explained by Lebedeva *et al.* to be due to the blocking of the active centers by Al-Cl groups. Indeed, beginning from an Al/Ni ratio of two to three (the optimum of the activity), the Ni/Si ratio on the surface is halved.

We noted on an earlier page that the decoration or encapsulation model is widely employed for explaining the chemisorption and catalytic effects of strong metal-support interaction. Figure 95 presents a typical relation between the change in the catalytic activity of Rh/TiO₂ in reactions of ethane hydrogenolysis and the hydrogenation of benzene and the temperature and duration of reduction, and also the dynamics of regenerating the initial activity under the effect of oxygen [610, 630]. The example of the reaction of hydrogenation of benzene, which is not so sensitive to SMSI, shows that both the drop in and the regeneration of the activity depend on the duration of treatment in H_2 or O_2. This can be associated with the diffusion nature of the process, namely, the migration of TiO_x to the surface of the Rh particles, and then with the oxidation of the TiO_x and the transformation of the two-dimensional layer into three-dimensional crystallites of TiO_2 [636]. The relation between the changes in the catalytic activity of this system in the dehydrogenation of cyclohexane and the diffusion processes was indicated by Haller *et al.* [653]. They found that the drop in the reaction rate is proportional to $\sqrt{t}$, where t is the duration of reduction.

At the same time, we would again like to draw attention to the presence of local electron interaction between Me and Ti^{3+}, which has been proved for highly dispersed catalysts by XPS, XAES, SIMS, and EXAFS [630, 682]. The increase in the electron density on the metallic particles and the filling of the d vacancies will unfavorably affect the proceeding of the donor reactions such as hydrogenolysis and hydrogenation-dehydrogenation. The correlation between the electron deficiency of Pt and the rate of transformation of C_2H_6 shown in the preceding section confirms these assumptions.

Investigations of the hydrogenation of CO on Group VIII metals supported on TiO_2 or containing TiO_2 as an additive have aroused the greatest interest and discussion in the literature [636]. Here complicated changes in the activity and selectivity are observed that are associated both with the use of TiO_2 and with the change in the conditions of reduction, of the reaction, etc. Although the data diverge somewhat, the following can be considered as finally established:

(a) quite a few metals on TiO_2 have a higher catalytic activity in the hydrogenation of CO in comparison with other supports. This relates especially to metals with a low activity such as Pt and Os;

(b) supporting on TiO_2 not only changes the activity, but also the selectivity, which generally shifts in the direction of more high-molecular products, alkenes, etc.;

(c) in the state of SMSI, there may occur both suppression and growth in the activity and additional changes in the selectivity.

An interesting example of the catalytic behavior of Ru/TiO_2 in the hydrogenation of CO is given in Table 26 [609]. At 300 °C, the catalyst exhibits a high activity and the usual selectivity in the formation of alkenes. Next the activity gradually diminishes with increasing T_{red}, while the selectivity with respect to alkenes grows sharply and reaches an unusually high value of 90% (mass) at 500 °C. Already with quite mild oxidation (100 °C), the initial activity is practically restored with the retaining of a somewhat higher selectivity. Figure 96 shows the range of the change in the selectivities with respect to

Table 26. Influence of the Treatment Conditions on the Catalytic Properties of 2% Ru/TiO$_2$

Conditions of consecutive treatments	W, mol CO/g$_{Ru}$.s	Selectivity, %							CO/Ru
		CH$_4$	C$_2$H$_4$	C$_2$H$_6$	C$_3$H$_6$	C$_3$H$_8$	C$_4$H$_8$	C$_4$H$_{10}$	
300°C, H$_2$	11.5	34.7	24.2	16.0	14.6	2.5	6.0	1.0	0.11
400°C, H$_2$	4.1	13.7	38.1	12.0	36.3	—	—	—	0.10
500°C, H$_2$	3.8	7.6	88.6	—	3.8	—	—	—	0.10
25°C, O$_2$ + 300°C, H$_2$	3.1	12.3	71.7	—	10.9	1.2	2.6	1.1	0.10
100°C, O$_2$ + 300°C, H$_2$	13.5	9.3	24.0	8.1	33.2	5.7	14.7	4.6	0.11

C$_1$, C$_2$, and C$_3$ hydrocarbons (C$_2$ and C$_3$ are mainly represented by alkenes) depending on the overall rate of hydrogenation [609]. The points on the curves correspond to various degrees of the state of oxidation-reduction of Ru/TiO$_2$ achieved by the change in T_{red}, the duration of reduction, the degree of purification of H$_2$, etc. We can see that conditions can be found when the catalytic system is in an "intermediate" state of metal-support interaction, having a high activity and a sufficiently high selectivity. Investigation of Ru/TiO$_2$ by XPS has shown that beginning from T_{red} = 400 °C, the Ru is partly encapsulated, and, moreover, a small negative charge is observed for it as for Rh [609].

Many aspects of the kinetics and mechanism of hydrogenation of CO on model and real systems of the Me/TiO$_2$ type have been discussed in detail in an excellent monograph [636]. This is why we would like to treat here only the aspects that have become clearer owing to very recent studies not dealt with in the monograph. It must be stressed once more that the reaction of hydrogenation of CO on Me/TiO$_2$ is characterized not only by the change in the activity, but also in the selectivity of the process. The latter is pronounced the most when the catalyst passes into the state of strong or intermediate metal-support interaction. It is difficult to explain this fact only from the viewpoint of the decoration or encapsulation model. The data given in Chap. 4 and here for Pt, Rh, and Ru on TiO$_2$, in our opinion point quite convincingly to the increase in the electron density on metals in the state of SMSI. By analogy with the effect of electropositive promoters (alkali metals),

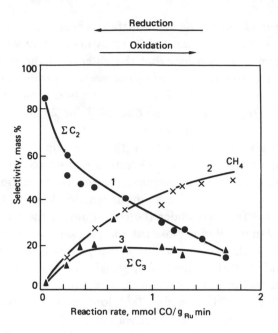

Fig. 96. Range of the changes in the selectivity with respect to C_1-C_3 hydrocarbons and the rate of hydrogenation of CO depending on the state of reduction and oxidation of the surface of Ru/TiO$_2$:

$1 - \Sigma C_2$; $2 - CH_4$; $3 - \Sigma C_3$

this should increase the contribution of $\alpha - \pi^*$ back donation and also the rate of CO dissociation and the concentration of the reactive carbon [754]. A correlation between the negative shift in the Ni $2p_{3/2}$ spectra of nickel catalysts (boride, skeleton Ni, phosphide, and formate) and the ratio C_2^+/CH_4 has been found by Okamoto et al. [757].

Recent data on the kinetics of the reaction of CO and H_2 on pure metals (Pt, Rh) and on the same surfaces coated with various amounts of TiO$_x$ confirm this assumption [758, 759]. The previously expressed by Andersson et al. [656] and Haller et al. [658] doubts that a state of SMSI will hardly be retained under the conditions of the reaction in the presence of H_2O and CO_2 are partly annulled by the data obtained for TiO$_x$/Rh [759]. Levin et al. [759] consider that in the course of the reaction Ti^{3+} is constantly regenerated:

1. Ti^{4+}—O—Me + H (a) $\rightleftarrows$ Ti^{4+}—OH (Me = Rh or Ti^{4+})
2. Ti^{4+}—OH + H (a) $\rightleftarrows$ Ti^{3+} + H_2O

The presence of oxygen (TiO$_x$) on the metal surface is considered as a positive factor in connection with the possibility of its ready detachment with the formation of water. Levin *et al.* consider that the absence of the influence of T_{red} on the catalytic properties is explained by the reaction medium itself forming a definite amount of Ti^{3+} that facilitates the formation of an active center or directly enters one. With a view to the data of Table 26 and Fig. 96, we can agree with this conclusion only partly.

Still another contradiction consists in that the suppression of the chemisorption of H$_2$ or CO at room temperature is observed at TiO$_x$/Me coverages corresponding to the maximal activity in the hydrogenation of CO. Proceeding from an analysis of the change in the reaction kinetics, this effect can be explained by the fact that under the conditions of the reaction, the TiO$_x$ layer having a high mobility is loosened and provides access for the adsorption of CO [758, 759]. As a result, the active centers include atoms of a metal, Ti^{3+}, O, and C [758]. It apparently remains to ascertain the extent to which such a mechanism is suitable for real systems, or whether three-dimensional crystallites of TiO$_2$ containing Ti^{3+} ions form at the phase interface with metallic clusters, as was assumed earlier [656].

Consequently, the SMSI models whose proofs were found by EES and IS techniques make it possible to interpret the involved behavior of metal-supported systems in various reactions and, primarily, in the hydrogenation of CO. At the same time, there is a large lack of clarity as regards the mechanism of the influence of SMSI or intermediate metal-support interaction. The employment of a combination of XPS, SIMS, and EXAFS seems promising for revealing the local structure of real catalysts under the reaction conditions, namely, for appraising the concentrations of Ti^{3+} and the size of the metal particles that may disintegrate in a CO atmosphere [636]. Numerous questions are raised by reactions of hydrocarbons, e.g. alkane conversions, including their hydrogenolysis, that proceed on the surface of metals which do not virtually adsorb hydrogen at room temperature [760].

The catalytic properties of supported bimetal systems are determined by many factors, among which the EES and IS techniques can be used to study, primarily, the states of the components and their ratio on the surface. These techniques are less sensitive to local interaction between the components, which can be appraised indirectly, or by employing other techniques. Let us consider several examples of the comparative studies of supported bimetal systems and their activity. The closeness of the catalytic effect of Co-zeolites and Ni-Co-zeolites in the NO reduction reaction is explained using XPS data by the fact that the Co under the reaction conditions segregates on the surface with the formation of the oxide [475]. Moreover, similarity was found in the change in the surface composition of an unsupported Co-Ni alloy and a supported Co-Ni catalyst in the adsorption of NO, which confirmed the formation of a dispersion alloy.

Detailed studies of the surface composition and reactivity of single crystal Pt-Rh alloys by AES, XPS, FEM, and TDS in vacuum, and also in the presence of NO, O_2, CO, and H_2 enabled Wolf *et al.* [761] to arrive at a number of important conclusions on the surface composition and mechanism of action of real three-way Pt-Rh catalysts for completing the combustion of motor vehicle exhaust gases. As in unsupported alloys, the surface composition (Pt/Rh ratio) in Pt-Rh/SiO_2 catalysts is determined by the bulk composition, as well as by the reaction medium. In lean fuel mixtures (enriched in air), the surface becomes highly enriched in rhodium that partly forms the inactive oxide Rh_2O_3. In this case, the catalytic properties of the alloy particles remind one of the properties of rhodium, although its oxidation in the presence of platinum is hindered. In rich mixtures (a reducing atmosphere), the surface becomes enriched in platinum. The surface composition is additionally affected by impurities in the fuel (P, S, Si, B). Another similarity of unsupported and dispersion alloys is their reactivity with respect to various molecules present in the reaction mixture. The reactivity is determined to a considerable extent by the properties of the individual metals: inhibition of the reactions of CO + O_2 and CO + NO at low temperatures by carbon monoxide reflects the properties of Pt, as well as the high activity of the catalysts in an excess of O_2 due to the weak Pt—O bonds. The rate of dissociation of NO depends on the Pt/Rh ratio on the surface. Consequently, depending on the conditions of the process, the properties of Pt-Rh catalysts may be close to those of the first or second component, while synergism of their effect is practically not observed.

Unlike this catalyst, Co-Ru on γ-Al_2O_3 and on SiO_2 exhibits synergism of the catalytic effect in the reaction of hydrogenation of CO: in alloying, an increase in the rate of the reaction and the selectivity with respect to C_2^+ hydrocarbons is observed [358]. A comparison of these data and of the state of Co and Ru under the reaction conditions in a massive and dispersed alloy shows that the presence of ruthenium increases the concentration of the activated hydrogen, and also accelerates the dissociation of the CO.

Another system — Pd-Cu/zeolite — exhibits a high activity in the synthesis of allyl acetate from C_3H_6, CH_3COOH, and O_2 [475], the copper having a clearly promoting effect [Fig. 97]. There can be seen to be an optimal activity and ratio of the components on the surface. The reaction medium substantially

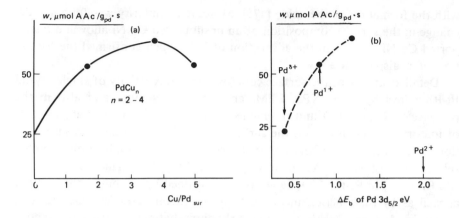

Fig. 97. Correlation between the composition of Pd-Cu clusters (a), the state of Pd (b), and the rate of oxidative acetoxylation of propylene on the catalyst Pd-Cu/DM [475]

changes the state of the Pd and Cu: the copper is oxidized to Cu^+ and Cu^{2+}, while the palladium passes over into the state $Pd^{\delta+}$ (where $\delta^+ \leqslant 1$), correlation of the activity and charge on the Pd being observed. But if the Pd was oxidized to Pd^{2+}, the activity was suppressed completely. These data show that the $PdCu_n$ clusters ($n = 2$-4) can be considered as the precursors of the active centers. Bonds of the type of Pd—O—Cu also form in the course of the reaction, which is confirmed by SIMS data.

The role of the copper probably consists in facilitating the transition of the Pd to an intermediate active state. It is quite possible that in addition to a redox mechanism, a contribution to the reaction on Pd-Cu/DM is introduced by the mechanism proposed for the homogeneous clusters L_mPd_n [762]. The formation of Pd-molecular oxygen adducts is the first stage of this mechanism. The high reactivity of Pd^+ with respect to oxygen was shown by data of EPR in the zeolite Pd/DM [582]. In the zeolite PdCuNaY, no Pd-Cu alloy forms, and the promoting effect of the copper manifests itself only slightly [475].

The effect of Cr on the properties of Pt in zeolites also depends on the degree of interaction of the two metals [540] (Table 27). In a catalyst prepared from $[Pt(NH_3)_4]Cl_2$ and CrO_3 at a low pH (No. 1), the reduction of the Cr to the zero-valence state was detected by XPS. Judging from the data of electron microscopy, the presence of metals on the outer surface was not observed in the reduced catalyst. We can thus assume that the Pt and Cr are localized

Table 27. Catalytic Properties of Pt/H-ZSM-5 and Pt-Cr/H-ZSM-5 in Reactions of Hydrogenolysis of C_2H_6 and C_3H_8

Catalyst	C₂H₆ hydrogenolysis					C₃H₈ hydrogenolysis				
	air + H₂			H₂		H₂			air + H₂	
	$w \times 10^3$ mol/$g_{Pt}\cdot$s 400 °C	Turnover frequency, s⁻¹ 400 °C	E_a, kJ/mol 300–400 °C	$w \times 10^3$, mol/$g_{Pt}\cdot$s 400 °C	Turnover frequency, s⁻¹ 400 °C	E_a, kJ/mol 300–450 °C	Turnover frequency, s⁻¹ 400 °C	$w \times 10^3$, mol/$g_{Pt}\cdot$s 350 °C	E_a, kJ/mol 300–400 °C	$S_{C_2H_6}^{a}$, %
0.5%Pt/H-ZSM-5	7.2	1.74[b] 2.31[c]	134	0.25	0.30[b]	199	—	—	—	—
	1.2[d]	—	134	—	—	—	—	2.9	76	92.5
1.0%Pt/H-ZSM-5	3.3	1.58[c]	133	0.19	0.17[c]	188	—	—	—	—
1.75%Pt/H-ZSM-5	2.4	1.54[c]	125	0.14	0.34[c]	120	—	—	—	—
0.5%Pt-0.75%Cr/ H-ZSM-5, No. 1	0.36[d]	—	232	—	—	—	—	0.28	120	65.7
0.5%Pt-0.75%Cr/ H-ZSM-5, No. 2	0.45[d]	—	160	—	—	—	—	0.55	132	90.3

[a] The selectivity has been determined as $(C_2H_6/CH_4 + C_2H_6) \times 100\%$.

[b] The number of surface Pt atoms has been calculated from TEM.

[c] The number of surface Pt atoms has been calculated from the chemisorption of CO.

[d] The reaction temperature is 350°C.

in the zeolite channels and interact with each other, possibly, to form an alloy. In a catalyst prepared by impregnation $[Pt(NH_3)_4]Cl_2$ and $Cr(NO_3)_3$ at a high pH (No. 2), the Cr is not reduced to Cr^0 because the chromium is chiefly on the outer surface in the form of aggregates of Cr(III) oxide [540]. Examination of Table 27 shows that the suppression of the hydrogenolysis of ethane and propane as well as the change in the selectivity in the second reaction are more noticeable for the first catalyst.

The same conclusions were made for Pt-Cr and Pt-Cr-Re/Al_2O_3 reforming catalysts. It is exactly the reduction of Cr and the formation of the alloy Pt-Cr [464, 713] that suppress the destructive reactions in the reforming process and improve the stability of the catalytic effect. XPS, EXAFS, and other techniques show that the second component can be only partly reduced or not reduced at all. An essential condition for the manifestation of the ligand effect is its direct interaction with the active metal (the systems Pt-Re, Pt-Sn, Pt-Cr, Pt-In), which not only results in stabilization of the highly dispersed Pt, but also in a change in its electron state. To ensure such interaction and a high activity in the dehydrogenation of the higher alkanes [763] on Pt-Sn and Pt-In/Al_2O_3, it is necessary to preliminarily calcine the catalysts. The calcination leads to electron interaction of the platinum and promoter; in direct reduction, the promoters migrate to the surface and block the active centers.

The examples given in this chapter show what kind of correlations can be established between the characteristics of a surface and catalytic properties. Some of them have a quite general nature, e.g. the dependences of the activity in the hydrogenation of CO on the degree of reduction of the metals or the surface composition. For systems such as CuO-ZnO or WO_3/Al_2O_3, XPS or XAES data make it possible to appraise the turnover frequency of the reaction on centers differing in their state. At the same time in many cases, the relation between the activity and selectivity on the one hand and the state of the active elements or their surface concentration on the other is of a qualitative nature. This is due to the fact that the activity of centers such as highly dispersed metal or bimetal particles depends on many factors whose contribution can meanwhile be appraised only qualitatively. Of importance is the fact that by employing methods of electron and ion spectroscopy, one can form or appreciably extend the observations on the surface structure of an active catalyst.

The use of EES and IS techniques also makes it possible to give specific recommendations on the optimization of the ways of preparing active catalysts and their treatment. For example, to a large extent investigations by XPS made it possible to optimize the conditions of stabilization of highly dispersed and active particles of metals in high silica zeolites, choose the optimal conditions of sulfidizing aluminocobalt-molybdenum catalysts, etc. Also important is the studying of industrial catalysts during or after their service. SAM [764] makes it possible to estimate the distribution of C and S on the surface of industrial catalysts and how it affects the activity. An attempt to appraise the content of the carbon deposit and its distribution on the outer surface and in the channels of zeolites by XPS has been undertaken by Sexton *et al.* [765]. The distribution of the metals (Ni and V) deposited from heavy petroleum fractions on industrial cracking catalysts has been studied by scanning SIMS [720]. Both metals are deposited on the outer surface, but in the course of multifold "cracking-regeneration" cycles the vanadium migrates into the bulk and interacts with the matrix and zeolite. Conversely, the concentration of the nickel on the outer surface increases. These data together with those described in Chap. 4 can be used to optimize the way of preparing a cracking catalyst (e.g. by varying the SiO_2/Al_2O_3 ratio in the matrix, the grain size), and also the way of depositing the passivators.

It is quite possible that in a number of cases the diagnosis of the surface of catalysts by XPS and other "on-line" techniques, like that conducted in the analysis of materials of microelectronics, may increase the service life of catalytic systems.

At the same time, the fundamental concepts of catalysis that in the long run have a decisive influence on the level of development of the technology of catalytic processes may evidently be worked out only by the detailed studying of the structure and reactivity of the surface of both real and model systems.

Conclusion

During the period of preparation of the manuscript, which took about a year, we noted two clear trends in the studying of catalysts by modern physical methods. One of them consists in the development of most up-to-date techniques for studying the structure of catalysts [766]. Nonstructural techniques such as high resolution electron microscopy, scanning electron microscopy, tunnel scanning microscopy, and multinuclear high resolution NMR in solid bodies are being involved more and more in structure studies. Owing to the widespread application of new sources (synchrotron radiation, a pulse source of neutrons), classical methods such as X-ray diffraction, electron diffraction, and neutron diffraction are obtaining a new impetus. The methods of absorption X-ray spectroscopy — EXAFS and XANES are gaining ground. We can also name absolutely novel original procedures employing synchrotron radiation or a pulse source of neutrons [766]:

— large-angle Rutherford scattering;
— a chemically sensitive structural image;
— microscopy of atomic forces;
— incoherent inelastic scattering of neutrons;
— high-resolution powder diffraction;
— microdiffraction on single crystals.

A merit of these methods is the possibility of *in situ* studies, while a drawback is the need of special, sometimes unique, sources and monochromators, and also the fact that most of these methods yield information averaged over the bulk.

Hence the second trend becomes clear, namely, the continuing growth of investigations of catalysts with the use of physical methods of studying a surface. It is sufficient to mention a bi-annual review of the literature on surface studying [767]. Surface studying remains important not only for such systems as metals, which are sharply nonuniform in the properties of the surface and bulk, but also for catalysts which by J. Thomas's classification relate to homogeneous systems [768] (zeolites, stratified clays, and some oxides). The examples contained in the monograph give us the right to state that a description of the structure and catalytic behavior of zeolite systems is possible only when two groups of methods are used — surface and bulk ones. The publications that have appeared in the last two years only confirm the impetous expansion of the application of electron and ion spectroscopy techniques in catalysis.

In conclusion, we would like to express our profound gratitude to all the co-authors of our works who were cited in the monograph, and first of all to G. V. Antoshin, who participated in many investigations. We are grateful to A. Yu. Stakheev, V. Grünert, and S. Yu. Panov, whose methical developments ensured the required experimental level of our studies. We are especially grateful to O. P. Tkachenko and N. S. Telegina, who are not only co-authors of some studies, but also took on the strenuous task of designing the manuscript.

REFERENCES

1. Somorjai, G. A. *Proc. 8th Int. Congr. Catal.*, Vol. 1, West Berlin, p. 113 (1984).
2. Boreskov, G. K. and Savchenko, V. I. *Proc. 7th Int. Congr. Catal.*, Tokyo, p. 655 (1980).
3. Siegbahn, K., Nordling, C., Fahlman, A., Nordberg, R., Hamrin, K., Hedman, J., Johansson, G., Bergmark, T., Karlsson, S. E., Lindgren, I., and Lindberg, B. *ESCA: Atomic, Molecular and Solid State Structure Studied by Means of Electron Spectroscopy.* Uppsala: Almqvist and Wiksells (1967).
4. Siegbahn, K., Nordling, C., Johansson, G., Hedman, J., Heden, P. F., Hamrin, K., Gelius, U., Bergmark, T., Werme, L. O., Manne, R., and Baer, Y. *ESCA Applied to Free Molecules.* Amsterdam: North Holland Publ. Co. (1969).
5. Siegbahn, K. *Usp. Fiz. Nauk*, **138**, 2: 223 (1982).
6. Czanderna, A. W. (ed.). *Methods of Surface Analysis.* New York: Elsevier (1975).
7. Briggs, D. and Seah, M. P. (eds.). *Practical Surface Analysis by Auger and X-Ray Photoelectron Spectroscopy.* New York: Wiley (1983).
8. Fiermans, L., Vennik, J., and Dekeyser, W. (eds.). *Electron and Ion Spectroscopy of Solids.* New York: Plenum Press (1978).
9. Nefedov, V. I. and Cherepin, V. T. *Fizicheskie metody issledovaniya poverkhnosti tverdykh tel* (Physical Methods of Studying the Surface of Solids). Moscow: Nauka (1983).
10. Woodruff, D. P. and Delchar, T. A. *Modern Techniques of Surface Science.* Cambridge University Press (1986).
11. Siegbahn, K. In: Meisel, A. and Finster, J. (eds.). *X84.* Leipzig: Karl-Marx-Universität, p. 114 (1984).
12. Wagner, C. D. and Joshi, A. *Surface and Interface Anal.*, 6: 215 (1984).
13. Turner, N. H. *Anal. Chem.*, **58**: 153 (1986).
14. Baker, A. D. In: Brundle, C. R. and Baker, A. D. (eds.). *Electron Spectroscopy: Theory, Techniques, and Applications*, Vol. 1, New York: Academic Press, p. 1 (1977).
15. Canesson, P. *J. Microsc. Spectros. Electron.*, **1**: 429 (1976).
16. Vedrine, C. R. *J. Microsc. Spectros. Electron.*, **1**: 285 (1976).
17. Brinen, J. C. *J. Electron Spec.*, **5**: 377 (1974).
18. Brundle, C. R. *J. Electron Spec.*, **5**: 291 (1974).
19. Gomoyunova, M. V. *Zh. Teor. Fiz.*, **46**: 673 (1977).
20. Joyner, R. W. *Surf. Sci.*, **63**: 291 (1977).
21. Delgass, W. N. In: Delgass, W. N., Haller, G. L., Kellerman, R., and Lunsford, J. H., (eds.). *Spectroscopy in Heterogeneous Catalysis*, New York: Academic Press, p. 267 (1979).
22. Defossé, C. In: Delannay, F. (ed.). *Characterization of Heterogeneous Catalysts.* New York: Marcel Dekker, p. 225 (1984).
23. Barr, T. L., In: Briggs, D. and Seah, M. P., (eds.) *Practical Surface Analysis by Auger and X-Ray Photoelectron Spectroscopy.* New York: Wiley, p. 283 (1983).
24. Thomas, J. M. and Lambert, R. M. (eds.). *Characterization of Catalysts.* New York: Wiley-Interscience (1980).

25. Minachev, Kh. M., Antoshin, G. V., and Shpiro, E. S. In: Krylov, O. V., and Shilanova, M. D. (eds.). *Novye metody issledovaniya poverkhnosti katalizatorov (Problemy kinetiki i kataliza)* [New Methods of Studying Catalyst Surfaces (Problems of Kinetics and Catalysis)]. Moscow: Nauka, Vol. 16, p. 189 (1975).

26. Minachev, Kh. M., Antoshin, G. V., and Shpiro, E. S. *Uspekhi Khimii*, **47**: 2097 (1978).

27. Minachev, Kh. M., Antoshin, G. V., and Shpiro, E. S. *Fotoelektronnaya spektroskopiya i ee primenenie v katalize* (Photoelectron Spectroscopy and Its Use in Catalysis). Moscow: Nauka (1981); Shpiro, E. S., Antoshin, G. V., and Minachev, Kh. M., In: Kallo, D. and Minachev, Kh. M. (eds.). *Catalysis on Zeolites*. Budapest: Akademiai Kiado, p. 46 (1988).

28. Anstermann, R. L., Denley, D. R., Hart, D. W., Himelfarb, P. B., Irwin, R. M., Narayana, M., Szentirmay, R., Tang, S. C., and Yeates, R. S. *Anal. Chem.*, **59**: 68 (1987).

29. Benninghoven, A. *Surf. Sci.*, **53**: 596 (1975); Vickerman, J. C. In: Clark, R. J. H. and Hester, R. E. (eds.). *Spectroscopy of Surfaces*. New York: Wiley, p. 155 (1988).

30. Ryazanov, M. I. and Tilinin, I. S. *Issledovanie poverkhnosti po obratnomu rasseyaniyu chastits* (Investigating a Surface by Particle Backscattering). Moscow: Energoizdat (1985).

31. Nemoshkalenko, V. V. and Aleshin, V. G. *Elektronnaya spektroskopiya kristallov* (Electron Spectroscopy of Crystals). Kiev: Naukov Dumka (1976).

32. Carlson, T. A. *Photoelectron and Auger Spectroscopy*. New York: Plenum Press (1975).

33. Baker, A. D. and Betteridge, D. *Photoelectron Spectroscopy*. Oxford: Pergamon Press (1972).

34. Nefedov, V. I. *Rentgenoelektronnaya spektroskopiya khimicheskikh soedinenii* (X-Ray Electron Spectroscopy of Chemical Compounds). Moscow: Khimiya (1984).

35. Williams, A. R. and Lang, N. D. *Phys. Rev. Lett.*, **40**: 954 (1978).

36. Shirley, D. A. *J. Vac. Sci. Tech.*, **13**: 280 (1976).

37. Watson, R. E. and Perlman, M. L., *Phys. Rev.*, **B13**: 2358 (1976).

38. Broughton, J. Q. and Perry, D. L. *J. Electron Spec.*, **16**: 45 (1979).

39. Wagner, C. D., Gale, L. H., and Raymond, R. H. *Anal. Chem.*, **51**: 466 (1979).

40. Nefedov, V. I., Gati, D., Dzhurinsky, B. F., Sergushin, N. P., and Salin, Yu. N. *Zh. Neorg. Khim.*, **20**: 2307 (1975).

41. Jolly, W. L. *J. Amer. Chem. Soc.*, **92**: 3260 (1970).

42. Plummer, E. W., Chen, C. T., Ford, W. K., Eberhardt, W., Messmer, R. P., and Freund, H. J. *Surf. Sci.*, **158**: 58 (1985).

43. Carver, J. C., Schweitzer, G. K., and Carlson, T. A. *J. Chem. Phys.*, **57**: 973 (1972).

44. Allen, G. C. and Tucker, P. *Inorg. Chim. Acta.*, **16**: 41 (1976).

45. Rosencwaig, A. and Wertheim, G. K. *J. Electron Spec.*, **1**: 493 (1973).

46. Kim, K. S. and Davis, R. E. *J. Electron Spec.*, **1**: 251 (1973).

47. Frost, D. C., Ishitani, A., and McDowell, C. A. *Mol. Phys.*, **24**: 861 (1972).

48. Campagna, M., Wertheim, G. K., and Bucher, E. *Struct. and Bonding*, **30**: 92 (1976).

49. Brisk, M. A. and Baker, A. D. *J. Electron Spec.*, **7**: 95 (1975).
50. Ioffe, M. S. and Borodko, Yu. G. *J. Electron Spec.*, **11**: 235 (1977).
51. Signorelli, A. J. and Hayes, G. F. *Phys. Rev.*, **B8**: 81 (1973).
52. Teterin, Yu. A., Baranov, A. V., and Kulakov, V. M. *Koordinatsionnaya Khimiya*, **4**: 1860 (1978).
53. Pireaux, J. J., Caudano, R., and Verbist, J. *J. Electron Spec.*, **5**: 267 (1974).
54. Barr, T. L. *Applications of Surf. Sci.*, **15**: 1 (1983).
55. Carlson, T. A. and McGuire, G. E. *J. Electron Spec.*, **1**: 161 (1973).
56. Wagner, C. D. *Anal. Chem.*, **44**: 1050 (1972).
57. Fraser, W. A., Florio, J. V., Delgass, W. N., and Robertson, W. D. *Surf. Sci.*, **36**: 661 (1973).
58. Reilman, P. F., Msezane, A., and Manson, S. T. *J. Electron Spec.*, **8**: 389 (1976).
59. Nefedov, V. I., Sergushin, N. P., Band, I. M., and Trzhaskovskaya, M. B. *J. Electron Spec.*, **2**: 383 (1973).
60. Nefedov, V. I., Sergushin, N. P., Salyn, Y. V., Band, I. M., and Trzhaskovskaya, M. B. *J. Electron Spec.*, **7**: 175 (1975).
61. Scofield, J. H. *J. Electron Spec.*, **8**: 129 (1976).
62. Evans, S., Pritchard, R. G., and Thomas, J. M. *J. Electron Spec.*, **14**: 341 (1978).
63. Szajman, J., Jenkin, J. G., Leckey, R. C. G., and Liesegang, J. *J. Electron Spec.*, **19**: 393 (1980).
64. Klasson, M., Berndtsson, A., Hedman, J., Nilsson, R., Nyholm, R., and Nordling, C. *J. Electron Spec.*, **3**: 427 (1974).
65. Lindau, I. and Spicer, W. E. *J. Electron Spec.*, **3**: 409 (1974).
66. Powell, C. J. *Surf. Sci.*, **44**: 29 (1974).
67. Penn, D. R. *J. Electron Spec.*, **9**: 29 (1976).
68. Seah, M. P. and Dench, W. A. *Surface and Interface Anal.*, **1**: 2 (1979).
69. Quinn, J. J. *Phys. Rev.*, **126**: 1453 (1962).
70. Chang, C. C. *Surf. Sci.*, **48**: 9 (1975).
71. Tokutaka, H. Nishimori, K., and Hayashi, H. *Surf. Sci.*, **149**: 349 (1985).
72. Tanuma, S., Powell, C. J., and Penn, D. R. *Surf. Sci.*, **192**: L849 (1987).
73. Baschenko, O. A. and Nefedov, V. I. *J. Electron Spec.*, **17**: 405 (1979).
74. Clark, D. T., Dilks, A., Shuttleworth, D., and Thomas, H. R. *J. Electron Spec.*, **14**: 247 (1978).
75. Ebel, M. F., Ebel, H., and Hirokawa, K. *Spectrochimica Acta*, **7B**: 461 (1982).
76. Ebel, H., Ebel, M. F., Wernisch, J., and Jablonski, A. *Surface and Interface Anal.*, **6**: 140 (1984).
77. Jablonski, A., Mrozek, P., Gergely, G., Menyhard, M., and Sueyok, A. *Surface and Interface Anal.*, **6**: 291 (1984).
78. Jablonski, A. *Surf. Sci.*, **188**: 164 (1987).
79. Wagner, C. D. *Anal. Chem.*, **49**: 1282 (1977).
80. Demuth, J. E. and Eastman, D. E. *J. Vac. Sci. Tech.*, **13**: 283 (1976).
81. Boronin, A. I., Vishevsky, A. A., Bukhtiyarov, V. I., Stefanov, P. K., and Savchenko, V. I. *Tez. dokl. VI Seminar sots. stran po elektronnoi spektroskopii* (Abstracts of Reports to 6th Seminar of Socialist Countries on Electron Spectroscopy). Liblice, CSSR, p. 10 (1986).
82. Wandelt, K. *J. Vac. Sci. Tech.*, **A2**: 802 (1984).
83. Spicer, W. E. In: Fiermans, L., Vennik, J., and Dekeyser, W. (eds.). *Electron and*

Ion Spectroscopy of Solids. New York: Plenum Press, p. 54 (1978).

84. Hofmann, P., Bare, S. R., Richardson, N. V., and King, D. A. *Surf. Sci.*, **133**: L459 (1983).

85. Allyn, C. L., Gustafsson, T., and Plummer, E. W. *Chem. Phys. Letters*, **47**: 127 (1977).

86. Weng, S. L., Plummer, E. W., and Gustafsson, T. *Phys. Rev.*, **B18**: 1572 (1978).

87. Dose, V. *Applications of Surf. Sci.*, **22/23**: 338 (1985).

88. Reihe, B., Schlitter, R. R., and Neff, H. *Surf. Sci.*, **152/153**: 231 (1985).

89. Smith, N. V. *Applications of Surf. Sci.*, **22/23**: 349 (1985).

90. Gallon, T. In: Fiermans, L., Vennik, J., and Dekeyser, W. (eds.). *Electron and Ion Spectroscopy of Solids.* New York: Plenum Press, p. 230 (1978).

91. Borodko, Yu. G. and Mikhailov, G. M. In: *Khimicheskaya svyaz' i stroenie molekul* (The Chemical Bond and Structure of Molecules). Moscow: Nauka, p. 123 (1984).

92. Mikhailov, G. M., Gutsev, G. L., and Borodko, Yu. G. *Chem. Phys. Letters*, **96**: 70 (1983).

93. Fleish, T. H. and Mains, G. L. *Applications of Surf. Sci.*, **10**: 51 (1982).

94. Cini, M. *Solid State Comm.*, **20**: 605 (1977).

95. Cini, M. *Phys. Rev..*, **B17**: 2788 (1978).

96. Sawatzky, G. A. *Phys. Rev. Letters*, **39**: 504 (1977).

97. Sawatzky, G. A. and Lenselink, A. *Phys. Rev.*, **B21**: 1790 (1980).

98. Fuggle, J. C., Hillebrecht, F. U., Zeller, R., Zolnierek, Z., Bennett, P. A., and Freiburg, C. *Phys. Rev.*, **B27**: 2194 (1983).

99. Rubtsov, V. I., Shulga, Yu. M., Gutsev, G. L., Borodko, Yu. G., and Trusov, L. I. *Metallofizika*, **9**: 96 (1987).

100. Shulga, Yu. M., Rubtsov, V. I., Gutsev, G. L., and Borodko, Yu. G. *Poverkhnost'*, **7**: 86 (1987).

101. Rubtsov, V. I., Shulga, Yu. M., Gutsev, G. L., *et al.*, *Poverkhnost'*, **9**: 57 (1988).

102. Jablonski, A. *Surface and Interface Anal.*, **1**: 122 (1979).

103. Palmberg, P. W. In: Shirley, D. A. (ed.). *Electron Spectroscopy Proceedings of the International Conference on Electron Spectroscopy, Asilomar, California, Sept. 1971.* Amsterdam: North-Holland Publ., p. 835 (1972).

104. Holloway, P. H. *Surf. Sci.*, **66**: 479 (1977).

105. Chorkendorff, I. *J. Electron Spec.*, **32**: 1 (1983).

106. Pauling, R. and Szajman, J. *J. Electron Spec.*, **43**: 37 (1987).

107. Streubel, P. and Berndt, H. *Surface and Interface Anal.*, **6**: 48 (1984).

108. Jablonski, A. *Surf. Sci.*, **124**: 39 (1983).

109. McHugh, J. A. and Sheffield, J. C. *J. Appl. Phys.*, **35**: 512 (1964).

110. McHugh, J. A. In: Czanderna, A. W. (ed.). *Methods of Surface Analysis.* New York: Elsevier, p. 223 (1975).

111. Fogel, Ya. M. *Uspekhi Fiz. Nauk.*, **91**: 75 (1967).

112. Fogel, Ya. M., Nadykto, B. T., Rybalko, V. F., Shvachko, V. I., and Korobchaskaya, I. E. *Kinetika i Kataliz*, **5**: 496 (1964).

113. Werner, H. W. In: Fiermans, L., Vennik, J., and Dekeyser, W. (eds.). *Electron and Ion Spectroscopy of Solids.* New York: Plenum Press (1978).

114. Sigmund, P. *Phys. Rev..*, **184**: 383 (1969).

115. Wehner, G. K. In: Czanderna, A. W. (ed.). *Methods of Surface Analysis*. New York: Elsevier, p. 5 (1975).
116. Tompkins, H. G. *J. Vac. Sci. Tech.*, **16**: 778 (1979).
117. Nefedov, V. I., Chulkov, N. G., and Lunin, V. V. *Surface and Interface Anal.*, **2**: 207 (1980).
118. Kelly, R. *Surface and Interface Anal.*, **7**: 1 (1985).
119. Betz, G. *Surf. Sci.*, **92**: 283 (1980).
120. Lam, N. Q., Hoff, H. A., Wildersich, H., and Rehn, L. E. *Surf. Sci.*, **149**: 517 (1980).
121. Andersen, H. H. In: Matic, H. (ed.). *Physics of Ionized Gases*. Beograd, p. 421 (1980).
122. Kim, K. S. and Winograd, N. *Surf. Sci.*, **43**: 625 (1974).
123. Christensen, N. E. and Seraphin, B. O. *Phys. Rev.*, **B4**: 3321 (1971).
124. Andersen, C. A. and Hinthorne, J. R. *Anal. Chem.*, **45**: 1421 (1973).
125. Schroer, J. M., Rhodin, T. N., and Bradey, R. C. *Surf. Sci.*, **34**: 571 (1973).
126. Ashton, A. G., Elliot, I. E., Dwyer, J., Fitch, F. R., and Machado, F. J. *Proc. 8th Int. Congr. Catal.*, Vol. IV. West Berlin, p. 531 (1984).
127. Buck, T. M. In: Czanderna, A. W. (ed.). *Methods of Surface Analysis*. New York: Elsevier, p. 75 (1975).
128. Panin, B. V. *Zh. Eksp. i Teor. Fiz.*, **42**: 313 (1962).
129. Mashkova, E. S. and Molchanov, V. A. *Dokl. Akad. Nauk SSSR*, **146**: 585 (1962).
130. Smith, D. P. *J. Appl. Phys.*, **38**: 340 (1967).
131. Baun, W. L. *Surface and Interface Anal.*, **3**: 243 (1981).
132. Horrell, B. A. and Cocke, D. L. *Catal. Rev. Sci. Eng.*, **29**: 447 (1987).
133. Woodruff, D. P. *Nucl. Instr. and Meth.*, **194**: 639 (1982).
134. Baun, W. L. *Applications of Surf. Sci.*, **1**: 81 (1977).
135. Powell, C. J. *Applications of Surf. Sci.*, **4**: 492 (1980).
136. Swartzfager, D. G. *Anal. Chem.*, **56**: 55 (1984).
137. Nelson, G. C. *SAND 79-0712*, June, 1979.
138. Brongersma, H. H. and Theeten, J. B. *Surf. Sci.*, **54**: 519 (1976).
139. Woodruff, D. P. and Godfrey, D. J. *Solid State Comm.*, **34**: 679 (1980).
140. Brongersma, H. H. and Buck, T. M. *Nucl. Instrum. and Meth.*, **149**: 569 (1978).
141. Ibach, H. and Mills, D. A. *Electron Energy Loss Spectroscopy*. New York: Academic Press (1982).
142. Lagarde, P. and Dexpert, H. *Adv. Phys.*, **33**: 567 (1984).
143. Stöhr, J., Jaeger, R., and Brennan, J. *Surf. Sci.*, **117**: 503 (1982).
144. Stöhr, J., Gland, J. L., Eberhardt, W., Datka, D., Madix, R. J., Settle, F., and Koestner, R. J. *Phys. Rev. Letters*, **51**: 2414 (1983).
145. Prince, K. C., Holub-Krappe, E., Paolucel, G., Horn, K., and Woodruff, D. P. In: Meisel, A. and Finster, J. *X84*. Leipzig: Karl-Marx-Universität, p. 479 (1984).
146. Riviere, J. C. In: Briggs, D. and Seah, M. P. (eds.). *Practical Surface Analysis by Auger and X-Ray Photoelectron Spectroscopy*. New York: Wiley, p. 17 (1983).
147. Hirsch, D. and Leonhardt, G. *Surf. Sci.*, **89**: 660 (1979).
148. Nizovsky, A. I., Savchenko, V. I., and Dyplyakin, V. K. In: 7th Seminar of Socialist Countries on Electron Spectroscopy, Burgas, p. 63 (1988).

149. Gelius, U., Asplund, L., Basilier, E., Hedman, S., Helenelund, K., and Siegbahn, K. *Nucl. Instrum. and Meth.*, **B1**: 85 (1984).
150. Allison, D. A., and Anater, T. F. *J. Electron Spec.*, **43**: 245 (1987).
151. Beamson, G., Porter, H. Q., and Turner, D. W. *Nature*, **290**: 556 (1981).
152. Gurker, N., Ebel, M. F., and Ebel, H. *Surface and Interface Anal.*, **5**: 13 (1983).
153. Smith, G. C. and Seah, M. P. *J. Electron Spec.*, **42**: 359 (1987).
154. Bertrand, P. A., Kalinovski, W. J., Tribble, L. S., and Tolentino, L. U. *Rev. Sci. Instr.*, **54**: 387 (1984).
155. Allison, D. A. and Anater, T. F. *J. Electron Spec.*, **43**: 243 (1987).
156. Hellinga, G. J. A., Ottevanger, H., Boelns, S. W., Knibbeler, C. L. C. M., and Brongersma, H. H. *Surf. Sci.*, **162**: 913 (1985).
157. Shimizu, R. and Kurokawa, A. *Surf. Sci.*, **176**: 653 (1986).
158. Seah, M. P. *J. Vac. Sci. Tech.*, **17**: 16 (1980).
159. Wirth, A. and Thompson, S. P. *Proc. 11th Int. Cong. on X-Ray Optics and Microanalysis*, (1988).
160. Fadley, C. S. In: Brundle, C. R. and Baker, A. D. (eds.). *Electron Spectroscopy. Theory, Techniques and Applications*, Vol. 2. London: Academic Press, p. 9 (1978).
161. Anthony, M. T. and Seah, M. P. *Surface and Interface Anal.*, **6**: 95, 107 (1984).
162. Huchital, D. A. and McKeon, R. T. *Appl. Phys. Letters*, **20**: 158 (1972).
163. Lewis, R. T. and Kelly, M. A., *J. Electron Spec.*, **20**: 105 (1980).
164. Windawi, H. *J. Electron Spec.*, **22**: 373 (1981).
165. Swift, P., Shuttleworth, D., and Seah, M. P. In: Briggs, D. and Seah, M. P. (eds.). *Practical Surface Analysis by Auger and X-Ray Photoelectron Spectroscopy.* New York: Wiley, p. 437 (1983).
166. Edgell, M. J., Paynter, R. W., and Castle, J. E. *Surface and Interface Anal.*, **8**: 113 (1986).
167. Swift, P. *Surface and Interface Anal.*, **4**: 47 (1982).
168. Bird, R. J. and Swift, P. *J. Electron Spec.*, **21**: 227 (1980).
169. Nefedov, V. I., Salyn, Ya. N., Leonhardt, G., and Scheife, R. *J. Electron Spec.*, **10**: 121 (1977).
170. Morokawa, Y. *et al. J. Electron Spec.*, **14**: 129 (1978).
171. Kohiki, S. *Applications of Surf. Sci.*, **17**: 497 (1984).
172. Madey, T. E., Wagner, C. D., and Joshi, A. *J. Electron Spec.*, **10**: 359 (1977).
173. Nefedov, V. I. *J. Electron Spec.*, **25**: 29 (1982).
174. Powell, C. J., Erickson, N. E., and Madey, T. E. *J. Electron Spec.*, **17**: 361 (1979).
175. Nelson, G. C. *Surface and Interface Anal.*, **6**: 144 (1984).
176. Muilenberg, G. E. (ed.). *Handbook of X-Ray Photoelectron Spectroscopy.* Eden Prairie, Minn.: Perkin-Elmer Corp., Physical Electronics Div. (1978).
177. Wagner, C. D., Riggs, W. N., Davis, L. E., Moulder, J. F., and Muilenberg, G. E. (eds.). *Handbook of X-Ray Photoelectron Spectroscopy.* Eden Prairie, Minn.: Perkin-Elmer Corp., Physical Electronics Div. (1979).
178. Wagner, C. D. In: Briggs, D. and Seah, M. P. (eds.). *Practical Surface Analysis by Auger and X-Ray Photoelectron Spectroscopy.* New York: Wiley, p. 477 (1983).
179. Briggs, D. (ed.). *Handbook of X-Ray and Ultraviolet Photoelectron Spectroscopy.* London: Heyden & Sons (1977).

180. Seah, M. P. In: Briggs, D. and Seah, M. P. (eds.). *Practical Surface Analysis by Auger and X-Ray Photoelectron Spectroscopy.* New York: Wiley, p. 247 (1983).
181. Hanke, W., Ebel, M. F., Jablonski, A., and Hirokawa, K. *J. Electron Spec.*, **40**: 241 (1986).
182. Anthony, M. T. and Seah, M. P. *DMA Report* (D): 311, March (1982).
183. Seah, M. P. and Anthony, M. T. *Surface and Interface Anal.*, **6**: 230, 242 (1984).
184. Wagner, C. D., Davis, L. E., Zeller, M. V., Taylor, J. A., Raymond, R. H., and Gale, L. H. *Surface and Interface Anal.*, **2**: 211 (1981).
185. Kerkhof, F. P. J. and Moulijn, J. A., *J. Phys. Chem.*, **83**: 1612 (1979).
186. Miller, A. C. *Surface and Interface Anal.*, **8**: 47 (1988).
187. ASTM-42. *Surface and Interface Anal.*, **10**: 48 (1987).
188. Proctor, A. and Sherwood, P. M. A. *Anal. Chem.*, **52**: 2315 (1980).
189. Shirley, D. A. *Phys. Rev.*, **B5**: 4709 (1972).
190. Bishop, H. E. *Surface and Interface Anal.*, **3**: 272 (1981).
191. Carley, A. F. and Joyner, R. W. *J. Electron Spec.*, **16**: 1 (1979).
192. Chornik, B., Sopiret, R., and LeGressus, C. *J. Electron Spec.*, **42**: 329 (1987).
193. Doniach, S. and Sunjic, M. *J. Phys.*, **C3**: 285 (1970).
194. Hughes, A. E. and Sexton, B. A. *J. Electron Spec.*, **46**: 31 (1988).
195. Grünert, W., Feldhaus, R., Anders, K., Shpiro, E. S., Antoshin, G. V., and Minachev, Kh. M. *J. Electron Spec.*, **40**: 187 (1986).
196. Bragin, O. V., Shpiro, E. S., Preobrazhensky, A. V., Isaev, S. A., Vasina, T. V., Dyusenbina, B. B., Antoshin, G. V., and Minachev, Kh. M. *Appl. Catal.*, **27**: 219 (1986).
197. Minachev, Kh. M., Shpiro, E. S., Tkachenko, O. P., and Antoshin, G. V. *Izv. AN SSSR, Ser. Khim.*, 5 (1984).
198. Grünert, W., Stakheev, A. Yu., Shpiro, E. S., Feldhaus, R., Anders, K., and Minachev, Kh. M. In: *7th Seminar of Socialist Countries on Electron Spectroscopy, Bourgas, 1988, Abstracts*, p. 40 (1988).
199. Haber, J., Stoch, J., and Ungier, L. *J. Solid State Chem.*, **19**: 113 (1978).
200. Haber, J., Marczewski, W., Stoch, J., and Ungier, L. *Ber. Bunsengls.*, **79**: 970 (1975).
201. Broclawik, E., Foti, E. A., and Smith, V. H. *J. Catal.*, **51**: 380 (1978).
202. Patterson, T. A., Carver, J. C., Leyden, D. C., and Hercules, D. M. *J. Phys. Chem.*, **80**: 1700 (1976).
203. Nikishenko, S. B., Slinkin, A. A., Shpiro, E. S., Antoshin, G. V., and Minachev, Kh. M. *Kinetika i Kataliz*, **20**: 524 (1979).
204. Grünert, W., Shpiro, E. S., Feldhaus, R., Anders, K., Antoshin, G. V., and Minachev, Kh. M. *J. Catal.*, **107**: 522 (1987).
205. Shpiro, E. S., Avaev, V. I., Antoshin, G. V., Ryashentseva, M. A., and Minachev, Kh. M. *J. Catal.*, **55**: 402 (1978).
206. Braun, W. and Kuhlenbeck, H. *Surf. Sci.*, **180**: 279 (1987).
207. Sundberg, P., Larson, R., and Folkesson, B. *J. Electron Spec.*, **46**: 19 (1988).
208. Okamoto, Y., Konishi, Y., Fubino, K., Imanaka, T., and Teranishi, S. *Proc. 8th Int. Congr. Catal.*, Vol. 5, West Berlin, p. 159 (1984).
209. Fleish, T. H. and Mieville, R. J. *J. Catal.*, **90**: 165 (1984).
210. Apai, G., Monnier, J. R., and Preuss, D. R. *J. Catal.*, **98**: 563 (1986).

211. Minachev, Kh. M., Antoshin, G. V., Shpiro, E. S., and Yusifov, Yu. A. In: Bond, G. C., Wells, P. B., and Tomkins, F. C. (eds.). *Proc. 6th Int. Congr. Catal.*, Vol. 2. London: Chem. Soc., p. 621 (1977).

212. Minachev, Kh. M., Antoshin, G. V., and Shpiro, E. S. *Izv. AN SSSR, Ser. Khim.*, 1012 (1974).

213. Narayana, M., Contarini, S., and Kevan, L. *J. Catal.*, **94**: 370 (1985).

214. Chin, R. L. and Hercules, D. M. *J. Phys. Chem.*, **86**: 360 (1982).

215. Chin, R. L. and Hercules, D. M. *J. Phys. Chem.*, **86**: 3077 (1982).

216. Zingg, D. S., Makovsky, L. E., Tischer, R. E., Brown, R. F., and Hercules, D. M. *J. Phys. Chem.*, **84**: 2898 (1980).

217. Strohmeir, B. R., Leyden, D. C., Field, R. S., and Hercules, D. M. *J. Catal.*, **94**: 514 (1985).

218. Stakheev, A. Yu., Shpiro, E. S., Lysenko, S. V., Karachanov, E. A., and Minachev, Kh. M. In: Viswanathan, B. and Meenakshisundaram, A. (eds.). *9th National Indian Symp. on Catal.*, Madras, Vol. 3, OR 10-1, OR 10-13 (1988).

219. Zhdan, P. A., Shepelin, A. P., Osipova, Z. G., and Sokolovskii, V. D. *J. Catal.*, **58**: 8 (1979).

220. Vinek, H., Noller, H., Latzel, J., and Ebel, M. *Z. Phys. Chem.* (GDR), **105**: 319 (1977).

221. Ascarelli, P. and Moretti, G. *Surface and Interface Anal.*, **7**: 8 (1982).

222. Minachev, Kh. M. and Shpiro, E. S. *Kinetika i Kataliz*, **27**: 824 (1986).

223. Fischer, T. E., Keleman, S. R., Wang, K. P., and Johnson, K. H. *Phys. Rev.*, **B20**: 3124 (1979).

224. Fuggle, J. C., Hillebrecht, F. U., Zeller, R., Zolnierek, Z., Bennett, P. A., and Freiburg, C. *Phys. Rev.*, **B27**: 2145 (1982).

225. Fuggle, J. C., Hillebrecht, F. U., Zeller, R., Zolnierek, Z., Bennett, P. A., and Freiburg, C. *Phys. Rev.*, **B27**: 2179 (1982).

226. Mason, M. G. *Phys. Rev.*, **B27**: 27 (1983).

227. Takasu, Y., Unwin, R., Tesche, B., and Bradshaw, A. W. *Surf. Sci.*, **77**: 219 (1978).

228. Kohiki, S. *Applications of Surf. Sci.*, **25**: 81 (1986).

229. Ascarelli, P., Cini, M., Missoni, G., and Nistico, N. *J. Phys.* (Paris), *Colloq.*, **2**: 125 (1977).

230. Yong, V., *J. Chem. Phys.*, **72**: 4247 (1980).

231. Bahl, M. K., Tsai, S. C., and Chung, Y. W. *Phys. Rev.*, **21**: 1344 (1980).

232. Wertheim, G. K., Dicenze, S. B., and Yongquist, S. E. *Phys. Rev. Letters.*, **51**: 2310 (1983).

233. Citrin, P. H. and Wertheim, G. K. *Phys. Rev.*, **B27**: 3176 (1983).

234. Mason, M. G. and Beatzold, R. C. *J. Chem. Phys.*, **64**: 271 (1976).

235. Unwin, R. and Bradshaw, A. M. *Chem. Phys. Letters*, **58**: 58 (1978).

236. Egelhoff, V. F., Jr. and Tibbets, G. G. *Phys. Rev.*, **B19**: 5028 (1979).

237. Hamilton, J. F., Apai, G., Lee, S.-T., and Mason, M. G. In: Bourdon, J. (ed.). *Growth and Properties of Metal Clusters*. Amsterdam: Elsevier, p. 387 (1980).

238. Beatzold, R. C., Mason, M. G., and Hamilton, J. F. *J. Chem. Phys.*, **72**: 366, 6820 (1980).

239. Oberli, L., Monot, R., Mathieu, H., Landolt, D., and Buttef, J. *Surf. Sci.*, **106**: 301 (1981).

240. Cheung, T. T. P. *Surf. Sci.*, **140**: 151 (1984).
241. Fritish, A. and Légaré, P. *Surf. Sci.*, **162**: 742 (1985).
242. Burkstrand, J. M. *J. Vac. Sci. Tech.*, **20**: 440 (1982).
243. Parmigiani, F., Kay, E., Bagus, P. S., and Nelin, C. J. *J. Electron Spec.*, **36**: 257 (1985).
244. Légaré, P., Sakisaha, Y., Brucker, C. F., and Rhodin, T. N. *Surf. Sci.*, **139**: 316 (1984).
245. Fritish, A. and Légaré, P. *Surf. Sci.*, **145**: L517 (1984).
246. Ferrer, S., Salmeron, M., Okai, G., Roubin, P., and Lecante, J. *Surf. Sci.*, **160**: 488 (1985).
247. Thomas, T. D. *J. Electron Spec.*, **20**: 117 (1980).
248. Dicenzo, S. B. and Wertheim, G. K. *J. Electron Spec.*, **43**: 7 (1987).
249. Bagus, P. S., Nelin, C. J., Kay, E., and Parmigiani, F. *J. Electron Spec.*, **43**: 13 (1987).
250. Bastl, Z., Pribyl, O., and Mikusik, P. *Czech. J. Phys.*, **B34**: 981 (1984).
251. Wertheim, G. K., Dicenzo, S. B., Buchanan, D. N. E., and Bennett, P. A. *Solid State Comm.*, **53**: 377 (1985).
252. Dicenzo, S. B. and Wertheim, G. K. *Comments Solid State Phys.*, **11**: 203 (1985).
253. Wertheim, G. K., Dicenzo, S. B., and Buchanan, D. N. E. *Phys. Rev.*, **B33**: 5384 (1986).
254. Schmidt-Ott, A., Schurtenberger, P., and Siegman, H. C. *Phys. Rev. Letters*, **45**: 1284 (1980).
255. Bagus, P. S., Nelin, C. J., and Bauschlicher, C. W. *Surf. Sci.*, **156**: 615 (1985).
256. Citrin, P. H., Wertheim, G. K., and Bear, Y. *Phys. Rev.*, **B27**: 3160 (1983).
257. Guillet, C., Lassailly, Y., Lecante, J., Yugnet, Y., and Vedrine, J. C. *Phys. Rev. Letters*, **43**: 789 (1979).
258. Egelhoff, W. F., Jr. *Phys. Rev. Letters*, **50**: 587 (1983).
259. Van der Veen, J. F., Himpsel, F. J., and Eastman, D. E. *Phys. Rev. Letters*, **44**: 189 (1980).
260. Beatzold, R. C., Apai, G., and Shustorovich, E. *Phys. Rev.*, **B26**: 4022 (1982).
261. Ducros, R. and Fusy, J. *J. Electron Spec.*, **42**: 305 (1987).
262. Van der Veen, J. F. and Eastman, D. E. *Solid State Comm.*, **39**: 1301 (1981).
263. King, T. S. and Donelley, R. G. *Surf. Sci.*, **151**: 374 (1985).
264. Defay, R., Prigogine, I., Billenan, A., and Everett, D. H. In: *Surface Tension and Adsorption*. New York, Longman and Green (1966).
265. Williams, F. J. and Nason, D. *Surf. Sci.*, **45**: 317 (1974).
266. Miedema, R. R. *Z. Mettallkünde*, **69**: 455 (1978).
267. Abraham, F. F. *Phys. Rev. Letters*, **46**: 546 (1981).
268. Seah, M. P. *J. Catal.*, **57**: 450 (1979).
269. Kelley, M., *J. Catal.*, **57**: 113 (1979).
270. Wynblatt, P. and Ku, R. C. *Surf. Sci.*, **65**: 511 (1977).
271. Sachtler, W. M. H. *Catal. Rev. Sci. Eng.*, **14**: 193 (1976).
272. Chelikowsky, J. R. *Surf. Sci. Letters*, **139**: L197 (1984).
273. Sundaram, V. S. and Wynblatt, P. *Surf. Sci.*, **52**: 569 (1975).
274. Aksyantsev, L. M., Shpiro, E. S., Antoshin, G. V., Shteinberg, A. S., and Minachev, Kh. M. *Poverkhnost'*, **3**: 89 (1987).

275. Aksyantsev, L. M., Shpiro, E. S., and Minachev, Kh. M. In: Shopov, D., Andreev, A., Palazov, A., and Petrov, L. (eds.). *Proc. 7th Int. Symp. Heterogeneous Catalysis, Sofia, 1987.* Sofia: Publ. House of the Bulgarian Acad. of Sciences, Part 1, p. 37 (1987).
276. Sachtler, W. M. H. *Applications of Surf. Sci.,* 19: 167 (1984).
277. Nelson, G. C. *Surf. Sci.,* 59: 310 (1976).
278. Varga, P. and Hetzendorf, G. *Surf. Sci.,* 162: 544 (1985).
279. Jablonski, A. *Quantitative Analysis of Solid Surface by Auger Electron Spectroscopy.* Warsaw (1982).
280. Ishimura, S., Shimizu, R., and Langeron, J. P. *Surf. Sci.,* 115: 259 (1982).
281. Shpiro, E. S., Telegina, N. S., Rudnyǐ, Yu., Panov, S. Yu., Tkachenko, O. P., Gryaznov, V. M., and Minachev, Kh. M. *Poverkhnost',* 12:38 (1987).
282. Bertolini, J. C. and Massardier, J. *Surf. Sci.,* 162: 342 (1985); Rosengren, A. and Johansson, B. *Phys. Rev.,* B23: 3852 (1980).
283. Feilman, P. J. *Phys. Rev.,* B27: 2531 (1983).
284. Angevine, P. J., Delgass, W. N., and Vartulli, J. C. In: Bond, G. C., Wells, P. B., and Tompkins, F. C. (eds.). *Proc. 6th Int. Congr. Catal.,* Vol. 2. London: The Chemical Society, p. 611 (1977).
285. Defossé, C. *J. Electron Spec.,* 23: 157 (1981).
286. Mukaida, K. and Araya, T. *J. Soc. Mater. Sci. Japan,* 29: 346 (1980).
287. Fung, S. C. *J. Catal.,* 58: 454 (1979).
288. Kuipers, H. P. C. E., van Lenven, H. C. E., and Visser, W. M., *Surf. Interf. Anal.,* 8, 235 (1986).
289. Garbassi, F., Bossi, A., and Petrini, G. *J. Mater. Sci.,* 15: 2559 (1980).
290. Kalaguine, S., Adnot, A., and Lemay, G. *J. Phys. Chem.,* 91: 2886 (1987).
291. Minachev, Kh. M., Antoshin, G. V., and Shpiro, E. S. *Izv. AN SSSR, Ser. Khim.,* 5: 1012 (1974).
292. Finster, J., Lorenz, P., and Meisch, A. *Surface and Interface Anal.,* 1: 179 (1979).
293. Cross, Y. M. and Pyke, D. R. *J. Catal.,* 58: 61 (1979).
294. Suib, S. L., Stucky, G. D., and Bealtner, R. J. *J. Catal.,* 65: 179 (1980).
295. Chin, R. L. and Hercules, D. M. *J. Phys. Chem.,* 88: 456 (1984).
296. Brinen, J. C., Avingnon, D. A., Meyers, E. A., Deng, P. T., and Behnken, D. W. *Surface and Interface Anal.,* 6: 295 (1985).
297. Derouane, E. G. *J. Amer. Chem. Soc.,* 248: 219 (1984).
298. Edmonds, T. and Mitchell, P. C. H. *J. Catal.,* 64: 491 (1980).
299. Lars, S., Anderson, S. L. T., and Jaras, S. *J. Catal.,* 64: 51 (1980).
300. Ott, G. L., Delgass, W. N., Winograd, N., and Baitinger, W. F. *J. Catal.,* 56: 174 (1979).
301. Brown, A., van der Berg, J. A., and Vickerman, J. C. In: *Proc. 8th Int. Congr. Catal.,* Vol. 4. West Berlin, p. 35 (1984).
302. Brown, A. and Vickerman, J. C. *Surf. Sci.,* 140: 261 (1984).
303. Priggs, S. and Baner, E. In: Benninghoven, A. *et al.* (eds.). SIMS II. Berlin: Springer, p. 133 (1979).
304. Garrison, B. J., Holland, S. P., and Winograd, N. In: Benninghoven, A. *et al.* (eds.). *SIMS II.* Berlin: Springer, p. 44 (1979).
305. Schleich, B., Schmeisser, D., and Göpel, W. *Surf. Sci.,* 191: 367 (1987).

306. Jablonski, A., Eder, S., and Wandelt, K. *Applications of Surf. Sci.*, **22/23**: 309 (1985).
307. Forgues, D., Ehrhardt, J. J., Abon, M., and Bertolini, J. C. *Surf. Sci.*, **194**: 149 (1988).
308. Alnot, M., Gorodetski, V., Cassuto, A., and Ehrhardt, J. J. *Surf. Sci.*, **162**: 886 (1985).
309. Armstrong, R. A. and Egelhoff, W. F. Jr. *Surf. Sci.*, **154**: 225 (1985).
310. Lai, S. Y. and Vickerman, J. C. *J. Catal.*, **90**: 337 (1984).
311. Hoffman, S., Maniv, T. S., and Folman, M. *Surf. Sci.*, **182**: 56 (1987).
312. Pavlovska, A. and Bauer, E. *Surf. Sci.*, **177**: 473 (1986).
313. Shulga, Yu. M., Rubtsov, V. I., and Borodko, Yu. G. *Metallofizika*, **8**: 22 (1986).
314. Göpel, W., Anderson, J. A., Frankel, D., Jachnig, M., Phillips, K., Schafer, J. A., and Rocker, G. *Surf. Sci.*, **139**: 333 (1984).
315. Cord, B. and Courth, S. *Surf. Sci.*, **162**: 14 (1985).
316. Komiyama, M., Ogino, Y., Akai, Y., and Goto, M. *J. Chem. Soc.., Faraday Trans. II*, **79**: 1719 (1983).
317. Foger, K. and Anderson, J. R. *Appl. Catal.*, **23**: 139 (1986).
318. Weiss, M. and Ertl, G. *Metal-Support and Metal-Additive Effects in Catalysis. Stud. Surf. Sci. Catal.*, Vol. 18 (1982).
319. Brinen, J. C., Graham, S. W., Hammond, J. S., and Paul, D. F. *Surface and Interface Anal.*, **6**: 68 (1984).
320. Smith, N. V. and Chiang, S. *Phys. Rev.*, **B19**: 5015 (1979).
321. Smith, N. V. *Phys. Rev.*, **B19**: 5019 (1979).
322. Dowden, D. A. *Research*, **1**: 239 (1948).
323. Dowden, D. A. *J. Chem. Soc.*, **1950**: 242 (1950).
324. Dowden, D. A. *Ind. Eng. Chem.*, **44**: 977 (1952).
325. Ponec, V. *Adv. Catal.*, **32**: 149 (1983).
326. Slinkin, A. A. *Itogi nauki i tekhniki. Ser. Kinetika i Kataliz* (Results of Science and Engineering. Series Kinetics and Catalysis), Vol. 10. Moscow: VINITI (1982).
327. Sinfelt, J. H., Lytle, F. W., and Via, G. H. *J. Chem. Phys.*, **70**: 2773 (1979).
328. Sinfelt, J. H., Via, G. H., and Lytle, F. W. *J. Chem. Phys.*, **75**: 5527 (1981).
329. Sinfelt, J. H., Via, G. H., and Lytle, F. W. *J. Chem. Soc.*, **72**: 4832 (1980).
330. Somorjai, G. A. *Adv. Catal.*, **26**: 1 (1977).
331. Savchenko, V. I. *Uspekhi Khim.*, **55**: 462 (1986).
332. Somorjai, G. A. *React. Kinet. Catal. Letters*, **35**, 37 (1987).
333. Boudart, M. *Adv. Catal.*, **20**: 153 (1969).
334. Boreskov, G. K. *ZhVKhO im. D. I. Mendeleeva*, **22**: 495 (1977).
335. Messmer, R. P., Knudson, S. K., Johnson, K. H., Diamond, J. B., and Yang, C. Y. *Phys. Rev.*, **B13**: 1396 (1976).
336. Dose, V. *Progr. Surf. Sci.*, **13**: 225 (1983).
337. Gartland, P. O. and Slagsvold, B. J. *Phys. Rev.*, **B12**: 4047 (1975).
338. Heimann, P., Neddermeyer, H., and Roloft, H. F. *J. Phys.*, **C10**: L17 (1977).
339. Garbe, J., Venus, D., Suger, S., Schneider, C., and Kirschner, J. *Surf. Sci.*, **178**: 342 (1986).
340. Steiner, P., Hüfner, S., Martensson, N., and Johanson, B. *Solid State Comm.*, **37**: 73 (1981).

341. Godfrey, D. J. and Somorjai, G. A. *Surf. Sci.*, **202**: 204 (1988).
342. Unger, W. and Baunack, S. *Surf. Sci.*, **184**: L361 (1987).
343. Tomanek, D., Mukherjee, S., Kumer, Y., and Benneman, K. H. *Surf. Sci.*, **114**: 11 (1982).
344. Cheung, T. T. P. *Surf. Sci.*, **177**: 493 (1986).
345. Cheung, T. T. P. *Surf. Sci.*, **177**: L887 (1986).
346. Hofflund, G. B., Asbury, D. A., Kirszesztein, P., and Laitinen, A. *Surf. Sci. Letters*, **161**: L583 (1985).
347. Unger, W. and Marton, D. *Surf. Sci.*, **166**: 262 (1986).
348. Hilaire, L., Guerrero, G. D., Légaré, P. J., Maire, G., and Krill, G. *Surf. Sci.*, **146**: 569 (1984).
349. Schevchik, H. J. and Bloch, D. *J. Phys. F: Metal Phys.*, **7**: 543 (1977).
350. Van Langeveld, A. D., Hendrickx, H. A. C M., and Niewenhuys, B. E. *Thin Solid Films*, **109**: 179 (1983).
351. Bardi, U., Ross, P. N., and Rovida, G. *Surf. Sci.*, **205**: L798 (1988).
352. Shpiro, E. S., Rudnyĭ, Yu., Tkachenko, O. P., Antoshin, G. V., Dyusenbina, B. B., and Minachev, Kh. M. *Izv. AN SSSR, Ser. Khim.*, 2466 (1981).
353. Sampath Kumar, T. S. and Hagde, M. S. *Applications of Surf. Sci.*, **20**: 290 (1985).
354. Conner, G. R. *J. Vac. Sci. Tech.*, **15**: 343 (1978).
355. Sampath Kumar, T. S. and Hegde, M. S. *Surf. Sci.*, **150**: L123 (1985).
356. Van Pruissen, O. G., Luerdam, G. C., Bijzeman, L. J., and Gens, J. W. *Applications of Surf. Sci.*, **29**: 86 (1987).
357. Hedge, M. S., Sampath Kumar, T. S., and Mallya, R. M. *Surf. Sci.*, **188**: 255 (1987).
358. Shpiro, E. S., Tkachenko, O. P., Belyatsky, V. I., Rudnyĭ, Yu., Telegina, N. S., Panov, S. Yu., Gryaznov, V. M., and Minachev, Kh. M., *Kinetika i Kataliz* (in print).
359. Bardi, U., Beard, B. C., and Ross, P. N. *J. Vac. Sci. Tech.*, **A6**: 665 (1988).
360. Abraham, F. F. and Brundle, C. R. *J. Vac. Sci. Tech.*, **18**: 361 (1981).
361. Viefhaus, H. and Rüsenberg, M. *Surf. Sci.*, **159**: 1 (1985).
362. Kang, H. J., Kawaton, E., and Shimizu, R. *Surf. Sci.*, **144**: 541 (1984).
363. Dawson, P. T. and Petrole, S. A. *Surf. Sci.*, **152/153**: 925 (1983).
364. Koshikawa, T. *Surf. Sci.*, **22/23**: 118 (1985).
365. Temmernan, L., Creenurs, C., van Hove, H., Neyen, S., Massardier, J., and Bertolini, J. C. *Surf. Sci.*, **178**: 888 (1986).
366. Peacock, D. C. *Applications of Surf. Sci.*, **26**: 306 (1986).
367. Aksyantsev, L. M., Shpiro, E. S., and Minachev, Kh. M. In: Krylov, O. V. (ed.). *Trudy IV Vsesoyuznoĭ konferentsii po mekhanizmu kataliticheskikh reaktsiĭ* (Proceedings of the 4th All-Union Conference on the Mechanism of Catalytic Reactions). Moscow, Nauka, p. 366 (1986).
368. Sanchez, J. M. and Moran-Lopez, J. L. *Surf. Sci.*, **157**: L397 (1985).
369. Segaud, J. P., Blanc, E., Lanroz, C., and Baudiong, R. *Surf. Sci.*, **206**: 297 (1988).
370. Kok, G. A., Noordermeyer, A., and Niewenhuys, B. E. *Surf. Sci.*, **152/153**: 505 (1985).
371. Noordermeyer, A., Kok, G. A., and Niewenhuys, B. E. *Surf. Sci.*, **165**: 375 (1986).
372. Sonderricker, D., Jona, F., and Marcus, P. M. *Phys. Rev.*, **B34**: 6770 (1986).

373. Vanezia, A. M. and Loxton, C. M. *Surf. Sci.*, **194**: 1 (1988).
374. Kishi, K. *Surf. Sci.*, **192**: 210 (1987).
375. Rotemund, H. H., Penba, V., De Lonisc, L. A., and Brundle, C. R. *J. Vac. Sci. Tech.*, **A5**: 1198 (1987).
376. Niewenhuys, B. E. *Thin Metal Films and Gas Chemisorption.* Amsterdam: North Holland Publ. Co. (1987).
377. Frick, B. and Jacobi, K. *Surf. Sci.*, **178**: 907 (1986).
378. Sakakini, B., Dunhill, N., Harendt, C., Steeples, B., and Vickerman, J. C. *Surf. Sci.*, **189/190**: 211 (1987).
379. Sakakini, B., Swift, A. J., Vickerman, J. C., Harendt, C., and Christmann, C. *J. Chem. Soc., Faraday Trans. I*, **83**: 1975 (1987).
380. Houston, J. E., Peden, C. H. F., Feibelman, P. J., and Haman, D. H. *Phys. Rev. Letters*, **56**: 375 (1986).
381. Dunhill, N. I., Sakakini, B., and Vickerman, J. C. In: *Proc. 9th Int. Congr. Catal.*, Vol. 3, Calgary, p. 1166 (1988).
382. Berlowitz, P. J., Houston, J. E., White, J. M., and Goodman, D. W. *Surf. Sci.*, **205**: 1 (1988).
383. Judd, R. W., Reichelt, M. A., and Lambert, R. M. *Surf. Sci.*, **198**: 26 (1988).
384. Qwen, G., Howkes, C. M., Lloyd, D., Jennings, J. R., Nix, R. M., and Lambert, R. M. *Appl. Catal.*, **33**: 405 (1987).
385. Nix, R. M., Judd, R. W., and Lambert, R. M. *Surf. Sci.*, **203**: 307 (1988).
386. Nix, R. M., Judd, R. W., and Lambert, R. M. *Surf. Sci.*, **205**: 59 (1988).
387. Kleshchevnikov, A. M., Boeva, O. A., Zhavoronkova, K. N., and Nefedov, V. I. *Poverkhnost'*, **6**: 98 (1983).
388. Nefedov, V. I., Vinogradov, A. P., Zhavoronkova, K. N., and Boeva, O. A. *Poverkhnost'*, **10**: 123 (1987).
389. Hildebrand, E. I. and Fasman, A. B. *Skeletnye katalizatory v organicheskoĭ khimii* (Skeleton Catalysts in Organic Chemistry). Alma-Ata: Nauka (1982).
390. Okamoto, Y., Nitta, Y., Imanak, T., and Teranishi, S. *J. Chem. Soc., Faraday Trans. I*, **76**: 998 (1980).
391. Klein, J. C. and Hercules, D. M. *Anal. Chem.*, **53**: 754 (1981).
392. Kelley, R. D., Candeba, G. A., Madey, T. E., Newbury, D. E., and Shene, R. R. *J. Catal.*, **80**: 235 (1983).
393. Baeva, G. N., Fasman, A. B., Shpiro, E. S., Antoshin, G. V., Pavlyukevich, L. V., Iskhanov, Zh. A., and Minachev, Kh. M. *Kinetika i Kataliz*, **24**: 1220 (1985).
394. Shpiro, E. S., Baeva, G. N., Fasman, A. B., and Minachev, Kh. M. *React. Kinet. Catal. Letters*, **33**: 499 (1986).
395. Mikhailenko, S. D., Fasman, A. B., Maksimova, N. A., Leongard, E. V., Shpiro, E. S., and Antoshin, G. V. *Appl. Catal.*, **19**: 141 (1984).
396. Fasman, A. B., Antoshin, G. V., Shpiro, E. S., Musabekova, G. S., and Almashev, B. K. *Kinetika i Kataliz*, **27**: 451 (1986).
397. Wallace, W. E. *Chem. Tech.*, **12**: 752 (1982).
398. Lunin, V. V. *Izvestiya AN SSSR, Ser. Neorg. Mater.*, **14**: 1593 (1978).
399. Lunin, V. V., Nefedov, V. I., Shchumadilov, E. K., Rakhamilov, B. Yu., and Chernavsky, P. A. *Dokl. AN SSSR*, **240**: 114 (1978).
400. Houalla, M., Dang, T. A., Kibby, C. L., Petrakis, L., and Hercules, D. M. *Applications of Surf. Sci.*, **19**: 414 (1984).

401. Dang, T. A., Petrakis, L., Kibby, C. L., and Hercules, D. M. *J. Catal.*, **88**: 26 (1984).
402. Khan Ashraf, Z., Chetina, O. V., Erivanskaya, L. A., Shpiro, E. S., Antoshin, G. V., and Lunin, V. V. *Neftekhimiya*, **23**: 61 (1983).
403. Khan Ashraf, Z., Antoshin, G. V., Shpiro, E. S., Erivanskaya, L. A., Lunin, V. V., and Minachev, Kh. M. *React. Kinet. Catal. Letters*, **23**: 221 (1983).
404. Khan Ashraf, Z., Shpiro, E. S., Antoshin, G. V., Erivanskaya, L. A., Lunin, V. V., and Minachev, Kh. M. *Poverkhnost'*, 1: 68 (1985).
405. Lunin, V. V., Khan Ashraf, Z., Erivanskaya, L. A., Shpiro, E. S., Antoshin, G. V., and Minachev, Kh. M. *Dokl. AN SSSR*, **279**: 393 (1984).
406. Yamashita, H., Funabiki, T., and Yoshida, S., *J. Chem. Soc., Faraday Trans. I*, **82**: 702, 1771 (1986); **83**: 2883 (1987).
407. Bertolini, J. C., Brissot, J., Le Mogne, T., Montes, H., Calvayrac, J., and Bigot, J. *Applications of Surf. Sci.*, **29**: 29 (1987).
408. Tomanek, D., Hauert, R., Oelhafen, P., Schloegl, R., and Güntherodt, H. J. *Surf. Sci. Letters*, **160**: L493 (1985).
409. Moruzi, V. L., Oelhafen, P., Williand, A. R., Lapka, R., Güntherodt, H. J., and Kübler, J. *Phys. Rev.*, **B27**: 2049 (1983).
410. Biwar, B. M. and Bernasek, S. L. *Applications of Surf. Sci.*, **25**: 41 (1986).
411. Benecke, W., Feller, H. G., and Ralek, M. *Z. Metallk.*, **75**: 625 (1984).
412. Shuzhong, D., Fanhua, X., and Jingfa, D. *J. Catal.*, **109**: 170 (1988).
413. Tureis, N., Verykios, Y., Khalid, S. M., Bunker, G., and Korszun, Z. P. *J. Catal.*, **109**: 143 (1988).
414. Boreskov, G. K. *Geterogennyǐ kataliz* (Heterogeneous Catalysis). Novosibirsk: Nauka (1986).
415. Krylov, O. V. and Kiselev, A. V. *Kataliz i adsorbtsiya na oksidakh perekhodnykh metallov* (Catalysis and Adsorption on Transition Metal Oxides). Moscow: Khimiya (1981).
416. Volkenstein, F. F. *The Electron Theory of Catalysis on Semiconductors*. New York: MacMillan (1953).
417. Chuang, Y. W., Lo, W., and Somorjai, G. A. *Surf. Sci.*, **64**: 588 (1977).
418. Egdell, R. G., Erickson, S., and Flavell, S. *Solid State Comm.*, **600**: 835 (1986).
419. Erickson, S. and Egdell, R. G. *Surf. Sci.*, **180**: 263 (1987).
420. De Frisart, E., Darville, J., and Gillets, J. M. *Surf. Sci.*, **126**: 518 (1983).
421. Egdell, R. G., Erickson, S., and Flawell, W. R. *Surf. Sci.*, **192**: 265 (1987).
422. Atanasoska, L., O'Grady, W. E., Atanasoski, R. T., and Pollak, F. H., *Surf. Sci.*, **202**: 142 (1988).
423. Cox, P. A., Goodenough, J. B., Tavaner, P. J., Telles, D., and Egdell, R. G. *J. Solid State Chem.*, **62**: 360 (1986).
424. Kurtz, R. J. and Henrich, V. E. *Surf. Sci.*, **129**: 345 (1983).
425. Henrich, V. E. *Rep. Prog. Phys.*, **48**: 1481 (1985).
426. Roberts, M. W. and Smart, R. S. C. *J. Chem. Soc., Faraday Trans. I*, **80**: 2957 (1984).
427. Szot, K., Hillebrecht, F. U., Sarma, D. D., Campagna, M., and Arend, H. *Appl. Phys. Lett.*, **48**: 490 (1986).
428. Rocker, G. and Göpel, W. *Surf. Sci.*, **175**: L675 (1986).
429. Tanabe, T., Tanaka, M., and Imoto, S. *Surf. Sci.*, **187**: 499 (1987).

430. Finster, J., Shulze, D., Bechstedt, E., and Meisel, A. *Surf. Sci.*, **152/153**: 1063 (1985).
431. Grunthaner, F. J. and Maserjian, J., In: *IEEE Trans. Nucl. Sci.*, NS-24(6): 2108 (1977).
432. Bechstadt, F. *Phys. Stat. Solidi.*, **B91**: 167 (1979).
433. Finster, J., Shulze, D., and Meisel, A. *Surf. Sci.*, **162**: 671 (1985).
434. Stevenson, D. A. and Binkowski, N. J. *J. Non-Crystal Solids*, **22**: 399 (1976).
435. Mullins, W. M. and Averbach, B. L. *Surf. Sci.*, **206**: 29 (1988).
436. Mullins, W. M. and Averbach, B. L. *Surf. Sci.*, **206**: 41 (1988).
437. Mullins, W. M. and Averbach, B. L. *Surf. Sci.*, **206**: 52 (1988).
438. Stoch, J. *Poverkhnost',* 10: 68 (1987).
439. Zazhigalov, V. L., Konovalova, N. D., Zaitsev, Yu. N., Beloussov, V. M., Stoch, J., and Krupa, R. *Poverkhnost',* 10: 55 (1987).
440. Velso, J. F. and Pantano, C. G. *Surface and Interface Anal.*, **7**: 228 (1985).
441. Martin, P. J. and Netterfield, R. P. *Surface and Interface Anal.*, **10**: 13 (1987).
442. Sarman, D. J., van der Berg, J. A., and Vickerman, J. C. *Surface and Interface Anal.*, **4**: 160 (1982).
443. Brown, A. and Vickerman, J. C. *Surface and Interface Anal.*, **6**: 1 (1984).
444. Unger, W. and Pritzkow, W. *Int. J. Mass Spectrometry, Ion Process.*, **75**: 15 (1987).
445. Thomas, J. M. and Tricker, M. J. *J. Chem. Soc., Faraday Trans. II*, **21**: 329 (1975).
446. Panzner, G., Egert, B., and Schmidt, H. P. *Surf. Sci.*, **151**: 400 (1985).
447. Chudinov, M. G., Kuznetsov, B. N., and Pak Giont Pil. *Fiziko-khimicheskie issledovaniya katalizatorov* (Physicochemical Studies of Catalysts). Moscow: Khimiya (1987).
448. Ocelli, M. L., Psaras, D., Suib, S. L., and Stencel, J. M. *Appl. Catal.*, **28**: 143 (1986).
449. Sexton, B. A., Hughes, A. E., and Turney, T. V. *J. Catal.*, **97**: 390 (1986).
450. Minachev, Kh. M., Avaev, V. I., Shpiro, E. S., and Ryashentseva, M. A. *Izv. AN SSSR, Ser. Khim.*, 2704 (1978).
451. Grünert, W., Shpiro, E. S., Feldhaus, R., Anders, K., and Minachev, Kh. M. *J. Catal.* **120**: 444 (1989).
452. Chappel, P. S. C., Kibel, M. H., and Baker, B. G. *J. Catal.*, **110**: 139 (1988).
453. Wachs, I. E., Chersich, C. C., and Hardenbergth, J. H. *Appl. Catal.*, **13**: 355 (1985).
454. Broclawik, E., Haber, J., and Ungier, L. *J. Phys. Chem. Solids*, **42**: 203 (1981).
455. Grünert, W., Shpiro, E. S., Feldhaus, R., Anders, K., Antoshin, G. V., and Minachev, Kh. M. *J. Catal.*, **100**: 138 (1986).
456. Cimino, A., DeAngelis, B. A., Gazzoli, D., and Valigi, M. Z. *Anorg. Allg. Chem.*, 460: 86 (1980); 471: 208 (1980).
457. Degruse, R., Landuyt, J., Vandenbroucke, L., and Vennik, J. *Surface and Interface Anal.*, **L1**: 168 (1982).
458. Menon, P. G. and Prasada Rao, T. S. R. *Catal. Rev. Sci. Eng.*, **20**: 97 (1979).
459. Cimino, A., Minelli, G., and DeAngelis, B. A. *J. Electron Spec.*, **13**: 291 (1978).
460. Garbassi, F. and Petrini, G. *J. Catal.*, **90**: 106 (1984).

461. Copperthwaite, R. G., Hack, H., Hutchings, G. J., and Seleshop, J. P. F. *Surf. Sci.*, **164**: L827 (1985).
462. Hess, D., Papp, H., and Baerns, M. *Ber. Bunsenges. Phys. Chem.*, **92**: 515 (1988).
463. Kuznetsova, L. I., Nguen Quang Huynh, Suzdorf, A. P., Beilin, L. A., Shpiro, E. S., and Minachev, Kh. M. *Kinetika i Kataliz* (1989) (in print).
464. Stocj, A. and Stoch, J. *Materials Chemistry*, **6**: 335 (1981).
465. Kluz, Z., Stoch, J., and Czeppl, T. *Z. für Phys. Chem.*, **134**: 125 (1983).
466. Angelov, S., Tyuliev, G., and Marinova, Ts. *Applications of Surf. Sci.*, **27**: 381 (1987).
467. Oku, M. and Hirokawa, K. *J. Solid State Chem.*, **30**: 45 (1979).
468. Suzuki, H., Yasuoka, S., Risseu, J., and Nakabayashi, H. *Kochi Tech. Coll. (Jap.)*, **29**: 38 (1988).
469. Kaushik, X. K., Prasada Rao, T. S. R, Jaday, B. L. S., and Chabre, M. S. *Applications of Surf. Sci.*, **32**: 93 (1988).
470. Satsuma, S., Hattori, A., Furuta, A., Miyamoto, A., Hattori, T., and Murakami, Y. *J. Phys. Chem.*, **92**: 2275 (1988).
471. Varsanyl, G., Bertoll, I., Mink, G., Ritu, F., and Revesz, M. *Kem. Közl.*, **66**: 1 (1986).
472. Davis, B. H. *Applications of Surf. Sci.*, **19**: 200 (1984).
473. Paparazzo, E. *J. Electron Spec.*, **43**: 97 (1987).
474. Minachev, Kh. M., Antoshin, G. V., Shpiro, E. S., and Isakov, Ya. I. *Izv. AN SSSR, Ser. Khim.*, **9**: 2131 (1973).
475. Minachev, Kh. M. and Shpiro, E. S. *React. Kinet. Catal. Letters*, **35**: 195 (1987).
476. Shpiro, E. S., Antoshin, G. V., and Minachev, Kh. M. In: Kallo, D. and Minachev, Kh. M. (eds.). *Catalysis on Zeolites*. Budapest: Akademiai Kiado, p. 43 (1988).
477. Dempsey, Z., *J. Catal.*, **49**: 115 (1977).
478. Jacobs, P. A. *Catal. Rev. Sci. Eng.*, **24**: 415 (1982).
479. Kazansky, V. B. *Stud. Surf. Sci. Catal.*, **18**: 61 (1984).
480. Wagner, C. D., Six, H. A., Jensene, W. T., and Taylor, J. A. *Applications of Surf. Sci.*, **9**: 203 (1981).
481. Barr, T. L. *J. Amer. Chem. Soc.*, **108**: 3178 (1986).
482. Okamoto, Y., Ogawa, M., Meazawa, R., and Imanaka, T. *J. Catal.*, **112**: 427 (1988).
483. Shpiro, E. S. Candidate's Dissertation. Moscow: Inst. of Organic Chem., USSR Acad. Sci. (1975).
484. Thomas, J. M. In: *Proc. 8th Int. Cong. Catal.*, Vol. 1. West Berlin, p. 31 (1984).
485. Stakheev, A. Yu., Prakash, V. L. N., Shpiro, E. S., and Minachev, Kh. M. *Kinetika i Kataliz*, **29**: (1988).
486. Kokotailo, C. T., Fyse, C. A., Kennedy, C. J., Gobbi, C. C., Stuobl, H., Paeztor, C. T., *et al.* In: Murokami, J., Iijima, A., and Ward, J. W. (eds.). *New Developments in Zeolite Science and Technology (Proc. 7th Int. Zeolite Conf., Tokyo, 1986)*. New York: Kodansha-Elsevier, p. 36 (1986).
487. Antoshin, G. V., Shpiro, E. S., and Minachev, Kh. M. In: *Trudy I Vsesoyuznoi konferentsii "Primenenie tsiolitov v katalize" (Transactions of 1st All-Union Conference "The Use of Zeolites in Catalysis")*. Moscow, p. 7 (1981).

488. Gabelica, Z., Deroune, B. D., and Blom, N. *Amer. Chem. Soc. Symp. Ser.*, **248**: 219 (1984).
489. Von Ballmoos, R. and Meier, W. M. *Nature*, **289**: 722 (1981).
490. Lyman, G. E., Betteridge, P. W., and Moran, E. F. *Amer. Chem. Soc. Symp. Ser.*, **218**: 199 (1983).
491. Auroux, A., Dexpert, H., Leclerq, C., and Vedrine, J. C. *Appl. Catal.*, **6**: 95 (1983).
492. Vedrine, J. C., Auroux, A., and Coudoner, G. *Amer. Chem. Soc. Symp. Ser.*, **248**: 253 (1984).
493. Chao, K.-G. *Zeolites*, **8**: 82 (1988).
494. Minachev, Kh. M., Shpiro, E. S., Mishin, I. V., Metkhe, T., and Antoshin, G. V. *Izv. AN SSSR, Ser. Khim.*, 2682 (1983).
495. Scherzer, J. *Amer. Chem. Soc. Symp. Ser.*, **248**: 157 (1984).
496. Meyers, B. L., Fleish, T. H., Ray, G. J., Miller, J. T., and Hall, J. B. *J. Catal.*, **110**: 82 (1988).
497. Fleish, T. H. *J. Catal.*, **99**: 117 (1986).
498. Gross, T., Lohse, U., Engelhardt, G., Richter, K.-H., and Patzelova, V. *Zeolites*, **4**: 25 (1984).
499. Kubelkova, L. *J. Chem. Soc., Faraday Trans. I*, **83**: 511 (1987).
500. Dwyer, J., Fitch, F. R., Qin, G., and Vickerman, C. *J. Phys. Chem.*, **86**: 4574 (1982).
501. Minachev, Kh. M., Isakov, Ya. I., Shpiro, E. S., Isakova, T. A., and Tkachenko, O. P. *Proc. 9th Int. Congr. Catal.* Vol. 2. Calgary, p. 461 (1988).
502. Minachev, Kh. M., Isakov, Ya. I., Shpiro, E. S., Isakova, T. A., and Tkachenko, O. P. *Kinetika i Kataliz*, **29**: 1413 (1988).
503. Rahman, A., Lemay, G., Adnot, A., and Kalaguine, S. *J. Catal.*, **112**: 453 (1988).
504. Minachev, Kh. M., Bondarenko, T. N., Dergachev, A. A., Kharson, M. S., Kondratyev, D. A., and Tkachenko, O. P. *Izv. AN SSSR, Ser. Khim.*, **12**: 2667 (1988).
505. Minachev, Kh. M., Bragin, O. V., Vasina, T. V., Dergachev, A. A., Shpiro, E. S., Sitnik, V. P., *et al. Dokl. Akad. Nauk SSSR*, **304**: 1391 (1989).
506. Yan Shi-Bin, She Ligin, Liu Xing-Yan, Li Xuan-Wen, Pang Li, Huang Hui-Zhong, and Zhon Yon. *J. Catal.*, **9**: 25 (1988).
507. Suib, S. L., Wintecki, A. M., and Kostapapa, S. *Proc. 7th Int. Zeolite Conf.*, Tokyo, p. 409 (1986).
508. Edgell, M. J., Mugford, S. A., and Castle, J. E. *J. Catal.*, **111**: 433 (1988).
509. Vedrine, J. C., Dufaux, M., Naccache, C., and Imelik, B., *J. Chem. Soc., Faraday Trans. I*, **74**: 44 (1970).
510. Tkachenko, O. P., Shpiro, E. S., Antoshin, G. V., and Minachev, Kh. M. *Izv. AN SSSR, Ser. Khim.*, **6**: 1249 (1980).
511. Minachev, Kh. M., Antoshin, G. V., Shpiro, E. S., and Yusifov, Yu. A. *Application of Zeolites in Catalysis*. Budapest; p. 153 (1979).
512. Suib, S. L., Coughlin, D. F., Otter, F. A., and Conopask, L. F. *J. Catal.*, **84**: 410 (1983).
513. Minachev, Kh. M., Shpiro, E. S., Tkachenko, O. P., and Antoshin, G. V. In: *Fifth Soviet-French Seminar on Catalysis*, Lille, p. 126 (1980).
514. Pedersen, L. A. and Lunsford, J. H. *J. Catal.*, **61**: 39 (1980).

515. Okamoto, Y., Ishida, N., Imanaka, T., and Teranishi, S. *J. Catal.*, **58**: 82 (1979).
516. Antoshin, G. V., Shpiro, E. S., and Minachev, Kh. M. *Tezisy doklada seminara Kataliz na tseolitakh, Liblitse, ChSSR* (Abstract of Report at Seminar "Catalysis on Zeolites", Liblice, CSSR) (1982).
517. Gustafson, B. L. and Lunsford, J. H. *J. Catal.*, **74**: 393 (1982).
518. Lunsford, J. H. and Treybing, D. S. *J. Catal.*, **68**: 192 (1981).
519. Gudkov, S. V., Romanovsky, B. V., Shpiro, E. S., Antoshin, G. V., and Minachev, Kh. M. *Izv. AN SSSR, Ser. Khim.*, 2448 (1980).
520. Gudkov, S. V., Shpiro, E. S., Romanovsky, B. V., Antoshin, G. V., and Minachev, Kh. M. In: *Katalizatory, soderzhashchie nanesennye kompleksy* (Catalysts Containing Supported Complexes), Vol. 1. Novosibirsk, p. 37 (1980).
521. Shpiro, E. S., Antoshin, G. V., Tkachenko, O. P., Gudkov, S. V., Romanovsky, B. V., and Minachev, Kh. M. *Stud. Surf. Sci. Catal.*, **18**: 31 (1984).
522. Strutz, J., Diegruber, H., Jaeger, N. J., and Möseler, R. *Zeolites*, **3**: 102 (1983).
523. Romanovsky, B. V., Zakharov, V. Yu., and Borisenkova, S. A. *USSR Inventor's Certificate No. 522, 752* (1976); Gabrielov, A. G., Zakharov, A. N., Romanovsky, B. V., Tkachenko, O. P., Shpiro, E. S., and Minachev, Kh. M. *Koordinatsionnaya Khimiya*, **14**: 821 (1988).
524. Meyer, C., Wöhrle, D., Mohl, M., and Schultz-Ekloff, G. *Zeolites*, **4**: 30 (1984).
525. Gabrielov, A. G., Tkachenko, O. P., Shpiro, E. S., and Romanovsky, B. V. In: *Trudy IV Vsesoyuznoĭ konferentsii Primenenie tseolitov v katalize* (Proceedings of 4th All-Union Conference "Application of Zeolites in Catalysis") (1989).
526. Macquet, J. P., Hillard, M. M., and Theophanidis, T. *J. Amer. Chem. Soc.*, **100**: 4741 (1977).
527. Shultz-Ekloff, G., Wöhrle, D., Iliev, V., Ignatzek, E., and Andreev, A. In: Karge, H. G. and Weitkamp, J. (eds.). *Zeolites as Catalysts, Sorbents, and Detergent Builders*. New York: Elsevier, p. 315 (1989).
528. Nijs, H. H., Jacobs, P. A., Verdonck, J. J., and Uytterhöven, J. V. *Stud. Surf. Sci. Catal.*, **4**: 479 (1980).
529. Anderson, S. L. T. and Scurrell, M. S. *J. Catal.*, **71**: 233 (1981).
530. Givens, K. E. and Dillard, J. G. *J. Catal.*, **86**: 108 (1984).
531. Shannon, R. D., Vedrine, J. C., Naccache, C., and Lefebure, F. *J. Catal.*, **88**: 431 (1984).
532. Shpiro, E. S., Baeva, G. N., Sass, A. S., Shvets, V. A., Fasman, A. B., Kazansky, V. B., and Minachev, K. M. *Kinetika i Kataliz*, **28**: 1432 (1987).
533. Narayana, M., Michalik, J., Contarini, S., and Kevan, L. *J. Phys. Chem.*, **89**: 3895 (1985).
534. Minachev, Kh. M., Baeva, G. N., Shpiro, E. S., Antoshin, G. V., and Fasman, A. B. *Kinetika i Kataliz*, **26**: 1265 (1985).
535. Ben Taarit, Y., Vedrine, J. C., Dutel, J. F., and Naccache, C. *J. Magnetic Res.*, **31**: 251 (1978).
536. Suib, S. L., McMahon, K. C., and Tan Min Li. *J. Catal.*, **84**: 20 (1984).
537. Wichterlova, B., Krajĉikova, L., Tvarûzková, Z., and Beran, S. *J. Chem. Soc., Faraday Trans. I*, **80**: 2639 (1984).
538. Shpiro, E. S., Tkachenko, O. P., Azbel, B. I., Goldshleger, N. F., Isakov, Ya. I., Khidekel, M. L., and Minachev, Kh. M. *Kinetika i Kataliz*, **28**: 1248 (1987).

539. Shpiro, E. S., Tuleuova, G. Zh., Zaikovski, V. I., Tkachenko, O. P., and Minachev, Kh. M. *Kinetika i Kataliz*, **30**: 939 (1989).

540. Shpiro, E. S., Tkachenko, O. P., Tuleuova, G. Zh., Zaikovski, V. I., Vasina, T. V., Bragin, O. V., and Minachev, Kh. M. *Stud. Surf. Sci. Catal.*, **46**: 143 (1989).

541. Sana, N. C. and Wold, E. E. *Appl. Catal.*, **13**: 101 (1984).

542. Melson, G. A. and Zuckerman, E. B. *Proc. 9th Int. Congr. Catal.*, Vol. 2, Calgari, p. 807 (1988).

543. Jaeger, N. J., Ryder, P., and Schultz-Ekloff, G. *Stud. Surf. Sci. Catal.*, **18**: 299 (1984).

544. Wichterlova, B., Tvarûzkova, Z., Krajĉikova, L., and Novakova, J. *Stud. Surf. Sci. Catal.*, **18**: 249 (1984).

545. Nguen Quang Huynh, Varivonchik, N. E., Shpiro, E. S., Lapidus, A. L., and Minachev, Kh. M. *Kinetika i Kataliz*, **28**: 717 (1987).

546. Nguen Quang Huynh, Varivonchik, N. E., Shpiro, E. S., Lapidus, A. L., and Minachev, Kh. M. *Proc. 6th Int. Symp. Heterog. Catal.*, Vol. 2, Sofia, p. 246 (1987).

547. Nguen Quang Huynh, Beilin, L. A., Shpiro, E. S., and Minachev, Kh. M. *Kinetika i Kataliz*, **30**: 694 (1989).

548. Beilin, L. A., Nguen Quang Huynh, Shpiro, E. S., and Minachev, Kh. M. *Trudy III Sovetsko-indiŭskogo seminara po katalizu* (Proceedings of 3rd Soviet-Indian Seminar on Catalysis). Baku, p. 134 (1988).

549. Strohmeir, B. R. and Hercules, D. M. *J. Phys. Chem.*, **88**: 4922 (1984).

550. Wu, M. and Hercules, D. M. *J. Phys. Chem.*, **83**: 2003 (1979).

551. Houalla, M., Kibby, C. L., Eddy, E. L., Petrakis, L., and Hercules, D. M. *J. Catal.*, **83**: 50 (1983).

552. Antoshin, G. V., Shpiro, E. S., Belyatsky, V. N., Tkachenko, O. P., and Minachev, Kh. M. *Izv. AN SSSR, Ser. Khim.*, 1913 (1983).

553. Shpiro, E. S., Antoshin, G. V., Tkachenko, O. P., Belyatsky, V. N., Panov, S. Yu., and Minachev, Kh. M. *Dokl. Akad. Nauk SSSR*, **278**: 155 (1984).

554. Houalla, M., Delannay, F., Matsuura, I., and Delmon, B. *J. Chem. Soc., Faraday Trans. I*, **76**: 2128 (1980).

555. Stakheev, A. Yu., Shpiro, E. S., and Minachev, Kh. M. *Poverkhnost'*, 10: 141 (1987).

556. Yuffa, A. Ya., Stakheev, A. Yu., Shpiro, E. S., Slinkin, A. A., Kucherov, A. V., Fedorovskaya, E. A., Antoshin, G. V., and Minachev, Kh. M. *Appl. Catal.*, **39**: 153 (1988).

557. Houalla, M., Delennay, F., and Delmon, B. *J. Chem. Soc., Faraday Trans. I*, **76**: 1766 (1980).

558. Stakheev, A. Yu., Grigorian, A. A., Shpiro, E. S., Lysenko, S. V., Korolkov, N. S., Karakhanov, E. A., and Minachev, Kh. M. *Izv. AN SSSR, Ser. Khim.*, 2684 (1988).

559. Stakheev, A. Yu., Grigorian, A. A., Shpiro, E. S., Lysenko, S. V., Korolkov, N. S., Karakhanov, E. A., and Minachev, Kh. M. *Izv. AN SSSR, Ser. Khim.*, : 2678 (1988).

560. Okamoto, Y., Tomioka, H., Katoh, Y., Imanaka, T., and Teranishi, S. *J. Phys. Chem.*, **84**: 1833 (1980).

561. Stencel, J. M., Diehli, J. R., D'Este, J. R., Makovsky, L. E., Rodrigo, L., Marcinkowska, K. *et al. J. Phys. Chem.*, **90**: 4739 (1986).

562. Henker, M., Wendlandt, K.-P., Shpiro, E. S., and Tkachenko, O. P. *Appl. Catal.* (to be published).

563. Ghosh, A. K., Tanaka, K., and Toyoshima, I. *J. Catal.*, **109**: 221 (1987).

564. Masuyama, Y., Tomatsu, Y., Ishida, K., Kurusu, Y., and Secawa, K. *J. Catal.*, **114**: 347 (1988).

565. Tanaka, K., Sasaki, M., and Toyoshima, I. *J. Phys. Chem.*, **92**: 4730 (1988).

566. Scohart, P. O., Amin, A., Defosse, C., and Rouxhet, P. G. *J. Phys. Chem.*, **85**: 1406 (1981).

567. Carver, J. C., Wachs, I. E., and Murrell, L. L. *J. Catal.*, **100**: 500 (1986).

568. Abart, J., Delgado, E., Ertl, G., Jeziorowski, H., Knozinger, H., Thiele, N., and Wang, X. Zh. *Appl. Catal.*, **2**: 155 (1982).

569. Jesiorowski, H., Knozinger, H., Taglauer, E., and Vogdt, G. *J. Catal.*, **80**: 286 (1983).

570. Stencel, J. M., Makovsky, L. E., Diehl, J. R., and Sarkus, T. A. *J. Catal.*, **95**: 414 (1985).

571. Kasztlian, S., Grimblot, J., and Bonnelle, J. *J. Phys. Chem.*, **91**: 1503 (1987).

572. Delannay, F., Haeussler, E. N., and Delmon, B. *J. Catal.*, **66**: 469 (1980).

573. Horrell, B. A., Cocke, D. L., Sparrow, G. K., and Murray, J. *J. Catal.*, **95**: 309 (1985).

574. Maerawa, A., Kitamura, M., Okamoto, Y., and Imanaka, S. *Bull. Chem. Soc. Japan*, **61**: 2295 (1988).

575. Brown, J. R. and Ternan, M. *Ing. Eng. Chem. Prod. and Dev.*, **23**: 557 (1984).

576. Kazansky, V. B. *Kinetika i Kataliz*, **8**: 1125 (1967).

577. Slovetskaya, K. I. and Rubinshtein, A. M. *Kinetika i Kataliz*, **7**: 342 (1966).

578. Grünert, W., Shpiro, E. S., Antoshin, G. V., Feldhaus, R., Anders, K., and Minachev, Kh. M. *Dokl. Akad. Nauk SSSR*, **284**: 372 (1986).

579. Okamoto, Y., Fujii, M., Imanaka, T., and Teranishi, S. *Bull. Chem. Soc. Japan*, **49**: 859 (1976).

580. Merryfield, R., McDaniel, M., and Parks, G. *J. Catal.*, **77**: 348 (1982).

581. Cimino, A., Cordischi, D., De Rossi, S., Ferraria, G., Garroli, D., Indovina, V., Minelli, G., Occhiuzzi, M., and Valigi, M. *Proc. 9th Int. Congr. Catal.*, Vol. 3, Calgari, p. 1465 (1988).

582. Sachtler, W. M. H. *J. Molec. Catal.*, **25**: 1 (1984).

583. Ryashentseva, M. A. and Minachev, Kh. M. *Renii i ego soedineniya v katalize* (Rhenium and Its Compounds in Catalysis). Moscow: Nauka (1983).

584. Edreva-Kordjieva, R. M. and Andreev, A. A. *J. Catal.*, **94**: 97 (1985).

585. Edreva-Korjieva, R. M. and Andreev, A. A. *J. Catal.*, **97**: 321 (1986).

586. Duquette, L. G., Cieslinski, R. G., Jung, C. W., and Garrou, P. E. *J. Catal.*, **90**: 362 (1984).

587. Klier, K. *Adv. in Catal.*, **31**: 243 (1982).

588. Petrini, G. and Garbassi, F. *J. Catal.*, **90**: 113 (1984).

589. Turcheninov, A. L., Shpiro, E. S., Nekrasov, N. V., Yakerson, V. I., Sobolevsky, V. S., and Minachev, Kh. M. *Dokl. Akad. Nauk SSSR*, **296**: 165 (1987).

590. Turcheninov, A. L., Shpiro, E. S., Yakerson, V. I., Sobolevsky, V. S., Golosman, E. Z., Kiperman, S. L., and Minachev, Kh. M. *Kinetika i Kataliz* (in print).

591. Chen, H., Wang, S., Liao, J., Tsai, J., Zhang, H., and Tsai, K. *Proc. 9th Int. Congr. Catal.*, Vol. 2, Calgari, p. 537 (1988).

592. Fleish, T. H. and Mieville, R. J. *J. Catal.*, **97**: 284 (1986).

593. Okamoto, Y., Fukino, K., Imanaka, T., and Teranishi, S. *J. Phys. Chem.*, **87**: 3747 (1984).

594. Monnier, J. R., Apai, G., and Hanrahan, M. J. *J. Catal.*, **88**: 523 (1984).

595. Chinchen, G. C. and Waugh, K. G. *J. Catal.*, **97**: 280 (1986).

596. Dianqin, L., Qiming, Z., and Jinlu, L. *Proc. 9th Int. Congr. Catal.*, Vol. 2, Calgari, p.577 (1988).

597. Vlaic, G., Bart, J. C. J., Cavigiolo, W., and Pianzola, B. *J. Catal.*, **96**: 314 (1985).

598. Yoshida, S., Tanaka, T., Nishimura, Y., Mizutani, H., and Funabiki, T. *Proc. 9th Int. Congr. Catal.*, Vol. 3, Calgari, p. 1473 (1988).

599. Sonson, B., Rebenstorf, B., Larson, R., Andersson, S., and Lars, T. *J. Chem. Soc., Faraday Trans. I*, **84**: 1897 (1988).

600. Weng, L.-T., Zhou, B., Yasse, B., Doumain, B., Ruiz, P., and Delmon, B. *Proc. 9th Int. Congr. Catal.*, Vol. 4, Calgari, p. 1609 (1988).

601. Zamaraev, K. I. and Kochubei, D. I. *Kinetika i Kataliz*, **27**: 1031 (1986).

602. Ratnasamy, P. and Leonard, A. J. *Catal. Rev. Sci. Eng.*, **6**: 293 (1972).

603. Yacaman, M. J. *Appl. Catal.*, **13**: 1 (1984).

604. Kuzminsky, M. B. and Bagaturiants, A. A. *Itogi nauki i tekhniki. Ser. "Kinetika i Kataliz"* (Results of Science and Engineering. Series "Kinetics and Catalysis"), Vol. 10. Moscow: VINITI, p. 99 (1980).

605. Castner, D. G. and Santilli, D. S. *Amer. Chem. Soc. Symp. Ser.*, **249**: 39 (1984).

606. Vedrine, J. C., Hollinger, C., and Duc Minh Tran. *J. Phys. Chem.*, **82**: 1515 (1978).

607. Zakumbaeva, G. D., Shpiro, E. S., Beketaeva, L. A., Dyusenbina, B. B., Aitmatambetova, S. Z., Uvaliev, T. Yu., Khisametdiev, A. M., Antoshin, G. V., and Minachev, Kh. M. *Kinetika i Kataliz*, **23**: 941 (1982).

608. Tsisun, E. I., Nefedov, V. I., Shpiro, E. S., Antoshin, G. V., and Minachev, Kh. M. *React. Kinet. Catal. Letters*, **24**: 37 (1984).

609. Tkachenko, O. P., Shpiro, E. S., Grünert, W., Beilin, L. A., and Minachev, Kh. M. *Izv. AN SSSR, Ser. Khim.*, **10**: 2390 (1987).

610. Shpiro, E. S., Dyusenbina, B. B., Tkachenko, O. P., Antoshin, G. V., and Minachev, Kh. M. *J. Catal.*, **110**: 262 (1988).

611. Shyu, J. Z. and Otto, K. *Applications of Surf. Sci.*, **32**: 246 (1988).

612. Podkletnova, N. M., Shpiro, E. S., Kogan, S. B., Minachev, Kh. M., and Bursian, N. R. *Kinetika i Kataliz*, **28**: 712 (1987).

613. Bursian, N. R., Kogan, S. B., Podkletnova, N. M., Oranshaya, O. M., Glordievsky, Y. Yu., Lastovkin, G. A., Shpiro, E. S., and Minachev, Kh. M. *Proc. 9th Int. Congr. Catal.*, Vol. 3, Calgari, p. 1314 (1988).

614. Antoshin, G. V., Shpiro, E. S., Tkachenko, O. P., Nikishenko, S. B., Ryashentseva, M. A., Avaev, V. I., and Minachev, Kh. M. *Proc. 7th Int. Congr. Catal.*, Tokyo, p. 302 (1981).

615. Minachev, Kh. M., Avaev, V. I., Dmitriev, R. V., and Ryashentseva, M. A. *Izv AN SSSR, Ser. Khim.*, 1456 (1984).

616. Chen, H. W., White, J. M., and Ekerdt, J. G. *J. Catal.*, **99**: 293 (1986).

617. Di Castro, V., Furlani, C., Gargano, M., and Rossi, M. *Applications of Surf. Sci.*, **28**: 270 (1987).
618. Gallezot, P. *Proc. 5th Int. Symp. on Relations between Homog. and Heterog. Catal.*, Vol. 3, Novosibirsk, p. 2 (1986).
619. Whetten, R. L., Cox, D. M., Trevor, D. J., and Kaldor, A. *Surf. Sci.*, **156**: 8 (1985).
620. Beatzold, R. C. *Adv. Catal.*, **25**: 1 (1976).
621. Beatzold, R. C. *Surf. Sci.*, **106**: 241 (1981).
622. Dalla Betta, R. A. *Proc. 5th Int. Congr. Catal.*, Vol. 2, Amsterdam, p. 429 (1973).
623. Primet, M. and Ben Taarit, Y. *J. Phys. Chem.*, **81**: 1317 (1977).
624. Dalla Betta, R. A., Boudart, M., Gallezot, P., and Weber, R. S. *J. Catal.*, **69**: 514 (1981).
625. Schwab, G. M. *Adv. Catal.*, **27**: 1 (1978).
626. Slinkin, A. A. and Fedorovskaya, E. A. *Uspekhi Khim.*, **40**: 1857 (1971).
627. Tauster, S. J., Fung, S. C., and Garten, R. L. *J. Amer. Chem. Soc.*, **100**: 170 (1978).
628. Fritish, A. and Légaré, P. *Surf. Sci.*, **184**: L355 (1987).
629. Katrib, A., Petit, C., Légaré, P., Hilaire, L., and Maire, G. *J. Phys. Chem.*, **92**: 3527 (1988).
630. Shpiro, E. S., Dyusenbina, B. B., Tkachenko, O. P., Antoshin, G. V., and Minachev, Kh. M. *Kinetika i Kataliz*, **27**: 638 (1986).
631. Semikolenov, V. A., Likholobov, V. A., Zhdan, P. A., Nizovsky, A. I., Shepelin, A. P., Moroz, E. M., Bogdanov, S. V., and Yermakov, Yu. I. *Kinetika i Kataliz*, **22**: 1247 (1981).
632. Miyake, H., Doi, Y., and Kazuo, S. Shokubai, **29**: 450 (1987). *C.A.*, **108**: 188850c (1988).
633. Sauer, J., Haberlandt, H., and Shirmer, W. *Stud. Surf. Sci. Catal.*, **18**: 313 (1984).
634. Imelik, B. (ed.). *Metal-Support and Metal-Additive Effects in Catalysis*. Amsterdam: Elsevier (1982).
635. Baker, R. T. K., Tauster, S. J., and Dumesic, J. A. (eds.). *Metal-Support Interactions. Sympos. Div. Petr. Chem.*, ACS (1986).
636. Stevenson, S. A., Raupp, G. B., Dumesic, J. A., Tauster, S. J., and Baker, R. T. K. (eds.). *Metal-Support Interaction in Catalysis*. New York: Van Nostrand-Reinhold (1987).
637. Haller, G. L. and Resasco, D. E. *Advances in Catalysis*. New York: Academic Press (to be published).
638. Horsley, J. A. *J. Amer. Chem. Soc.*, **101**: 2870 (1979).
639. Sexton, B. A., Hughes, A. E., and Foger, K. *J. Catal.*, **77**: 85 (1982).
640. Fung, S. C. *J. Catal.*, **76**: 225 (1982).
641. Chien S.-H., Shelimov, B. N., Resasco, D., Lee, E. H., and Haller, G. L. *J. Catal.*, **77**: 301 (1982).
642. Koudelka, M., Monnier, A., Sanchez, J., and Augustynski, J. *J. Molec. Catal.*, **25**: 295 (1984).
643. Huizinga, T., Van't Blick, H. F. J., Yis, J. C., and Prins, R. *Surf. Sci.*, **135**: 580 (1983).
644. Short, D. R., Mansocr, A. N., Cook, J. W., Sayers, D. E., Jr., and Katzer, D. A. *J. Catal.*, **82**: 299 (1983).
645. Wehner, P. S., Tustin, G. C., and Gustafsson, B. J. *J. Catal.*, **88**: 246 (1984).

646. Meriaudeau, P., Ellesard, O. H., Dufaux, M., and Naccache, C. *J. Catal.*, **75**: 243 (1982).
647. Herrmann, J. M. *J. Catal.*, **84**: 404 (1984).
648. Baker, R. T. K., Prestridge, E. B., and Garten, R. L., *J. Catal.*, **59**: 293 (1979).
649. Singh, A. K., Pande, N. K., and Bell, A. T. *J. Catal.*, **94**: 422 (1985).
650. Sadeghi, H. R. and Henrich, V. E. *J. Catal.*, **87**: 279 (1984).
651. Belton, D. N., Sun, Y. M., and White, J. M. *J. Phys. Chem.*, **88**: 172 (1984).
652. Spencer, M. S. *J. Catal.*, **93**: 216 (1985).
653. Resasco, D. E. and Haller, G. L. *J. Catal.*, **82**: 279 (1983).
654. Chung, Y.-W., Xiong, G., and Kao, C.-C. *J. Catal.*, **85**: 237 (1984).
655. Vannice, M. A. and Garten, R. L. *J. Catal.*, **63**: 255 (1980).
656. Andersson, J. B. F., Bracy, J. D., Burch, R., and Flambard, A. R. *Proc. 8th Int. Congr. Catal.*, Vol. 5, West Berlin, p. 111 (1984).
657. Van't Blick, H. F. J., Vis, J. C., Huizinga, T., and Prins, R. *Appl. Catal.*, **19**: 405 (1985).
658. Haller, G. L., Henrich, V. E., McMillan, M., Resasco, D. E., Sadeghi, H. R., and Sakellson, S. *Proc. 8th Int. Congr. Catal.*, Vol. 4, West Berlin, p. 135 (1984).
659. Belton, D. N., Sun, Y.-M., and White, J. M. *J. Amer. Chem. Soc.*, **106**: 3059 (1984).
660. Ko, C. S. and Gorte, R. J. *J. Catal.*, **90**: 59 (1984).
661. Greenlief, C. M., White, J. M., Ko, C. S., and Gorte, R. J. *J. Phys. Chem.*, **89**: 5025 (1985).
662. Gorte, R. J., Altman, E., Corallo, G. R., Davidson, M. R., Asbury, D. A., and Hofflund, G. B. *Surf. Sci.*, **188**: 327 (1988).
663. Levin, M. E. *et al. Surf. Sci.*, **169**: 123 (1986).
664. Takatani, S. and Chung, Y. W. *J. Catal.*, **90**: 75 (1984).
665. Hofflund, G. B., Grogan, A. L., Jr., and Asbury, D. A. *J. Catal.*, **109**: 226 (1989).
666. Tamura, K., Bardi, U., and Nihel, J. *Surf. Sci*, **197**: L281 (1988).
667. Levin, M. E., Salmeron, M., Bell, A. T., and Somorjai, G. A. *Surf. Sci.*, **195**: 429 (1988).
668. Baker, R. T. K., Kim, K. S., Emerson, A. B., and Dumesic, J. A. *J. Phys. Chem.*, **90**: 860 (1986).
669. Dwyer, D. J., Robbins, J. I., Cameron, S. D., and Dutash, N. *ACS Symp. Ser.*, **298**: 21 (1986).
670. Tang Sheng, Xiong Guoxing, and Wang Hongli. *Cuihua Xuebao*, **8**: 234 (1987).
671. Tang Sheng, Xiong Guoxing, and Wang Hongli. *J. Catal.*, **111**: 136 (1988).
672. Horsley, J. A. *J. Catal.*, **88**: 549 (1984).
673. Horsley, J. A. and Lytle, F. M. *Preprints of ACS Symp. Div. Petr. Chem.*, p. 133 (1986).
674. Henrich, V. E. *J. Catal.*, **88**: 519 (1984).
675. Koboyashi, H. and Yamaguchi, M. *Proc. 9th Int. Congr. Catal.*, Vol. 3, Calgary, p. 1098 (1988).
676. Shpiro, E. S., Dyusenbina, B. B., Antoshin, G. V., Tkachenko, O. P., and Minachev, Kh. M. *Kinetika i Kataliz*, **25**: 1505 (1984).
677. Dyusenbina, B. B., Tkachenko, O. P., Shpiro, E. S., and Minachev, Kh. M. In: *IV Vsesoyuz. konfer. po mekhanizmu katal. reaktsii* (4th All-Union Conference on the Mechanism of Catalytic Reactions). Moscow, p. 337 (1986).

678. Okada, O., Ipponmatsu, M., Kawai, M., and Aika, K. *Chem. Letters*, 1041 (1984).

679. Chen Yanxin, Chen Yixuan, Li Wenzhao, and Sheng Shishan. *Cuihua Xuebao*, **8**: 337 (1987).

680. Andera, V. *7th Seminar of Socialist Countries on Electron Spectroscopy, Abstracts*. Bourgas, p. 11 (1988).

681. Sadeghi, H. R. and Heinrich, V. E. *J. Catal.*, **109**: 1 (1988).

682. Sakellson, S., McMillan, M., and Haller, G. L. *J. Phys. Chem.*, **90**: 1733 (1986).

683. Sadeghi, H. R., Resasco, D. E., Heinrich, V. E., and Haller, G. L. *J. Catal.*, **104**: 252 (1987).

684. Blasco, M. T., Conesa, S. C., Soria, S., Gonzalez-Elipe, A. R., Munuera, G., Rojo, S. M., and Saur, S. *J. Phys. Chem.*, **92**: 4685 (1988).

685. Gonzalez-Elipe, A. R., Munuera, G., Espinos, J. P., Soria, S., Conesa, S. C., and Sanz, J. *Proc. 9th Int. Congr. Catal.*, Vol. 3, Calgari, p. 1392 (1988).

686. Fuentes, S., Vazquez, A., Perez, J. C., and Yacaman, M. J. *J. Catal.*, **99**: 492 (1986).

687. Kelly, M., Short, D. R., and Swartzfager, D. C. *J. Molec. Catal.*, **20**: 235 (1983).

688. Tkachenko, O. P. *6-oĭ seminar sots. stran po elektronnoĭ spektroskopii* (6th Seminar of Socialist Countries on Electron Spectroscopy). Liblice, p. 84 (1986).

689. Kunimori, K., Abe H., Yamaguchi, E., Matsue, S., and Uchijima, T. *Proc. 8th Int. Congr. Catal.*, Vol. 5, West Berlin, p. 251 (1984).

690. Viswanathan, B., Tanaka, K., and Toyoshima, I. *Langmur*, **2**: 113 (1986).

691. Yermakov, Yu. I. and Ryndin, Yu. A. In: *5th Int. Symp. on Relations between Homog. and Heterog. Catal.*, Vol. 3, Part 2, Novosibirsk, p. 42 (1986).

692. Ryndin, Yu. A., Nogin, Yu. N., Chuvilin, A. L., Pashis, A. V., Zverev, Yu. B., and Ermakov, Yu. I. *Appl. Catal.*, **26**: 327 (1986).

693. Fleisch, T. H., Hicks, R. F., and Bell, A. T. *J. Catal.*, **87**: 398 (1984).

694. Bastl, Z. In: *7th Seminar of Socialist Countries on Electron Spectroscopy, Abstracts*. Bourgas, p. 1 (1988).

695. DeKoven, B. M. and Hagans, P. L. *Applications of Surf. Sci.*, **27**: 199 (1986).

696. Belyi, A. A., Chigladze, L. G., Rusakov, A. L., Shalikiani, M. O., Tabidze, A. S., Gagarin, S. G. *et al. Izv. AN SSSR, Ser. Khim.*, 2153 (1987).

697. Sinfelt, J. H., Garter, J. L., and Yates, D. J. C. *J. Catal.*, **24**: 283 (1972).

698. Dalmon, J. A., Primet, M., Martin, G. A., and Imelik, B. *Surf. Sci.*, **59**: 95 (1975).

699. McDonald, M. A. and Boudart, M. *J. Phys. Chem.*, **88**: 2185 (1984).

700. Gonzales, R. D. *Applications of Surf. Sci.*, **19**: 181 (1984).

701. Niemantsverdier, J. A. C, Kaamvan, A. C., Flipse, G. F., and van der Kzaam, A. M. *J. Catal.*, **96**: 58 (1985).

702. Bommannavar, A. S., Montano, P. A., and Yacaman, M. J. *J. Catal.*, **156**: 426 (1985).

703. Meitzner, G., Via, G. H., Lytle, F. W., and Sinfelt, J. H. *J. Chem. Phys.*, **78**: 2553 (1983).

704. Onuferko, J. H., Short, D. R., and Kelly, M. J. *Applications of Surf. Sci.*, **19**: 227 (1984).

705. Ioffe, M. S., Kuznetsov, B. N., Ryndin, Yu. A., and Ermakov, Yu. I. *Proc. 6th Int. Congr. Catal.*, Vol. 1, London, p. 131 (1977).

706. Budge, J. K., Lücke, B. F., Scott, J. P., and Gates, B. C. *Proc. 8th Int. Congr. Catal.*, Vol. 5, West Berlin, p. 89 (1984).
707. Bursian, N. R., Zharkov, B. B., Kogan, S. B., and Lastovkin, G. A. *Proc. 8th Int. Congr. Catal.*, Vol. 2, West Berlin, p. 48 (1984).
708. Lieske, H. and Völter, J. J. *J. Catal.*, **90**: 96 (1984).
709. Burch, R. and Garba, L. C. *J. Catal.*, **71**: 348 (1981).
710. Sexton, B. A., Hughes, A. E., and Foger, K. *J. Catal.*, **88**: 466 (1984).
711. Baronetti, G. T., de Miguel, S. R., Scelra, O. A., and Castro, A. A. *Appl. Catal.*, **24**: 109 (1986).
712. Kirlin, P. S., Stromeir, B. R., and Gates, B. C. *J. Catal.*, **98**: 308 (1986).
713. Engels, S., Lausch, H., Mahlow, P., Anders, K., Feldhaus, R., Vieweg, H.-G., Peplinski, B., Wilde, M., and Kzaak, P. *Chem Techn.*, **37**: 157 (1985).
714. Shyu, J. Z. and Otto, K. *J. Catal.*, **115**: 16 (1989).
715. Stoch, J., Ho Quy Dao, and Czeppl, T. *Bull. Pol. Acad. Sci.*, **35**: 387 (1987).
716. Kakhtashvili, G. N., Marshanova, E. N., Mishchenko, Yu. A., Sparzov, I. I., Kolotyrkin, I. Ya., Zhdan, P. A., and Gelbstein, A. I. *Kinetika i Kataliz*, **29**: 831 (1988).
717. Karakhanov, E. A., Kontsevaya, A. I., and Lysenko, S. V. *Neftekhimiya*, **25**: 435 (1985).
718. Parks, G. D., Schaffer, A. M., Dreiling, M. J., and Shiblom, C. M. *Symp. Surf. Sci. Petr. Chem.*, *ACS*, p. 334 (1980).
719. Anderson, S. L. T., Lundin, S. T., Yäras, S., and Otterstedt, J. E., *Appl. Catal.*, **9**: 317 (1984).
720. Kugler, E. L., and Leta, D. P. *J. Catal.*, **109**: 387 (1987).
721. Chukin, G. D., Zhdan, P. A., Dagurov, V. G., and Gertsik, B. M. *Khimiya i tekhnologiya topliv i masel* (The Chemistry and Technology of Fuels and Oils) (1986).
722. Ocelli, M. L. and Stencel, J. M. In: Karge, H. G. and Weitkamp, J. (eds.). *Zeolites as Catalysts, Sorbents, and Detergent Builders.* New York: Elsevier, p. 127 (1989).
723. Minachev, Kh. M., Kazansky, V. B., Dergachev, A. A., Kustov, A. M., and Bondarenko, T. N. *Dokl. AN SSSR*, **303**: 412 (1988).
724. Cai Guangyu, Li Xiyao, Zhu Shan, Wang Qingxia, and Chen Guoguan. In: Murakami, J., Iijima, A., and Ward, J. W. (eds.). *Abstracts of 7th Zeolite Int. Conf., Tokyo, 1986.* Tokyo: Kodansha, 3D-12 (1986).
725. Yakerson, V. I., Vasina, T. V., Lafer, L. I., Sitnik, V. P., Dykh, Zh. P., Mokhov, A. V. *et al. Dokl. Akad. Nauk SSSR*, **307**, 4: 923 (1989).
726. Lee, G. and Ponec, V. *Catal. Rev.*, **29**: 183 (1987).
727. Kucherov, A. V. and Slinkin, A. A. *Zeolites*, **7**: 38 (1987).
728. Ustinova, T. S., Shpiro, E. S., Smirnov, V. S., Gryaznov, V. M., Antoshin, G. V., and Minachev, Kh. M. *Izv. AN SSSR, Ser. Khim.*, 441 (1976).
729. Shpiro, E. S., Ustinova, T. S., Smirnov, V. S., Gryaznov, V. M., Antoshin, G. V., and Minachev, Kh. M. *Izv. AN SSSR, Ser. Khim.*, 763 (1978).
730. Kirsch, G., Schwab, E., Wicke, E., and Zöchner, H. *Proc. 8th Int. Congr. Catal*, Vol. 4, West Berlin, p. 209 (1984).
731. Tatarchuk, B. J. and Dumesic, J. A. *J. Catal.*, **70**: 323 (1981).

732. Zhavoronkova, K. N., Peshkov, A. V., Boeva, O. A., *Proceed. 9th Congr. Catal.,* Calgary, 1330 (1988).
733. Baeva, G. N., Mikhailenko, S. D., Shpiro, E. S., and Fasman, A. B. *Kinetika i Kataliz* (in print).
734. Birkenstock, U., Holm, R., Reinfandt, B., and Storp, S. *J. Catal.,* **93**: 55 (1985).
735. Maerawa, A., Kitamura, M., Okamoto, Y., and Imanaka, T. *Bull. Chem. Soc., Japan,* **61**: 2295 (1988).
736. Kondratyev, S. I., Nikishenko, S. B., Antoshin, G. V., Slinkin, A. A., Fedorovskaya, E. L., and Minachev, Kh. M. *Kinetika i Kataliz,* **25**: 1169 (1984).
737. Turcheninov, A. L., Nekrasov, N. V., Gaidai, N. A., Kostyukovsky, M. I., Yakerson, V. I., and Golosman, E. Z. *Kinetika i Kataliz,* **28**: 366 (1987).
738. Subbotin, A. N., Yakerson, V. I., Beilin, L. A., and Shpiro, E. S. *Kinetika i Kataliz* (in print).
739. Rappè, A. K. and Goddard III, W. A. *J. Amer. Chem. Soc.,* **104**: 448 (1982).
740. Lombardo, E. A., Houalla, M., and Hall, W. K. *J. Catal.,* **51**: 256 (1978).
741. Grünert, W., Saffert, W., Feldhaus, R., Anders, K. *J. Catal.,* **9**: 149 (1986).
742. Cristopher, J. and Swamy, C. S. *Faraday Discuss. Chem. Soc.,* Univer. of Liverpool, April 11-13, 1989, Abstract No. 33.
743. Steiner, P., Courths, R., Kinsinger, V., Sander, I., Siegwart, B., Huefner, S., and Politis, C. *Appl. Phys.,* **A44**: 75 (1987).
744. Laggenberg, P. M., Copperthwaite, R. G., Sallschop, J. P. P., and Wits, P. O. *Faraday Discuss. Chem. Soc.,* Univ. of Liverpool, April 11-13, 1989, Abstract No. 7.
745. Chukin, G. D., Khusid, B. I., Zhdan, P. A., Moskovskaya, I. F., Sinitsyna, O. A., Skarzov, I. I., and Nefedov, B. K. *Kinetika i Kataliz,* **29**: 231 (1988).
746. Kazansky, V. B., Kustov, L. M., and Khodakov, A. Yu. In: Jacobs, P. A. and van Santen, R. A. (eds.). *Zeolites: Facts, Figures, Future.* Elsevier, Stud. Surf. Sci. Catal., **49**, part B: 1173 (1989).
747. Inui, T., Ishihara, Y., Kamachi, K., and Matsuda, H. In: Jacobs, P. A. and van Santen, R. A. (eds.). *Zeolites: Facts, Figures, Future.* Elsevier, Stud. Surf. Sci. Catal., **49**, part B: 1183 (1989).
748. Tkachenko, O. P., Vasina, T. V., Shpiro, E. S., Bragin, O. V., Preobrazhensky, A. V., and Minachev, Kh. M. *Dokl. AN SSSR,* **304**: 1391 (1989).
749. Gaziev, R. G., Berman, A. D., Yanovsky, M. I., Yelinek, A. V., Krylov, O. V., Tkachenko, O. P., and Shpiro, E. S. *Kinetika i Kataliz,* **29**: 424 (1988).
750. Joyner, R., Shpiro, E. S., Pudney, P., and Minachev, Kh. M. *J. Catalysis* (submitted).
751. Bragin, O. V., Shpiro, E. S., Preobrazhensky, A. V., Vasina, T. V., Antoshin, G. V., and Minachev, Kh. M. *Izv. AN SSSR, Ser. Khim.,* 1256 (1980).
752. Abon, M., Billy, J., Bertolini, J. C., and Tardy, B. *Surf. Sci.,* **167**: 1 (1986).
753. Nguen Quang Huynh, Beylin, L. A., Tkachenko, O. P., and Shpiro, E. S. In: Jansen, J. C., Moscou, L., and Post, M. F. M. (eds.). *Zeolites for Nineties, Rec. Res. Reports 8th Int. Zeolite Conf.,* Amsterdam, 1989, p. 375.
754. Snell, R. *Catal. Rev.,* **29**: 361 (1987).
755. Joyner, R. W. *Vacuum,* **38**: 309 (1988).
756. Lebedeva, N. N., Stakheev, A. Yu., Shpiro, E. S., and Yuffa, A. Ya. *Kinetika i Kataliz,* **30**: 212 (1989).

757. Okamoto, Y., Matsunaga, E., Imanaka, T., and Teranishi, Sh. *Chem. Lett.*, 565 (1983).

758. Demmin, R. A. and Gorte, R. J. *J. Catal.*, **105**: 373 (1987).

759. Levin, M. E., Salmeron, M., Bell, A. T., and Somorjai, G. A. *J. Catal.*, **106**: 401 (1987).

760. Anderson, J. B. F., Burch, R., and Cairns, J. A. *J. Catal.*, **107**: 351 (1987); *Ibid.*, p. 364.

761. Wolf, R. M., Siera, J., van Delft, F. C., and Nieuwenhuys, B. E. *Faraday Discuss. Chem. Soc.*, **87**, paper 226 (1987).

762. Moiseev, I. I. In: Bondar, I. (ed.). *Itogi nauki i tekhniki. Ser. kinetika i kataliz* (Results of Science and Engineering. Series Kinetics and Catalysis). Moscow: VINITI, **13**: 147 (1984).

763. Bursian, N. R. *et al.*, *Proc. 8th Int. Congr. Catal.*, Vol. 2. West Berlin, p. 481 (1984).

764. Brinen, J. C., Graham, S. W., Hammond, J. C., and Paul, D. F. *Surface and Interface Anal.*, **6**: 68 (1984).

765. Sexton, B. A., Hughes, A. E., and Bibby, D. M. *J. Catal.*, **109**: 126 (1988).

766. Thomas, J. M. In: Jacobs, P. A. and van Santen, R. A. (eds.). *Zeolites: Facts, Figures, Future*. Elsevier, Stud. Surf. Sci. Catal., **49**, part A: 3 (1989).

767. Fulghum, J. E., McGuire, G. E., Musselman, I. H., Nemanich, R. J., White, J. M., Chopra, D. R., and Chourasia, A. R. *Anal. Chem.*, 61: 243R (1989).

768. Thomas, J. M. *Angew. Chem. Int. Ed. Engl.*, **27**: 1673 (1988).

SUBJECT INDEX

AUTHOR INDEX